T0231456

Welding Thermal Processes and Weld Pool Behaviors

Welding Thermal Processes and Weld Pool Behaviors

Chuan Song Wu

CRC Press
Taylor & Francis Group
Boca Raton   London   New York

CRC Press is an imprint of the
Taylor & Francis Group, an **informa** business

The original Chinese edition Copyright © 2008 by China Machine Press.

Taylor & Francis
6000 Broken Sound Parkway NW, Suite 300
Boca Raton, FL 33487-2742

© 2011 by Taylor and Francis Group, LLC
Taylor & Francis is an Informa business

No claim to original U.S. Government works

International Standard Book Number: 978-7-111-21962-0 (Hardback)

This book contains information obtained from authentic and highly regarded sources. Reasonable efforts have been made to publish reliable data and information, but the author and publisher cannot assume responsibility for the validity of all materials or the consequences of their use. The authors and publishers have attempted to trace the copyright holders of all material reproduced in this publication and apologize to copyright holders if permission to publish in this form has not been obtained. If any copyright material has not been acknowledged please write and let us know so we may rectify in any future reprint.

Except as permitted under U.S. Copyright Law, no part of this book may be reprinted, reproduced, transmitted, or utilized in any form by any electronic, mechanical, or other means, now known or hereafter invented, including photocopying, microfilming, and recording, or in any information storage or retrieval system, without written permission from the publishers.

For permission to photocopy or use material electronically from this work, please access www.copyright. com (http://www.copyright.com/) or contact the Copyright Clearance Center, Inc. (CCC), 222 Rosewood Drive, Danvers, MA 01923, 978-750-8400. CCC is a not-for-profit organization that provides licenses and registration for a variety of users. For organizations that have been granted a photocopy license by the CCC, a separate system of payment has been arranged.

Trademark Notice: Product or corporate names may be trademarks or registered trademarks, and are used only for identification and explanation without intent to infringe.

Library of Congress Cataloging-in-Publication Data

Wu, Chuan Song.
   Welding thermal processes and weld pool behaviors / Chuan Song Wu.
      p. cm.
   "A CRC title."
   Includes bibliographical references and index.
   ISBN 978-7-111-21962-0 (hardcover : alk. paper)
   1. Welding. 2. Physical metallurgy. 3. Heat--Transmission. I. Title.

TS227.2.W82 2011
671.5'2--dc22   2010027931

Visit the Taylor & Francis Web site at
http://www.taylorandfrancis.com

and the CRC Press Web site at
http://www.crcpress.com

Contents

Preface

Fusion welding, such as arc welding, laser beam welding, electron beam welding, and the like, is one category of advanced manufacturing processes widely used in industry. In a typical fusion welding process of metals, a heat source (arc or beam) is applied locally to the interfaces of the two metals to be joined so that they will be bridged by the liquid metal melted from them, and the liquid metal melted from the filler metal if any, and be joined together as a weld after the liquid metal solidifies. The locally melted metal forms a molten pool (weld pool) on the base metal. Complex processes and phenomena occur due to the heating/melting and cooling/solidifying inside the weld pool; if not appropriately controlled, these may produce adverse effects on the properties of the weld and degrade the base metal properties in the heat-affected zone. The weld thermal process and molten pool behaviors have decisive influences on the weld quality and productivity. Therefore, accurate analysis of the weld thermal process and weld pool behaviors is of critical significance for deep insight into weld metallurgy, stress/strain and distortion of the weldment, and welding process control and optimization. With rapid development of computer science and technology, numerical simulation of the weld thermal process and molten pool behaviors becomes a research frontier that has made materials welding evolve from almost an empirical art to an activity combining the most advanced tools of various basic and applied sciences.

For 30 years, I have been conducting investigations on numerical simulation of the weld thermal process and molten pool behaviors. In 1990, my book *Numerical Analysis of Weld Thermal Processes* was published. Since then, great progress has been made in this field, and my team has achieved many new results. Thus, I feel it necessary to write a new book to introduce the state of the art in modeling and simulation of the weld thermal process and molten pool behaviors.

This book discourses on the theories, models, algorithms, and study cases of the weld thermal process and molten pool behaviors in typical fusion welding processes; these lay the foundation for weld metallurgy analysis, residual stress and distortion analysis, and welding process control and optimization. The book is organized into 11 chapters. Chapter 1 introduces the concepts, physical meanings, and characteristics of the weld thermal process and molten pool behaviors. Chapter 2 discusses the category, distribution modes, and features of welding heat sources as well as weld thermal efficiency and melting efficiency. Chapters 3 through 5 disclose, respectively, the analytic approaches of weld thermal process calculation and dimensionless formulae, the finite difference method of weld thermal conduction, and finite element analysis of weld thermal conduction. Then, Chapter 6 introduces the numerical analysis method and examples of surface deformation, fluid flow, and heat transfer in tungsten-inert gas weld pools. The emphasis in Chapters 7 and 8 is on modeling and simulation of metal transfer and weld pool behaviors in gas metal arc welding. Chapter 9 deals with analysis of temperature profiles and keyhole shape in plasma arc welding, and Chapter 10 is concerned with the vision-based measuring of weld pool size and geometry. Finally, Chapter 11 discusses the transport mechanisms in welding arcs and anode boundary layers.

The research results introduced in this book were achieved by my research team at the laboratory of materials joining of Shandong University in China. I greatly appreciate the contributions of my colleagues and graduate students to the research and development projects of the laboratory. I also wish to express sincere gratitude to my family, friends, students, and colleagues, who have shown me genuine love, support, and help.

ChuanSong Wu

1

Introduction

During the fusion welding process, the heat input from the heat source (electric arc, electron or laser beam) is employed to heat the workpiece so that melting takes place in a local region, and a molten pool forms on the workpiece. Solidification occurs at the tail of the weld pool, and a weld bead is produced. For the points adjacent to the weld, their temperature rises to a level sufficient to cause variations of metallurgy and mechanical properties. Thus, the heat-affected zone (HAZ) appears around the weld. The deposition mode of heat input transferred to the workpiece, the heat flow and temperature distribution on the workpiece, and the physical phenomena of melting and solidification have decisive influence on the weld quality; these physical mechanisms are the most important theoretical basis of welding process analysis. To obtain an understanding of the problem, the characteristics of the weld thermal process and weld pool behaviors are briefly introduced first, and then a literature review is given in this chapter.

1.1 The Characteristics of the Weld Thermal Process

In welding, the welded metal experiences heating, melting (or reaching a thermal plastic condition), and subsequent cooling phases due to heat deposition and transfer; this is usually referred to as the *weld thermal process*. The weld thermal process, which accompanies the whole welding operation, is one of the main factors affecting and determining both the weld quality and welding productivity through the following ways:

1. The shape and dimension of the weld pool depend on the level and distribution of thermal energy deposited on the base metal.
2. There is a strong correlation between the extent of metallurgical reactions in the weld pool and thermal action as well as the existing duration of the weld pool.
3. The variations of heating and cooling parameters have an effect on solidification and phase transformation of the molten pool and microstructure evolution in the HAZ. Thus, thermal action during

welding plays a critical role in determining the microstructure and properties in both the weld and the HAZ.

4. Nonuniform heating and cooling processes experienced at different regions on the weldment result in nonuniform stress conditions, which cause thermal stress or strain and distortion.

5. Various kinds of cracks and other metallurgical defects may occur due to the combined influence of metallurgy, stress, and metal structure factors under thermal actions.

6. The heat input and transfer efficiency determine the melting rate of both base metal and electrode, which affects the welding productivity.

The weld thermal process is much more complicated than that in general heat treatment. It has the following main characteristics:

1. *Partial localization.* The workpiece is not heated as a whole body in welding. Only those points within and adjacent to the thermal deposition area are heated, so both heating and cooling are severely nonuniform.

2. *Heat source motion.* During welding, the heat source is moving with respect to the workpiece, and the thermal deposition domain is continuously changing. As the heat source approaches some point, its temperature rises quickly, while it drops quickly when the heat source moves far away from it.

3. *Transient nature.* Under the action of a highly centralized heat source, the heating rate is very high (above 1500°C/s in arc welding). A large amount of thermal energy is transferred onto the workpiece from the heat source within a short time. The localized heating and moving heat source make the cooling rate high as well.

4. *Hybrid heat transfer mechanism.* The molten metal flows strongly inside the weld pool. Convection and conduction dominate inside and outside the weld pool, respectively. The weld pool also produces evaporation. There exist heat losses due to convection and radiation on the workpiece surfaces. Therefore, the weld thermal process is concerned with hybrid heat transfer mechanisms.

The characteristics mentioned make the weld thermal process complicated. However, it is of great significance to understand the weld thermal process and its variations under different process conditions because of its critical influence on controlling the weld quality and raising welding productivity.

1.2 Weld Pool Behaviors

Based on heat transfer theory, the weld thermal process includes two parts. One is the convection-dominated heat transfer of overheated liquid metal inside the weld pool, and the other is the thermal conduction of solid in the HAZ and base metal. These two parts of heat transfer processes are closely related and interact with each other. The weld pool behaviors, referring to the weld pool geometry and size, fluid flow, heat transfer, and mass transfer in the pool, are one of the main factors affecting and determining the welding process and weld quality. To analyze and calculate the weld thermal process with more accuracy, it is essential to get deep insight into the fluid dynamics and heat transfer inside the weld pool.

During arc welding processes, the fluid flow in the weld pool is driven mainly by the following forces: surface tension gradient, electromagnetic force, buoyancy, and impinging force.

1. *Surface tension gradient.* Every liquid has a surface tension that depends on the temperature of that liquid (for a uniform and constant composition). Because a temperature gradient exists on the weld pool surface, a gradient of surface tension also exists. Surface tension gradient, as one of the main driving forces to make liquid metal flow in the pool, can cause the molten metal to be drawn along the surface from the region of lower surface tension to the region of higher surface tension. For molten steel, surface tension generally decreases as temperature increases; that is, the temperature coefficient of surface tension $\partial\gamma/\partial T$ is negative. At the pool center, temperature is higher and surface tension is lower, while near the boundary, temperature is lower and surface tension is higher. Thus, the surface tension gradient on the pool surface drives the molten metal to flow outward along the surface and to flow upward at the pool center [Figure 1.1(a), 1.2(b)]. However, if some surface active elements (sulfur, oxygen, selenium) are added into the weld pool, the temperature coefficient of surface tension $\partial\gamma/\partial T$ becomes positive.[1] Thus, temperature is higher and surface tension is also higher at the pool center, while temperature is lower and surface tension is also lower near the boundary. The surface tension gradient on the pool surface drives the molten metal to flow inward along the surface and to flow downward at the pool center [Figure 1.1(b)]. This flow pattern brings more heat to the bottom of the pool so that the penetration is deeper. Therefore, the sign and magnitude of the temperature coefficient of surface tension can change the flow direction of molten metal in the weld pool and affect the temperature distribution and fusion zone shape, as shown in Figure 1.1.

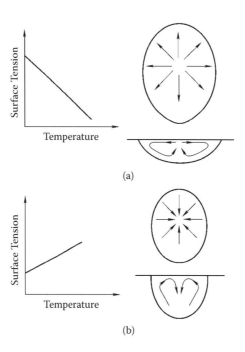

FIGURE 1.1
The effect of the temperature coefficient of surface tension on flow patterns in the weld pool.

2. *Electromagnetic force.* There is a divergence of current when it flows from the conduction spot into the weld pool, so that the interaction between the current and its magnetic field results in electromagnetic force (Lorentz force). Electromagnetic force plays an important role in driving the molten metal to flow in the pool. It pushes the molten metal to flow downward at the pool center, return to the pool surface along the pool edge, and then move inward from the edge to the center [Figure 1.2(c)].

3. *Buoyancy.* Buoyancy force results from the density variation of molten metal due to the gradient of temperature or composition in the weld pool. The density is lower at the region of higher temperature; it is larger at the region of lower temperature. Under buoyancy action, the overheated molten metal rises to the pool surface, and cooler metal is pushed downward to the pool bottom. Compared to the actions from surface tension gradient and electromagnetic forces, buoyancy force plays a less-important role [Figure 1.2(a)].

4. *Impinging force.* This is the shear force exerted on the pool surface when the plasma jet impinges onto the surface.

With considering the driving forces mentioned, the fluid flow and heat transfer in the weld pool can be described mathematically by a group of

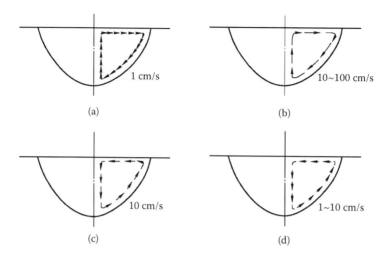

FIGURE 1.2
Schematic of the flow pattern induced by (a) only a buoyancy force; (b) only a surface tension gradient; (c) only an electromagnetic force; or (d) only an impinging force. (The arrows show flow directions as well as relative velocity by their lengths.)

partial differential equations, including thermal energy equation, momentum equation, and continuity equation. Although the treatment and solution of the problem are more complicated, the calculation accuracy can be effectively improved because this description and treatment fit the practical situation better.

1.3 The History and Evolution of an Analytic Solution of the Weld Thermal Process

An analytical solution of the weld thermal process was tried in the 1940s. Rosenthal[2,3] and Rykalin[4] provided a thermal conduction model of a moving heat source on solid plates and analytic formulae for welding temperature fields that became the classic theory of the weld thermal process: the Rosenthal-Rykalin formulae system. Grong[5] transformed the Rosenthal-Rykalin formulae into a dimensionless version and provided some calculation results in 1997. But, these analytic formulae were derived based on the following unreasonable assumptions: (1) The heat source is centralized at a point, line, or plane; (2) the materials remain at a solid state at any temperature without phase change; (3) the material properties are temperature independent; and (4) the weldment geometry is infinite (corresponding to the point and line heat source, it is considered as a semi-infinite body and infinite plate, respectively). There are big differences between these assumptions and practical cases of heat transfer during welding, which causes large error for the predicted temperature adjacent to the heat source. However, the positions near the heat source must be paid much attention because they are important for determining the fusion zone shape and size from the viewpoint of welding technology on one hand; on the other hand, the region heated above the phase change temperature must be emphasized from the viewpoint of metallurgy.

Because the Rosenthal-Rykalin formulae cannot give satisfactory results for temperature distributions in and near the weld pool, many researchers have been attempting to modify the unreasonable assumptions and improve the prediction accuracy since the 1950s. Even in 2004, some researchers still made modification of the Rosenthal-Rykalin formulae. The reason why people have been so insistent is that the analytic method can integrate the main influencing factors in a formula so that its physical meaning is comparatively clear, and the calculation process is simple and speedy.

In 1983, Eagar and Tsai made an important step in modifying the Rosenthal-Rykalin formulae.[6] They introduced the two-dimensional (2-D) Gaussian heat source into the Rosenthal-Rykalin formulae and obtained the welding temperature formula for a semi-infinite body under action of a moving Gaussian heat source. The prediction accuracy for the temperature field near the heat source was improved.

In 1999, Nguyen and coworkers applied the double-ellipsoid heat source to calculate the temperature on a semi-infinite body.[7] They first considered the volumetric distribution of the heat source. But, the multireflection method of a series of heat sources still had to be employed when their formula was used for finite-thickness plates.

In 2004, Nguyen and colleagues derived a set of formulae for the weld thermal process under the conditions of finite-thickness plates and a double-ellipsoid heat source.[8] The experimental data were used to calibrate the formulae, which can be directly employed to calculate the temperature field in general welding situations, and it is unnecessary to use the multireflection method of a series of heat sources any more.

In 2005, Wu and coworkers took into consideration both thermal losses on top and bottom surfaces of plates and a double-ellipsoid heat source and obtained the analytic formula of the welding temperature field that fit the practical cases better.[9]

1.4 Numerical Analysis of Welding Thermal Conduction

Due to the inherent drawbacks of the Rosenthal-Rykalin formulae, it is difficult to modify all the unreasonable assumptions completely. Only one or two assumptions can be avoided, but the problem cannot be solved from the very beginning of formulae derivation.

The development of modern computers promoted the application of numerical simulation in calculating the weld thermal process. A numerical simulation technique can be used to model the welding process by a group of partial differential equations describing the physical mechanisms in welding and their boundary conditions, which are solved numerically to obtain quantitative understanding of the welding process. The numerical approach is able to deal with various kinds of complex boundary conditions, intensity distribution of heat sources, and nonlinear problems, so it has incomparable advantages that the analytic approach does not possess. Thus, numerical analysis has been widely used to investigate the weld thermal process.

To conduct numerical simulation of weld pool behaviors and the thermal process completely and accurately, it is essential to consider the thermal convection inside the weld pool and the thermal conduction outside the weld pool simultaneously. However, it involves simultaneous solution of five partial differential equations and takes longer to calculate because of the complicated calculating process. To speed up the calculation for practical applications, people have attempted to simplify the whole calculation through just considering thermal conduction during welding without involving the convection heat transfer inside the weld pool.

Dilthey et al. developed a thermal conduction algorithm metal active gas arc welding—simulation (MAGSIM) for computer simulation of thin-sheet gas metal arc welding (GMAW) to calculate the weld shape and size and thermal cycles at any point on the plate.[10] The algorithm consists of two submodels; one simulates the process of wire heating and melting to describe the arc heat intensity distribution on the plate, another predicts the temperature field to calculate the weld pool, weld, and the HAZ shape and size. Also, MAGSIM can be used to determine the allowable variation range of weld geometry and to select the most suitable welding process parameters (welding current, arc voltage, welding speed) for specific conditions, including welded materials, plate thickness, wire diameter, and so on.

Radji et al. introduced a thermal conduction model of laser beam welding (LBW) with complex joint geometry and inhomogeneous material to predict the position, geometry, and dimension of transverse cross-section welds under different process parameters.[11]

Pardo and Weckman formulated a three-dimensional (3-D), steady-state thermal conduction model for the GMAW welding process.[12] The model included temperature-dependent material properties, a new finite element formulation for the inclusion of latent heat of fusion, a Gaussian distribution of heat flux from the arc, plus the effects of mass convection into the weld pool from the melted wire filler. The influence of weld pool convection on the pool shape was approximated using anisotropically enhanced thermal conductivity for the liquid phase. Weld bead width and reinforcement height were predicted using a unique iterative technique developed for this purpose.

Tekrival and Mazumder used a 3-D finite element model to analyze the thermal history of a weld joint produced by the GMAW process.[13] The finite element mesh is growing continuously in time to accommodate metal transfer in GMAW. The procedure of how to incorporate the growth of the mesh in the analysis was described. The metal transfer in GMAW can be simulated by using finite element analysis by adding elements at each time step corresponding to filler metal addition. A value of 2,300 K (melting point plus 550 K) was used as the droplet temperature.

Kumar and Bhaduri developed a 3-D finite element model for GMAW.[14] It assumed that the heat content of transferring droplets is distributed in a certain volume of the workpiece below the arc. Volumetric distribution of the heat content of transferring droplets is considered as an internal heat generation term, and the differences between penetration characteristics in two cases of globular and streaming conditions of metal transfer are analyzed. The heat generation term takes a cylindrical shape, with the radius equal to the droplet diameter and the height equal to the sum of the effective depth of cavity and drop diameter. The cavity is caused by the kinetic energy of the droplet.

1.5 Numerical Analysis of the Weld Thermal Process and Pool Behaviors in Tungsten Inert Gas Welding

The principal phenomena driving the weld pool formation and the directions of their actions on the heat and mass transport are schematically shown in Figure 1.3 for tungsten inert gas (TIG) welding. Various physical phenomena occur on the workpiece, such as melting, solidification, fluid flow, convection, thermal conduction, evaporation, volume expansion, and so on. The forces acting on the weld pool include the arc pressure, plasma gas drag, electromagnetic force, buoyancy, surface tension, and so on. There are distributed arc current density, heat flux, and pressure on the weld pool surface, which is deformed under the combined action of these factors. To get the weld pool shape and size and the temperature field on the workpiece, suitable boundary conditions must be selected to describe the process of TIG welding, and the conservation equations of mass, energy, and momentum as well as the equation of the pool surface deformation have to be solved simultaneously.

Developing a mathematical model of the TIG welding process is the basis of numerical simulation. According to the specific thermal processes, such models can be classified as steady, quasi-steady, and transient ones. Theoretically, for a stationary spot TIG welding process, the weld pool may keep an unchanged shape after initial expansion when the heat input to the workpiece from the arc equals that lost to the surroundings. Then, we say that the weld pool reaches its steady state. But, such so-called steady state

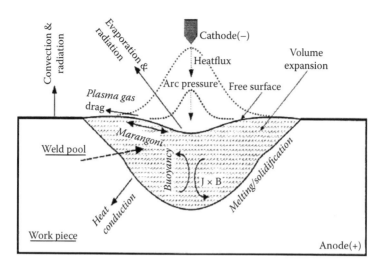

FIGURE 1.3
The physical mechanism of TIG weld pool behaviors.

is rarely encountered in practice. In most fusion welding processes, the heat source does not remain stationary. When the arc with constant power supply moves at a constant speed along the welding direction, after a short initial stage, both the weld pool shape and the temperature field will not vary with time if observed from a point located in the arc centerline. We call this situation the *quasi-steady state*, which means that the temperature distribution around the arc centerline moving at constant speed settles down to a steady form. In the quasi-steady state, although the weld pool keeps its shape and size, its position on the workpiece varies along the welding direction, so that the whole thermal process is still transient. It can be seen that the concept of quasi-steady state simplifies the transient thermal process. A transient model can simulate all stages of welding, including the initial period after arc ignition, weld pool expansion until quasi-steady state, dynamic response of the weld pool to variation of process parameters, and diminution of the weld pool after arc extinguishing. Generally, transient models can describe the practical situation in welding.

In 1984, Oreper and Szekely developed a mathematical model of fluid flow and heat transfer driven by surface tension gradient, buoyancy, and electromagnetic force in a stationary TIG weld pool.[15] Subsequently, a lot of investigation was conducted for numerical simulation of weld pool behaviors, and different quasi-steady or transient models were developed.[16–58] These models focused on different aspects in modeling of the weld pool. Generally, they can be classified into 2-D models for the stationary case, the 3-D quasi-steady state case for a moving arc, and the transient 3-D case for a moving arc. Tables 1.1 and 1.2 list the main features and shortcomings of these models.

TABLE 1.1

Two-Dimensional Model of Weld Pool for Stationary GTAW

Publication Year	Authors	Model Characteristics	Model Limitation	Refs.
1984	Oreper Szekely	Axisymmetric. Stream function and vorticity were used in the fluid flow calculation, including buoyancy, electromagnetic, and surface tension effects.	Flat free surface was assumed, nonpenetration.	15
1985	Kou Sun	Steady-state, and 2-D heat, fluid flow in stationary arc welds were computed, with the driving forces (buoyancy force, electromagnetic force, and surface tension gradient at the weld pool surface) considered.	2-D, flat free surface, nonpenetration.	16
1986 1987	Oreper Szekely Eagar	2-D, axisymmetric model was set up to calculate the unsteady-state laminar flow in the weld pool. The upper values of the temperature attainable at the free surface of the weld pool were bounded by a value 500°C below the boiling point of the metal. It was to postulate the impinging heat flux and current distribution as known quantities. Considered weld pool development in a range of metallic materials (steel, aluminum, and titanium) and examined the role played by material properties.	Flat free surface was assumed, nonpenetration.	17, 18
1989	Thompson Szekely	Transient heat transfer, fluid flow, and phase change in a weld pool with a deformed free surface were modeled using a 2-D coordinate transformation. Electromagnetic forces were neglected.	The free surface was assigned an initial position that remained fixed throughout the time interval examined. Nonpenetration.	19
1989 1990	Zacharia David DebRoy	$d\gamma/dT$ was calculated as a function of temperature and sulfur content. This allowed for a realistic simulation of the effect of the concentration of surface-active elements on the fluid flow and weld geometry. The predicted influences of surface-active elements and temperature distribution on weld pool geometry were verified by the transverse macrograph of the GTA weld indicating the fusion zone shape and size.	2-D, flat free surface, nonpenetration.	20, 21

Year	Authors	Description	Notes	Ref.
1990	Choo Szekely Westhoff	A first 2-D model was set up to describe the behavior of the arc weld pool interface, that is, the dynamic coupling between the welding arc and the weld pool surface, particularly the free surface deformation and the free surface instabilities.	The weld pool depression itself was not calculated from the model but reflects experimental measurements reported in the literature. Nonpenetration.	22
1990	Tsai Kou	A 2-D model was developed to predict heat transfer, fluid flow, and pool shape in stationary GTAW. Pool-surface-fitting orthogonal curvilinear coordinates were used.	Nonpenetration.	23
1991	Zacharia David Vitek	This evaluated the effect of weld pool evaporation and thermophysical properties on the development of the weld pool. An existing computational model was modified to include vaporization and temperature-dependent thermophysical properties.	2-D, flat free surface, nonpenetration.	24, 25
1991	Choo Szekely	This study examined the relative importance of the shear stress exerted by the plasma jet in GTAW operations to that due to surface tension shear. The calculations showed that, for low-current operation (i.e., below 200 A), gas shear is unlikely to play an important role.	2-D, flat free surface, nonpenetration.	26
1992	Choo Szekely	Special attention was given to vaporization from the free surface. It was found that heat losses due to vaporization do not play a major role in limiting the temperature at the free surface of weld pools. The computed results suggest that the free surface temperature is limited by thermocapillary motion.	2-D, flat free surface, nonpenetration.	27
1992	Choo Szekely David	Calculations are presented to describe the free surface temperature of weld pools for spot welding operations by combining a mathematical model of the welding arc and of the weld pool. Novel aspects were the heat and current fluxes falling on the free weld pool surface from the arc model; a realistic allowance for heat losses due to vaporization; and the temperature dependence of the surface tension.	2-D, flat free surface, nonpenetration.	28
1992	Knaoff Greif	The unsteady development of an axisymmetric stationary GTAW weld pool was numerically simulated. Moving grids were used to track the liquid-solid interface.	2-D, flat surface, nonpenetration.	29

(Continued)

TABLE 1.1 (Continued)

Two-Dimensional Model of Weld Pool for Stationary GTAW

Publication Year	Authors	Model Characteristics	Model Limitation	Refs.
1995	Domey Aidun Ahmadi	A series of numerical simulations was performed to study the effect of gravity on weld pool behavior. It was found that higher gravitational fields tended to enhance the convective flow within the weld pool and thus affect the heat transfer, the depth and width of the two-phase region, and the pool depth-to-width ratio.	2-D, flat surface, nonpenetration.	30
1997	Lee Na	The heat transfer and fluid flow in the molten pool during stationary GTAW were studied with an emphasis on the impact of geometric parameters on the welding arc, such as electrode bevel angle, arc length, and surface depression due to arc pressure action on the molten pool surface.	Nonpenetration.	31
1998	Chen David Zacharia	A numerical model was presented for a 2-D surface-tension-driven flow with two free surfaces. Focus was on the flow patterns with the presence of two surfaces and how they differed from those with one free surface.	Molten metal was contained in a cylinder without top and bottom walls. Although it is considered that the model approximately represents a full penetration spot weld, in fact it is quite different from the stationary full-penetration weld pool where the solid-liquid interface is calculated and moving with time, and the top radius of weld pool is larger than the bottom one.	32

Year	Authors	Description	Model	Ref.
2000	Ko Farson Choi	The dynamic behavior of stationary fully penetrated GTAW pools was investigated through numerical simulation. The effects of arc pressure, electromagnetic force, and surface tension gradients on surface depression, convection, and temperature distribution were calculated. During pool formation and growth, the fully penetrated molten pool sagged dramatically when the bottom pool diameter approached the top diameter.	2-D, stationary. The fully penetrated weld pool was simplified as a cylinder.	33
2001	Ko Choi Yoo	The effects of surface depression on pool convection and geometry in stationary GTAW were simulated numerically under direct current and pulsed-current conditions. Arc pressure was found to be a major factor in surface depression and in fluctuations of free surface and flow velocity.	2-D, nonpenetration.	34
2001	Wang Shi Tsai	A stationary GTAW weld pool of 304 stainless steel with different sulfur concentrations was studied. A parametric study showed that, depending on the sulfur concentration, one, two, or three vortexes may be found in the weld pool. These vortexes were caused by the interaction between the electromagnetic force and surface tension, which is a function of temperature and sulfur concentration, and had a significant effect on weld penetration. For given welding conditions, a minimum threshold sulfur concentration was required to create a single, clockwise vortex for deep penetration.	2-D, flat free surface, nonpenetration.	35
2001	Tanaka Terasaki Ushio	A unified numerical model of stationary TIG arc welding was developed to include melting of the anode, with inclusion of convective effects in the weld pool. 2-D distribution of temperature and velocity in the whole region of the TIG welding process, namely, tungsten cathode, arc plasma, workpiece, and weld pool, were predicted.	2-D, flat free surface, nonpenetration.	36

TABLE 1.2

Three-Dimensional Weld Pool Model for a Moving GTAW Arc

Publication Year	Authors	Model Characteristics	Model Limitation	Refs.
1986	Kou Wang	Computer simulation for 3-D convection in moving arc weld pools was described. Formulation of the electromagnetic force in the weld pool was presented. A 3.2-mm thick 6061 aluminum plate with welding parameters 110 A, 10 V, 5.5/mm was used.	Flat free surface, nonpenetration.	37, 38
1988 1989	Zacharia Eraslan Aidun	A transient, 3-D model was applied to the investigation of the nonautogenous, moving arc, GTAW process. The simulation results correctly predicted the development of the weld pool under moving arc conditions, with melting at the leading edge of the pool and weld solidification at the trailing edge.	The surface deformation of the weld pool was calculated by neglecting some higher-order terms. Nonpenetration.	39, 40
1988	Zacharia Eraslan Aidun	A transient, 3-D model (WELDER code) was developed to simulate the convection and heat transfer conditions of a GTAW weld pool. It treated for the first time the weld pool surface as a truly deformable free surface. A marked element technique was used to identify and keep track of the solid and liquid elements. The technique uses liquid-mass fraction conditions associated with the temperature conditions in the computational element to determine the solid-liquid interface.	Although it considered the melt-through weld pool, it ignored the deformation of the bottom surface of the weld pool.	41
1988	Wu et al.	A 3-D quasi-steady state model.	Flat pool surface, nonpenetration	42
1990	Wu Tsao	3-D fluid flow and heat transfer model of weld pool with a moving TIG arc.	Flat pool surface, non-penetration, quasi-steady state	43

TABLE 1.2 (Continued)

Three-Dimensional Weld Pool Model for a Moving GTAW Arc

Publication Year	Authors	Model Characteristics	Model Limitation	Refs.
1992 1993	Wu et al.	Full-penetration weld pool, considering free surface deformation in both sides.	Quasi-steady state.	44, 45
1994	Wu Dorn	3-D quasi-steady state weld pool with full penetration and surface deformation.	Quasi-steady state.	46
1995	Zacharia David Vitek	Transient 3-D convection model of weld pool.	Neglecting the surface depression at the back side of the weld pool.	47
1996	Cao and Wu	Full-penetration weld pool; focused on analysis of pool surface deformation.	Quasi-steady state.	48
1997	Joshi Dutta	3-D transient model.	Nonpenetration, without pool surface deformation.	49
1997	Wu et al	The arc current, heat flux, and pressure distributions taken from the model of welding arc physics.	Quasi-steady state.	50
1998	Cao Zhang Kovacevic	Full penetration and free top and bottom surfaces were incorporated in the model to simulate the welding process more practically.	The equation for calculating the arc pressure needs modification.	51
1997 1999	Wu et al.	Weld pool behaviors in pulsed current TIG welding.	Nonpenetration, without pool surface deformation.	52–54
2003	Hirata Asai Takenak, et al.	A 3-D numerical model, which predicted both fields of temperature and liquid metal flow in weld pool and the time-dependent penetration process by both spot and moving gas tungsten arc, was developed.	Weld pool surface was flat, nonpenetration.	55
2004 2005	Wu et al.	Transient 3-D model for weld pool geometry, surface deformation, heat and fluid flow in a moving full-penetration weld pool.		56–58

1.6 Numerical Simulation of the Weld Thermal Process and Pool Behaviors in Metal Inert Gas/Metal Active Gas Welding

In metal inert gas/metal active gas (MIG/MAG) welding, the wire itself acts as the electrode. The wire melts, droplets form at the end of the electrode, and metal transfer occurs. The droplets from the electrode are transferred to the weld pool, thereby providing filler besides simply heat to melt the substrates. There are various molten metal transfer modes, such as globular, spray, and short-circuiting transfer. The wire melting and metal transfer make the fluid flow and heat transfer phenomena in MIG/MAG weld pools more complicated than that in TIG weld pools, which can be seen from the following aspects:

1. With respect to different process parameters, there are different modes of metal transfer.
2. Due to the heat content and momentum delivered into the weld pool by metal transfer, the pool surface is more severely deformed, which changes the distribution modes of arc current density and heat flux.
3. The filler metal increases the volume of the molten pool, which affects the fluid flow and heat transfer in the pool and resultant weld formation.

Because of the complication of MIG/MAG weld pool behaviors, there are comparatively fewer models for MIG/MAG welding compared to those for TIG welding. But, MIG/MAG welding is one of the most widely used joining processes in industry, so more efforts have been made in this field. Table 1.3 lists the progress in modeling and simulation of MIG/MAG weld pools.

1.7 Numerical Models of Thermal Process in Plasma Arc Welding

Plasma arc welding (PAW) offers significant advantages over conventional gas tungsten arc welding (GTAW) in terms of penetration depth, joint preparation, and thermal distortion. The arc used in PAW is constricted by a small nozzle and has a much higher gas velocity and heat input intensity than that in GTAW. As the plasma arc impinges on the area where two workpieces are to be joined, it can melt material and create a molten liquid pool. Because of its high velocity and the associated momentum, the arc can penetrate through the molten pool and form a hole in the weld pool, which is usually referred to as a *keyhole*. Moving the welding torch and the associated

TABLE 1.3

MIG/MAG Weld Pool Models

Publication Year	Authors	Model Characteristics	Model Limitation	Refs.
1988	Tsao Wu	Stationary 2-D weld pool model considered the energy exchange between spray transfer droplets and the weld pool.	Nonpenetration, without considering reinforcement and pool surface deformation.	59
1992	Wu	3-D quasi-steady state model of fluid flow and heat transfer in a moving MIG weld pool. The heat content of droplets delivered into the pool was taken as an internal term in thermal energy equations.	Nonpenetration, without considering reinforcement and pool surface deformation.	60
1994	Kim Na	Computer simulation of 3-D heat transfer and fluid flow in GMAW was carried out. Molten surface deformation was considered.	Neither the droplet impact nor the droplet heat content was included; nonpenetration case.	61
1997	Wu et al	3-D GMAW model considered both the heat content and impact of transferring droplets.	Nonpenetration.	62, 63
1998	Wu Sun	Considered the effect of weld pool surface depression on the distribution variations of arc current density, heat flux, and pressure. The droplet impact was taken as a term in the equation of pool surface deformation.	Nonpenetration, quasi-steady state.	64, 65
1999	Fan Kovacevic	A transient 2-D model was developed to simulate the droplet formation, detachment, and transport phenomena associated with a weld pool into which molten droplets periodically were injected.	2-D, nonpenetration; no consideration of heat content of droplets.	66
1999	Ohring Lugt	Transient 3-D model. The addition of molten material was modeled by an impacting liquid metal spray on the weld pool. The mushy zone and the development of reinforcement without specific assumptions were included.	Nonpenetration.	67

(Continued)

TABLE 1.3 (Continued)

MIG/MAG Weld Pool Models

Publication Year	Authors	Model Characteristics	Model Limitation	Refs.
1999 2002	Sun Wu	Considered the reasonable distributions of arc heat flux on the severely depressed pool surface and droplet heat content inside the pool. The interaction between fluid flow field, temperature profile, and weld reinforcement was taken into account.	Nonpenetration, quasi-steady state.	68, 69
2001	Wang Tsai	A mathematical model and the associated numerical technique were developed to simulate, for the first time, the dynamic impinging process of filler droplets onto the weld pool in spot GMAW. Filler droplets driven by gravity, electromagnetic force, and plasma arc drag force, carrying mass, momentum, and thermal energy periodically impinge onto the base metal, leading to a liquid weld puddle. Transient weld pool shape and the complicated fluid flow in the weld pool caused mainly by the combined effect of droplet impinging momentum and surface tension were calculated.	2-D, nonpenetration.	70, 71
2003 2004	Kim Zhang DebRoy	A 3-D numerical heat transfer and fluid flow model was developed to examine the temperature profiles, velocity fields, weld pool shape and size, and the nature of the solidified weld bead geometry during GMAW of fillet joints.	Nonpenetration, without droplet impact.	72, 73

keyhole will cause the flow of the molten metal surrounding the keyhole to the rear region, where it resolidifies to form a weld bead. The keyhole mode of welding is the primary attribute of welding processes with high power density (PAW, laser welding, and electron beam welding) that makes them penetrate thicker pieces with a single pass. Compared to laser welding and electron beam welding, keyhole PAW is more cost-effective and more tolerant of joint preparation. The key issue for numerical analysis of the temperature field in keyhole PAW is how to develop a heat source model that reflects the thermophysical characteristics of the keyhole PAW process. Because of the complexity of the phenomena associated with the formation of a keyhole, only a limited number of theoretical studies treating the PAW process have been reported, each of varying degrees of approximation and each focusing on different aspects of the problem.

Hsu and Rubinsky[74] conducted the calculation for the 2-D (in the plane normal to the plasma jet flow axis) weld pool flow and temperature fields associated with steady travel PAW, but neglected surface tension effects and simplified the pool geometry by assuming flat upper and lower free surfaces, a keyhole of constant radius, and a nontapering solid-liquid phase boundary.

The most comprehensive model of PAW to date is Keanini and Rubinsky's quasi-steady state 3-D finite element simulation to calculate the weld pool shape and the fluid flow and temperature fields of the pool.[75] However, the model was based on quite a few simplifying assumptions: (1) The capillary surface shapes were determined by performing one Newton-Raphson iteration on the initial surface guess that was based on experimentally observed shapes, (i.e., the keyhole shape was assumed according to experimental results); (2) a conduction solution, based on a guessed capillary surface shape, was used to locate the solid-liquid phase boundary, so both weld pool convection and the difference between the guessed capillary surface and the final shape were neglected; (3) electromagnetic forces were neglected; (4) latent heat effects were not considered; and (5) plasma heat transfer was assumed to occur only on the keyhole portion of pool surface, and evaporation from the pool was neglected. Because of these somewhat oversimplified assumptions, the calculated weld pool shape, particularly the vertical cross sections at the rear portion of the weld pool, were not in agreement with the expected ones. The predicted weld pool shape seemed to have a much larger depression behind the keyhole. It is questionable that in such a case the back keyhole wall seemed to be inclined too seriously.

Fan and Kovacevic[76] developed a 2-D model to demonstrate the heat transfer and fluid flow in stationary keyhole PAW without including the plasma jet shear and the recoil pressure of the evaporating metal. Although the effect of vaporization was not dominant, however, the vapor pressure did have a significant effect on the PAW process.

For available heat source models, neither Gaussian nor double-ellipsoid modes of heat source is applicable to the keyhole PAW process. With considering the "digging action" of the arc plasma, Wu's group proposed a 3-D conical heat source model and its modification as the basis for the numerical analysis of temperature fields in the keyhole PAW process.[77,78] Further, a new heat source model for the quasi-steady state temperature field in keyhole PAW was developed to consider the "bugle-like" configuration of the keyhole and the "decline feature" of heat flux distribution of the plasma arc along the direction of the workpiece thickness. Based on this heat source model, finite element analysis of the temperature profile in keyhole PAW was conducted, and the weld geometry was determined. The results showed that the predicted location and locus of the melt line in the PAW weld cross section were in good agreement with experimental measurements.

1.8 Numerical Simulation of Metal Transfer in GMAW

Metal transfer is one of the key factors affecting the process stability and weld quality of GMAW. During the metal transfer process, the molten drop detaching from the filler wire is overheated, so it transmits some thermal energy to the weld pool. Besides, when the molten drop arrives at the weld pool at a certain speed, it impinges and applies some momentum to the pool. Therefore, metal transfer has immense effects on the welding thermal process and weld pool geometry and dimension. To improve the accuracy of numerical simulation of the welding thermal process and the geometry of the weld pool, it is necessary to simulate the dynamic process of metal transfer in GMAW.

The accuracy of numerical simulation results relies on the accuracy of the mathematical model describing the relevant physics phenomena. The mathematical modeling for metal transfer is always based on a certain theory of metal transfer. Therefore, theories of metal transfer are the basis for numerical simulation of metal transfer. So far, various theories have been proposed, including static force balance theory (SFBT), pinch instability theory (PIT), minimum energy theory, fluid dynamic theory, and "mass spring" theory.

1.8.1 SFBT and PIT

Of all the mentioned theories, SFBT and PIT are most widely used.[79–83] They have been used with reasonable results for predicting drop size, transfer frequency, and transfer mode.

1.8.1.1 Static Force Balance Theory

During GMAW, a molten drop detaches and moves into the weld pool through the combined action of gravity, electromagnetic force, plasma drag force, and surface tension. The gravity, electromagnetic force, and plasma drag force acting on the molten drop are detachment forces, and the surface tension is the holding force. The molten drop detaches once the summation of detachment forces exceeds the holding force. The balance between detachment forces and holding force is used as a detachment criterion.

1.8.1.2 Pinch Instability Theory

Pinch instability theory evolved from Rayleigh's liquid column instability model. A liquid column has a tendency to shrink and form several droplets because of the lower freedom energy of globular as compared with columnar

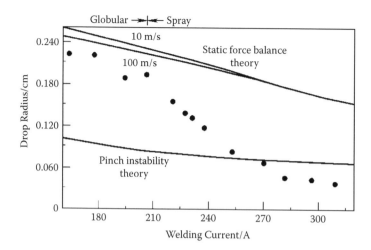

FIGURE 1.4
Comparison of predicted and measured drop size. (From Kim Y S and Eagar T W, *Weld. J.*, 72(6): 269s–277s, 1993.)

forms, resulting in oscillation of the liquid surface. Once the pressure difference between two points of a liquid column reaches a critical value, the liquid column breaks into droplets. The stability of the liquid column weakens with the increase in welding current or decrease in the radius of the liquid column.

Figure 1.4 shows the calculated drop size in different current values using SFBT and PIT. It can be seen that, in the range of globular transfer, the calculated drop sizes based on SFBT are in good agreement with experimental ones, while in the range of spray transfer, the predicted ones are much larger than the experimental ones. This is because the liquid metal at the tip of the wire under spray transfer takes a "pencil-tip" shape in spray transfer. The pencil-tip shaped liquid metal at the tip of the wire is very unstable. As shown in Figure 1.4, the spray transfer can be well predicted with PIT. High-speed photographic experiments revealed that, under the globular transfer mode, when the tip of filler metal melts, it takes a globular shape at once. Therefore, the PIT cannot describe globular transfer. Besides, both PIT and SFBT cannot explain the effect of wire extension on the metal transfer and the combination transfer of globular and spray transfer.[84] Therefore, each of the theories has its great limitation.

1.8.2 Minimum Energy Theory

Metal transfer is described not only from the viewpoint of forces but also from the viewpoint of energy. During metal transfer, the surface profile of a molten drop evolutes to minimize the total energy of the whole "arc-drop" system. Using Surface Evolver code as a platform of finite element analysis, Yang et al.

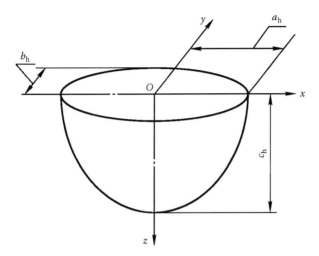

FIGURE 2.6
Schematic of semiellipsoid distribution.

$$= 4q_m \int_0^\infty \exp(-Ax^2)dx \int_0^\infty \exp(-By^2)dy \int_0^\infty \exp(-Cz^2)dz$$

$$= 4q_m \left(\frac{1}{\sqrt{A}} \frac{\sqrt{\pi}}{2} \right) \left(\frac{1}{\sqrt{B}} \frac{\sqrt{\pi}}{2} \right) \left(\frac{1}{\sqrt{C}} \frac{\sqrt{\pi}}{2} \right)$$

$$= \frac{q_m \pi \sqrt{\pi}}{2\sqrt{ABC}}$$

$$q_m = \frac{2Q\sqrt{ABC}}{\pi\sqrt{\pi}} \tag{2.33}$$

To evaluate the parameters A, B, C, the semiaxes of the ellipsoid a_h, b_h, c_h in the directions x, y, z, respectively, are defined such that the heat density falls to $0.05\,q_m$ at the surface of the ellipsoid. In the x-direction:

$$q(a_h,0,0) = q_m \exp(-Aa_h^2) = 0.05q_m, \ \exp(-Aa_h^2) = 0.05$$

$$A = \frac{3}{a_h^2} \tag{2.34}$$

Similarly,

$$B = \frac{3}{b_h^2} \tag{2.35}$$

developed an energy model for metal transfer that includes electromagnetic and gravity potential energy.[85] Based on the model, the equilibrium state and critical point of the molten metal drop were calculated, drop sizes at different current levels were predicted, and the effect of the applied mechanical force on the transfer behaviors was analyzed. The calculated results were in good agreement with the experimental ones. Using the minimum energy theory, Joo et al. calculated the geometry and volume of a molten metal drop.[86]

All the models mentioned are static models, which cannot describe the size and profile evolution of the molten drop. These evolutions include plenty of information about the metal transfer mechanism. Therefore, it is of great importance to simulate the dynamic process of metal transfer.

1.8.3 Fluid Dynamics Theory

Based on the fluid continuity equation and the momentum equation, Simpson et al. developed a one-dimensional dynamic model for metal transfer of GMAW.[87] Considering gravity, electromagnetic force, plasma drag, and surface tension, the time and current dependence of the pendant drop size and profile were calculated. Haider et al. developed dynamic models for globular and spray transfer, respectively, and calculated the metal transfer process using the volume of fluid (VOF) algorithm.[88] Kim et al. developed a 2-D model for metal transfer considering the unevenness and variation of drop temperature and calculated the velocity and temperature field, drop size, and transfer frequency by using VOF.[89,90] This model works well in simulating spray transfer but cannot be used to predict the transition current between globular and spray transfer. Using the VOF algorithm, Fan et al. calculated the heat transfer and fluid velocity field during the interaction between the molten drop and the molten pool. But, this kind of model is complex, and the calculation is time consuming.

1.8.4 Mass Spring Theory

The mass spring theory was proposed by Shaw.[91] It was first used to simulate the motion of water dripping from a faucet, then was used to simulate metal transfer by Watkins et al.[84] The calculated drop size and transfer frequency were very close to the experimental ones in the same welding conditions. Watkins et al. only analyzed the time dependence of displacement at a current approximate to the transitional current. They did not analyze the effect of drop displacement and oscillation velocity on the behaviors of globular and spray transfer.

Using the mass spring model, Jae et al. analyzed drop oscillation at different current values, calculated the displacement dependence of the spring constant, and found that the predicted drop sizes were in good agreement with the experimental ones.[92] But, they only analyzed the time dependence of displacement of a single drop on the hypothesis that the drop did not oscillate at the initial stage. The experiment revealed that, after detaching a drop, both the retained molten metal at the tip of the wire and the detached drop changed their profile, and the retained molten metal was subject to remarkable oscillation.[93]

Based on mass spring theory and the hypothesis that the molten drop takes the shape of an ellipse, Jones developed a dynamic model considering the spring force.[92] As described, this theory is far from perfect. The application of this theory is only limited to the simulation of metal transfer in conventional GMAW and has not yet been extended to the metal transfer in pulsed GMAW.

Wu's group has been working in the field of numerical simulation of metal transfer. Based on fluid dynamics, electromagnetic theory, and the VOF algorithm, a dynamic process model for metal transfer in GMAW was developed by taking into account the electromagnetic force, surface tension, and plasma drag force.[93,95] The formation, evolution, and detachment of the molten drop were calculated with this model. The calculated drop sizes and frequencies at different current values were in good agreement with experimental ones. The drop velocity fields at different stages were also calculated. The detaching mechanism was analyzed using the calculated velocity fields. An analytical model for metal transfer based on mass spring theory was also developed, and the dynamic processes of metal transfer in GMAW and pulsed GMAW were calculated using this model.[96,97]

1.9 Numerical Analysis of Deep-Penetration Laser Beam Welding

As deep-penetration LBW gets wider applications in manufacturing industry, it is more and more important to conduct optimization of the LBW process through numerical simulation. Compared to keyhole PAW, more theoretical models have been developed to explain the main physical mechanisms arising in deep-penetration LBW, but each focuses on one or more particular aspects with different simplifying assumptions because of the extreme complexity of the problem.

Andrews and Atthey[98] were the first to pay attention to vaporizing effects and formation of the keyhole in laser welding under assumption of non-ionized metal vapor. Dowden et al.[99] extended Andrews and Atthey's work, incorporated the absorption of radiation in the plasma, and assumed that the keyhole is everywhere circular in cross section. Kaplan used a formula for pure heat conduction, derived from Rosenthal's analytical model of a moving line source of heat,[100] to calculate the profile of the front and rear walls of the keyhole. The geometry of the keyhole wall in the vertical section is determined point by point by locally solving the energy balance at the wall. Lampa and Kaplan et al.[101] put forward a simplified version of Kaplan's model. With a real laser-weld cross section as reference, the usual value of thermal conductivity λ_1 was used to calculate the bottom width of the weld where the melt-solid interface is almost parallel, and an artificially high thermal conductivity ($\lambda_1 = 2.5\lambda_2$) was used to calculate the top width of the weld where lateral thermal transport is associated by thermocapillary flow. Lankalapalli et al.[102] developed a model for estimating penetration depth based on a 2-D heat conduction model and a conical keyhole assumption, which neglected latent heat effects and provided blind conical keyholes. For considering the melt flow around the keyhole, Colla et al.[103] developed a 2-D model to analyze the flow pattern of an ideal fluid flowing past a cylinder body (keyhole), disregarding the thermocapillary-driven convection, latent heat, and evaporation at the keyhole surface.

The first "self-consistent" model by Kroos et al.,[104] who investigated the stability of a cylindrical keyhole using the energy and pressure balance, made it possible to determine the radius of the keyhole as well as the pressure and temperature at the keyhole wall. A threshold for the laser power per thickness of the workpiece is found above which the formation of a stable keyhole commences. It should be noted, however, that in the model mentioned, the geometry of the keyhole was prescribed to be cylindrical, gas flow out of the keyhole and pressure effects due to a plasma were neglected, and with the energy flux density in Gaussian distribution inside the keyhole, the losses by heat conduction in the solid were calculated by solving the 2-D equation of heat conduction in the steady state.

The attempts to model laser welding as mentioned have characteristically studied the weld pool and keyhole with assumed shape in isolation from each other on the pragmatic ground that the physical processes involved in each region are complicated. Ducharme et al.[105] tried to develop an integrated model of laser welding of thin-sheet materials, taking account of the conditions in the keyhole as well as in the weld pool in an interactive manner. But, the variations of keyhole radius and temperature along the thickness of the metal were ignored, and the weld pool shape was calculated using an analytical solution of the equation of heat conduction. Solana et al.[106] constructed a model to determine the keyhole and weld pool geometry by setting the appropriate energy and pressure balances. However, the model could only account for blind keyholes and predict the maximum penetration depth for a semi-infinite plate. It neglected the hydrodynamics of the keyhole and the melting enthalpy of the liquid and took the temperature gradients along the keyhole axis as constant. Sudnik et al.[107,108] developed a model with characteristics that considered the keyhole, weld pool, and solid as a nonlinear thermodynamic continuum. The initial approximation to the keyhole surface was performed according to the boiling point of the metal, and the precise conditions were determined assuming a pressure equilibrium at the keyhole wall. The melt-solid boundary was given by the enthalpy values at the start and the completion of fusion. To predict the weld pool length sufficiently well, the model was enlarged by the heat transport caused by a recirculating flow in radial sections driven by the temperature-dependent surface tension and by the vapor friction at the capillary wall. But, it was concerned with only nonpenetration (blind keyhole) and quasi-steady state conditions. Pecharapa et al.[109] gave two simple expressions for the shapes of solid-liquid and liquid-vapor interfaces based on the Stefan condition for quasi-steady state laser welding and some oversimplified assumptions. Farson et al.[110] investigated the formation and stability of stationary laser weld keyholes, but the keyhole was blind, not penetrated.

2

Models of Welding Heat Sources

2.1 Introduction

There are two aspects for the physical model of welding heat sources: One is the amount of thermal energy acted on the base metal, and the other is how this energy is distributed on the base metal. Developing a suitable physical model of welding heat sources is a prerequisite for numerical analysis and simulation of the weld thermal process and weld pool behaviors. The two problems mentioned are discussed in this chapter.

2.2 Weld Thermal Efficiency and Melting Efficiency

For arc welding, electric power is transformed into thermal energy used to heat and melt wire (or electrode) and base metal. In gas metal arc welding, the wire melts, and molten droplets form at the wire tip. When droplets are detached and transferred into the weld pool, they deliver extra heat into the pool. For gas tungsten arc welding (GTAW), the tungsten electrode does not melt; only part of the arc heat is used to melt the base metal. The arc power is given by

$$Q_0 = IU_a \qquad (2.1)$$

where U_a is the arc voltage (V), I is the welding current (A), and Q_0 is the arc power (W), that is, the thermal energy released from the arc per unit time.

When energy or heat is transferred from an arc to the workpiece to produce a weld, the transfer is not perfect. Some energy is lost between the arc and the workpiece in fusion welding processes. The net energy for heating the workpiece may be written as

$$Q = \eta Q_0 = \eta IU_a \qquad (2.2)$$

TABLE 2.1

Weld Thermal Efficiencies for Arc Welding Processes

Arc Welding Process	η
Shielded metal arc welding (SMAW)	0.65–0.85
Submerged arc welding (SAW)	0.80–0.90
Gas metal arc welding (GMAW, CO_2)	0.75–0.90
Gas metal arc welding (GMAW, Ar)	0.70–0.80
Gas tungsten arc welding (GTAW)	0.65–0.70

where η is the weld thermal efficiency, which is defined as the ratio of the total energy that enters the plate being welded Q to the total arc power Q_0. The weld thermal efficiency η is dependent on the specific welding process, the process parameters, the consumables (wire, electrode, shielding gas), and so on. Table 2.1 lists some typical values of η for various arc welding processes.

If other conditions keep constant, η decreases in value as the arc length or voltage rises and increases in value as the welding current increases. It should be noted that the so-called thermal efficiency here only describes the net heat that reaches the workpiece. In fact, not all of this part of heat is used to melt the base metal and produce the weld. For the total heat transferred onto the workpiece, one part is used to melt the base metal and produce the weld; the remaining part is lost to the mass of the workpiece surrounding the region of energy deposition to form the heat-affected zone (HAZ). The value of η does not reflect the percentage of the former and latter parts for the total transferred heat.

By definition, the weld thermal efficiency is the ratio of the energy entering the plate Q to the total arc energy Q_0, that is,

$$\eta = \frac{Q}{Q_0} \tag{2.3}$$

$$\eta = \frac{Q_1 + Q_2}{Q_0} \tag{2.4}$$

$$Q = Q_1 + Q_2 \tag{2.5}$$

where Q_1 is the heat amount required to melt the weld metal in unit time ($T = T_m$, T_m is the melting point, including the latent heat of fusion), and Q_2 is the total heat consisting of the part making the liquid metal overheated ($T > T_m$) and the part conducted into the surroundings.

Equation (2.5) indicates that not all of the heat entering the base metal is totally used to melt the weld bead. Therefore, the melting efficiency η_m is defined as the ratio of the energy required to melt the weld bead (in liquid state, $T = T_m$) in unit time to the net heat input. It can be written as

$$\eta_m - \frac{Q_1}{Q_1 + Q_2}$$

(2.6)

Since

$$Q_1 = v_0 A_w \rho H_m$$

(2.7)

where v_0 is the welding speed, A_w is the weld area of a transverse cross section, ρ is the material density, and H_m is the heat content of molten metal defined by

$$H_m = C_p T_m + L_m$$

(2.8)

where C_p is the specific heat, T_m is the melting point, and L_m is the latent heat of fusion.

Substituting Eqs. (2.7) and (2.8) into Eq. (2.6), we obtain

$$\eta_m = \frac{v_0 A_w \rho (C_p T_m + L_m)}{Q}$$

(2.9)

Then, η can be calculated by the following equation:

$$\eta = \frac{Q}{Q_0} == \frac{v_0 A_w \rho H_m}{\eta_m} \cdot \frac{1}{IU_a}$$

(2.10)

From the viewpoint of calculating the weld thermal process, suitable and accurate selection of weld thermal efficiency η is a prerequisite for enhancing prediction accuracy. Many researchers investigated the approaches for determining η in different ways.[111–114] These approaches include calorimeter-based measurement, integrative calculation-measurement, and analysis of arc physics. However, there are big differences in η values from different literature.

2.2.1 Analysis of Arc Physics

Quigley et al. analyzed the heat transferred to the base metal from a tungsten inert gas (TIG) arc for a case with the following conditions: argon shielded arc, direct current electrode negative (DCEN), 2-mm tungsten diameter, 100-A current, 16 V, mild steel workpiece.[111] The various anode heating and cooling processes were considered. Heating of the anode is due to the deposition of potential and kinetic energy by electrons together with heat transferred by conduction and convection from the arc column.

2.2.1.1 Work Function

When electrons enter into a metal, they bring the work function absorbed from the cathode to the anode. For most metals, this value of work function V is around 4 V. A reasonable work function for steel is 4.2 V $\pm$ 0.3 V. For a 100-A arc, this energy deposited in the steel anode surface corresponds to a heat input of 420 $\pm$ 30 W.

It should be noted that if the arc polarity were reversed, the work function would produce cooling, rather than heating, of the workpiece, and the welding process would be correspondingly less efficient. The TIG process normally relies on electron cooling of the tungsten tip to attain higher power.

2.2.1.2 Electron Thermal Energy

Electrons flowing from the hot plasma to the cooler anode carry heat as well as current, and this extra transfer of heat is a type of Thomson effect. The electrons leaving the plasma possess random kinetic energy characterized by their temperature T_e (their mean energy is written as $3/2 k_e T_e$, where k_e is the Boltzmann constant, $k_e = 1.38 \times 10^{-23}$ J/K). By the time electrons reach the anode, their temperature has fallen to that of the anode (T_a). Since the total flux of electrons that must reach the anode per second to carry current I is I/e (e is the electronic charge), the rate of electron thermal energy transport is

$$\frac{I}{e} \cdot \frac{3}{2} k_e (T_e - T_a) = I V_{th} \qquad (2.11)$$

where V_{th} is an equivalent volt drop at the anode:

$$V_{th} = \frac{3}{2} k_e \frac{(T_e - T_a)}{e} \qquad (2.12)$$

Generally, $T_e \geq 7000$ K, $T_a \geq 2400$ K. Thus, a typical value of V_{th} is 0.58 V.

2.2.1.3 Anode Fall

The anode fall region covers the transition between the arc column and the anode. Across this transition there must be a rapid reduction in temperature, resulting in little thermal ionization and insufficient ion generation to maintain ion flow. A negative space charge is therefore formed that is responsible for a volt drop at the anode. An electron traveling across this region acquires a kinetic energy IV_a (V_a is the anode fall voltage). If the electron-neutral mean free path L_e is much greater than the thickness of the anode fall region S_a, the electrons do not lose any of the kinetic energy in collision with neutrals, and they deposit all of it in the anode. Based on the literature data ($V_a = 2$ V, $L_e = 3$ μm, $S_a = 0.6$ μm), the heat dissipated in the anode fall region is

$$IV_a = 2I \ (\text{W})$$

2.2.1.4 Conduction

The heat conducted through the gas to the anode is given by

$$P_{cd} = \lambda_g A_a \frac{dT}{dx} \tag{2.13}$$

where λ_g is the total thermal conductivity of the gas, A_a is the area of the arc at the anode, and dT/dx is the temperature gradient of the arc close to the anode. Substituting $\lambda_g = 0.1$ W/(m W), together with $A_a = 7.0 \times 10^{-6}$ m^2 and $dT/dx = 10^7$ K/m, the heat conducted from the gas to the anode workpiece under the present conditions is found to be

$$P_{cd} = 70 \ \text{W}$$

2.2.1.5 Convection

The heat transferred to the anode through convection of the plasma jet is estimated as

$$P_{cd} \approx 33 \pm 15 \ \text{W}$$

2.2.1.6 Radiation from the Arc to the Workpiece

The total radiated power input to the anode is unlikely to exceed 20 W, that is,

$$P_R = 20 \ \text{W}$$

2.2.1.7 Workpiece Cooling

To consider cooling effects on the workpiece, vaporization and radiation are two main processes involved. These two parts of heat loss from the anode are estimated as –33 W and –10 W, respectively.

$$P_E = -33 \ \text{W}, \ P_{ra} = -10 \ \text{W}.$$

TABLE 2.2

Total Heat Balance on the Anode

Total Arc Power (W)	1,600	Percentage
Work function IV (W)	420 ± 30	26.2
Thomson effect IV_{th} (W)	60 ± 25	3.8
Anode fall IV_a (W)	150 ± 50	9.4
Conduction P_{ed} (W)	70	4.4
Convection P_{CV} (W)	33 ± 15	2.1
Radiation P_R (W)	20	1.2
Evaporation P_E (W)	−33	2.1
Radiation from pool P_{ra} (W)	−10	0.6
Total power transfer (W)	710 ± 120	36.6–52.2

Source: From Quigley M B C, Richards P H, Swift-Hook D T, et al., *J. Phys. D. Appl. Phys.*, 6(18): 2250–2258, 1973.

In conclusion, the net heat input to the anode is given by

$$P_A = I(V_\varphi + V_a + V_{th}) + P_{cd} + P_{cv} + P_R - P_E - P_{ra} \qquad (2.14)$$

The various contributions are summarized in Table 2.2. It is seen that of a total arc power of 1,600 W, only 710 ± 120 W are calculated to be deposited in the anode. This represents an energy transfer efficiency of between 36.6% and 52.2%.

2.2.2 Integrative Calculation-Measurement

Guan et al. proposed the method of integrative calculation-measurement to estimate the value of weld thermal efficiency.[113] It combines the experimental measurement of the process parameters and weld geometry with calculation of the melting efficiency η_m and then determines the value of weld thermal efficiency η. The work was based on TIG welding of thin sheet, taking the arc as a line heat source. The estimated values of η for different materials are as follows:

Aluminum alloys: $\eta = 0.41$
Stainless steel: $\eta = 0.61$
Pure titanium: $\eta = 0.58$
Titanium alloy: $\eta = 0.62$
Mild steel: $\eta = 0.67$

2.2.3 Calorimetric Measurement

Giedt et al. used the experimental system shown in Figure 2.1 to determine the arc weld thermal efficiency.[114] A test workpiece was placed on the bottom

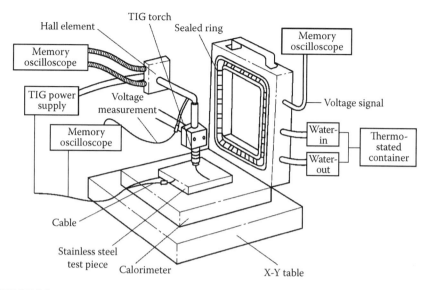

FIGURE 2.1
Schematic of experimental apparatus for GTAW thermal efficiency measurement. (From Giedt W H, Tallerico L N, and Fuerschbach P W, *Weld. J.*, 68(1): 28s–32s, 1989.)

of the calorimeter and connected to the cable lead as illustrated. An arc was initiated and a bead-on-plate weld formed as the workpiece and calorimeter were moved under the GTAW torch. The torch was quickly withdrawn and the calorimeter lid closed. The workpiece then was cooled as heat was transformed to cooling water flowing through passages under the outer surface of the calorimeter. The total heat transferred to the workpiece was determined by integrating the calorimeter signal over the time required for the workpiece to cool to its initial temperature.

The calorimeter used has internal dimensions of 150 × 150 × 75 mm. This instrument operates on the gradient layer principle. The temperature drop through a thin layer of material produced by the heat flow (Figure 2.2) was measured during formation of the hot and cold junctions of thermocouple circuits on the inner and outer surfaces of this layer. These circuits were connected in series to form a thermopile, which multiplies the thermoelectric output. The combination of the thermopile and gradient layer forms a heat rate meter based on the Seebeck thermoelectric effect. Since such heat rate meters are installed in all sides, the instrument is called a Seebeck envelop calorimeter. The heat rate meters are connected in series so that the calorimeter output is a single direct current (DC) signal. This signal is multiplied by a calibration constant, which accounts for the thickness and the thermal conductivity of the gradient layer material and the total wall area of the instrument.

Workpieces of 304L stainless steel 12.7 mm thick were mounted on the calorimeter base. With the lid open, a DCEN bead-on-plate weld 75 mm long was made. The lid was immediately closed over the workpiece, and

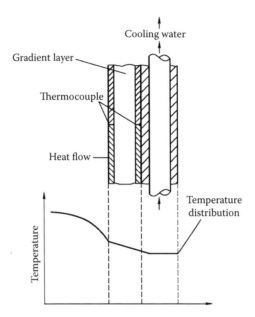

FIGURE 2.2
Operating principle of a gradient layer calorimeter. (From Giedt W H, Tallerico L N, and Fuerschbach P W, *Weld. J.*, 68(1): 28s–32s, 1989.)

the output voltage from the calorimeter was measured and recorded on a digital storage oscilloscope. Up to 6 hours was required for the workpiece to come to equilibrium with the constant-temperature cooling water flowing through the passages near the outer calorimeter surfaces. Integration of the output voltage over the cooling time multiplied by the calibration constant then yielded total energy transferred to the workpiece (listed as total heat input in Table 2.3).

Four test welds were made with a 3.2-mm diameter electrode and four test welds with a 2.4-mm diameter electrode to investigate the possible effect of electrode size. The included tip angles of these 2% thoriated tungsten electrodes were ground to 90°. The electrode-to-workpiece distance was set to 3.0 mm. Argon shielding gas flow rate was 15 L/min.

The welding current was measured with a Hall effect current transducer and recorded on a second digital oscilloscope for the entire weld duration. Arc voltage was also recorded with the same oscilloscope from a terminal located as close as possible to the electrode tip. The current and voltage waveforms were multiplied and integrated over the entire weld time using a data reduction feature of the oscilloscope. This result represents the true energy supplied to the arc and is reported as the total machine output in Table 2.3.

All welds were sectioned at 25 and 50 mm from the start to measure the penetration and the cross-sectional area. The cross-sectional area was

TABLE 2.3

Measured Data of Weld Thermal Efficiency in TIG

Arc Current (A)	Arc Voltage (V)	Rated Power Input (kW)	Total Machine Output (kJ)	Total Energy Entering the Plate (kJ)	Weld Thermal Efficiency	Weld Penetration (mm)	Cross-Sectional Area (mm^2)	Heat Used to Melt (kJ)	Melting Efficiency
100	9.77	0.977[b]	81.4	65.4	0.80	1.85	5.99	3.98	0.06
104	8.65	0.900[a]	79.6	64.7	0.81	1.72	4.54	3.66	0.06
152	10.2	1.54[b]	129	104	0.81	2.97	17.2	11.4	0.11
155	8.87	1.38[a]	122	103	0.84	3.00	16.9	11.2	0.11
200	10.3	2.95[b]	173	142	0.82	4.09	33.3	22.1	0.15
205	8.87	1.81[a]	162	131	0.81	3.71	28.6	19.0	0.15
204	9.60	1.96[a,c]	116	95.6	0.82	3.10	27.2	18.1	0.19

Source: From Giedt W H, Tallerico L N, and Fuerschbach P W, *Weld. J.,* 68(1); 28s–32s, 1989.

Welding speed $v_0 = 0.84$ mm/s.

[a] Electrode diameter = 2.4 mm.

[b] Electrode diameter = 3.2 mm.

[c] Welding speed = 1.27 mm/s.

measured from transverse weld micrographs with a planimeter. The average of the two areas was multiplied by the weld length to determine the total fused volume.

2.2.4 Determining η through Combining Theory Model with Temperature Measurement

In addition to the three methods to estimate η mentioned, there is another approach. First, develop a model for calculating the temperature field. With an initial estimated value of η, the model is used to calculate the temperature at some points on the workpiece. Then, compare the calculated temperature distribution with the measured one. If both do not match each other, an adjusted value of η is selected; the procedure is repeated until the calculated and measured temperature agree with each other within a prescribed range. Finally, the employed value of η is taken as the weld thermal efficiency for the situation.

If different models of welding temperature field are used, the values of η determined in this way are naturally different. Figure 2.3 shows the results obtained by Christensen et al.,[115] Nile et al.,[116] Eagar et al.,[6] and Glickstein and Friedman.[117,118]

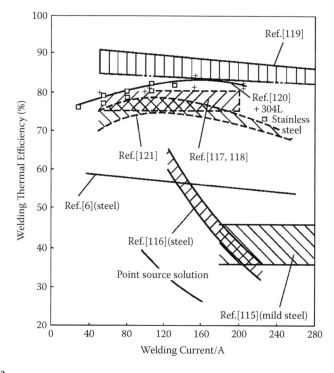

FIGURE 2.3
The weld thermal efficiency for TIG welding. (From Giedt W H, Tallerico L N, and Fuerschbach P W, *Weld. J.*, 68(1): 28s–32s, 1989.)

2.3 Deposition Mode of Welding Heat Sources

According to different action modes, welding heat sources can be treated as centralized, planar distribution, and volumetric distribution ones. If the examined point is far away from the weld centerline, the heat source can be approximated as a centralized point source. For arc welding, the heat flux from the arc is distributed in a spot area, which can be treated as a plane source. But, for high-density power beam welding, the larger depth-width ratio of the weld indicates that the heat intensity is distributed along the plate thickness direction, so that a volumetric distribution must be used.

2.4 Centralized Heat Source

The so-called centralized heat source takes the welding arc as a point or line source. The heat is released in an infinite medium either along a line (biaxial conduction) or at a point (triaxial conduction). A *line source* corresponds to a wide plate with finite thickness and infinite width and length. A *point source* corresponds to an infinite body. The analytical approaches of the weld thermal process (i.e., Rosenthal-Rykalin formulae) were based on the centralized sources. It is clear that centralized sources are a simplification of practical cases. For welding on the surface of a heavy slab, the arc can be seen as a point source, while it can be taken as a line source across the thickness for butt welding of a thin plate. If two long rods are welded in a butt joint, the heat source can be treated as a plane source across the cross section of the rods.

2.5 Planar-Distributed Heat Source

2.5.1 Gaussian Surface Flux Distribution

The transfer of energy from the arc to the workpiece is through a spot, that is, the heating spot on the workpiece surface. As shown in Figure 2.4, define the spot radius as $r_H = 0.5d_h$. The physical meaning of spot radius r_H is as follows: 95% of the transferred energy to the workpiece from the arc is distributed within a spot of radius r_H. The thermal flux within the spot is usually assumed to have a Gaussian distribution,

$$q(r) = q_m \exp(-Kr^2) \tag{2.15}$$

where $q(r)$ is the surface flux at radius r, q_m is the maximum flux at the center of the heat source, and K is the concentration coefficient. Since the total

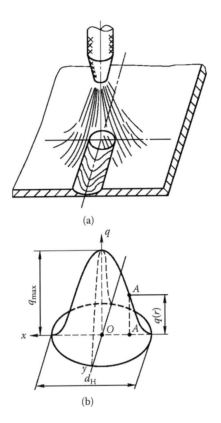

FIGURE 2.4
The heat flux distribution on a spot: (a) the arc action on the plate; (b) Gaussian distribution model.

energy on the workpiece surface is equal to the effective power Q of the arc, we have

$$Q = \int_0^\infty q(r)2\pi r\,dr = q_m \int_0^\infty \exp(-Kr^2)2\pi r\,dr = \frac{q_m\pi}{-K}\int_0^\infty \exp(-Kr^2)d(-Kr^2)$$

$$= \frac{q_m\pi}{K}$$

$$q_m = \frac{QK}{\pi} \tag{2.16}$$

Substituting Eq. (2.16) into Eq. (2.15), we obtain

$$q(r) = \frac{QK}{\pi}\exp(-Kr^2) \tag{2.17}$$

where $Q = \eta I U_a$ is the effective arc power.

The definition of the heating spot of radius r_H requires that

$$95\% Q = \int_0^{r_H} q(r)2\pi r dr \tag{2.18}$$

Substituting Eq. (2.17) into Eq. (2.18):

$$0.95Q = \int_0^{r_H} \frac{QK}{\pi} \exp(-Kr^2)2\pi r dr = Q\int_{r_H}^0 \exp(-Kr^2)d(-Kr^2)$$

$$= Q\exp(-Kr^2)\Big|_{r_H}^0 = Q[1-\exp(-Kr_H^2)]$$

After manipulation,

$$0.95 = 1-\exp(-Kr_H^2), \quad \exp(-Kr_H^2) = 0.05, \; -Kr_H^2 = \ln 0.05 = -3$$

$$K = \frac{3}{r_H^2} \tag{2.19}$$

Substituting Eq. (2.19) into Eq. (2.17), we obtain the form of Gaussian heat source usually used in the literature:

$$q(r) = \frac{3Q}{\pi r_H^2} \exp\left(-\frac{3r^2}{r_H^2}\right) \tag{2.20}$$

There is another form of Gaussian heat source in the literature:

$$q(r) = \frac{Q}{2\pi\sigma_q^2} \exp\left(-\frac{r^2}{2\sigma_q^2}\right) \tag{2.21}$$

where σ_q is the distribution parameter, and $q(r)$ is the heat flux at a point with a distance r away from the center.

To get the correlation of K and σ_q, we substitute Eq. (2.21) into Eq. (2.18):

$$0.95Q = \int_0^{r_H} \frac{Q}{2\pi\sigma_q^2} \exp\left(-\frac{r^2}{2\sigma_q^2}\right)2\pi r dr = Q\int_{r_H}^0 \exp\left(-\frac{r^2}{2\sigma_q^2}\right)d\left(-\frac{r^2}{2\sigma_q^2}\right)$$

$$= Q\exp\left(-\frac{r^2}{2\sigma_q^2}\right)\Big|_{r_H}^0 = Q\left[1-\exp\left(-\frac{r_H^2}{2\sigma_q^2}\right)\right]$$

After manipulation,

$$\exp\left(-\frac{r_H^2}{2\sigma_q^2}\right) = 0.05$$

$$r_H^2 = 6\sigma_q^2 \tag{2.22}$$

Thus, three parameters r_H, K, and σ_q describe the heat flux distribution in the heating spot with different concepts, and they have the following relationship:

$$\frac{1}{2\sigma_q^2} = K = \frac{3}{r_H^2} \tag{2.23}$$

2.5.2 Double-Ellipse Surface Distribution

A Gaussian heat source is a symmetric surface distribution with respect to the spot center, so it needs only one parameter (r_H or K or σ_q) to define the heat flux distribution. In fact, the heat flux distribution is asymmetric in the spot due to movement of the heat source along the welding direction. Because of the effect of welding speed, the heating region ahead of the arc is less than that behind the arc. The heating spot is not round but elliptical, and the ellipse size for the front and rear sections is different, as shown in Figure 2.5.

The heat flux inside the front half-ellipse ahead of the arc can be written as

$$q_f(x,y) = q_{mf} \exp(-Ax^2 - By^2) \tag{2.24}$$

where q_{mf} is the maximum value, and A and B are the elliptic distribution parameters.

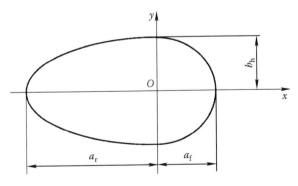

FIGURE 2.5
Schematic of double-ellipse distribution.

The heat flux inside the rear half-ellipse behind the arc can be written as

$$q_r(x,y) = q_{mr} \exp(-A_1 x^2 - B_1 y^2) \tag{2.25}$$

where q_{mr} is the maximum value, and A_1 and B_1 are the elliptic distribution parameters.

The total heat amount in the front section is determined as follows:

$$Q_f = 2\int_0^\infty\int_0^\infty q_{mf} \exp(-Ax^2 - By^2)dxdy = 2q_{mf}\int_0^\infty \exp(-Ax^2)dx\int_0^\infty \exp(-By^2)dy$$

Employing the improper integral,

$$\int_0^\infty \exp(-u^2)du = \frac{\sqrt{\pi}}{2}$$

$$\int_0^\infty \exp(-Ax^2)dx = \frac{1}{\sqrt{A}}\int_0^\infty \exp\left[-(\sqrt{A}x)^2\right]d(\sqrt{A}x) = \frac{1}{\sqrt{A}}\frac{\sqrt{\pi}}{2}$$

We have

$$Q_f = 2q_{mf}\left(\frac{1}{\sqrt{A}}\frac{\sqrt{\pi}}{2}\right)\left(\frac{1}{\sqrt{B}}\frac{\sqrt{\pi}}{2}\right) = q_{mf}\frac{\pi}{2\sqrt{AB}}$$

$$q_{mf} = Q_f\frac{2\sqrt{AB}}{\pi} \tag{2.26}$$

As shown in Figure 2.5, the semiaxes of the front half-ellipse are a_f and b_h. Assuming that 95% of the total thermal energy transferred to the workpiece is deposited in the double-ellipse region, we have

$$q_f(0,b_h) = q_{mf}\exp(-Bb_h^2) = 0.05q_{mf}$$

$$B = \frac{3}{b_h^2} \tag{2.27}$$

Similarly,

$$q_f(a_f,0) = q_{mf}\exp(-Aa_f^2) = 0.05q_{mf}$$

$$A = \frac{3}{a_f^2} \tag{2.28}$$

Substituting Eqs. (2.26), (2.27), and (2.28) into Eq. (2.24), we get the heat density distribution inside the front half-ellipse:

$$q_f(x,y) = \frac{6Q_f}{\pi a_f b_h} \exp\left(-\frac{3x^2}{a_f^2} - \frac{3y^2}{b_h^2}\right) \tag{2.29}$$

Similarly, we get the heat density distribution inside the rear half-ellipse:

$$q_r(x,y) = \frac{6Q_r}{\pi a_r b_h} \exp\left(-\frac{3x^2}{a_r^2} - \frac{3y^2}{b_h^2}\right) \tag{2.30}$$

where

$$Q = \eta I U_a = Q_f + Q_r, \quad Q_f = \frac{a_f}{a_f + a_r} Q, \quad Q_r = \frac{a_r}{a_f + a_r} Q \tag{2.31}$$

If $a_f = a_r = b_h = r_H$, then $Q_f = Q_r = Q/2$, and Eqs. (2.29) and (2.30) will be transformed to Eq. (2.20). That means that the double-ellipse distribution turns out to be a Gaussian distribution if the semiaxes of both front and rear ellipses are equal to each other.

2.6 Volumetric Heat Sources

For high power density welding, the heat density of the welding heat source acts not only on the workpiece surface, but also in the direction of thickness. It must be taken as a volumetric distribution. To consider its action along the workpiece thickness, an ellipsoid mode can be used.[122]

2.6.1 Semiellipsoid Power Density Distribution

As shown in Figure 2.6, the semiaxes of the semiellipsoid are set as a_h, b_h, c_h parallel to coordinate axes x, y, z, respectively. The distribution of the heat density in such a semiellipsoid with center at (0, 0, 0) can be written as

$$q(x,y,z) = q_m \exp(-Ax^2 - By^2 - Cz^2) \tag{2.32}$$

where q_m is the maximum value of the power density at the center of the ellipsoid, and A, B, C are the distribution parameters.

Since the heat density is distributed inside the semiellipsoid, we have

$$Q = \eta I U_a = 4 \int_0^\infty \int_0^\infty \int_0^\infty q(x,y,z) dx dy dz$$

$$C = \frac{3}{c_h^2} \tag{2.36}$$

Substituting Eqs. (2.33) to (2.36) into Eq. (2.32), we obtain the semiellipsoid distribution

$$q(x,y,z) = \frac{6\sqrt{3}Q}{a_h b_h c_h \pi \sqrt{\pi}} \exp\left(-\frac{3x^2}{a_h^2} - \frac{3y^2}{b_h^2} - \frac{3z^2}{c_h^2} \right) \tag{2.37}$$

2.6.2 Double-Ellipsoid Power Density Distribution

Considering the effect of welding speed, the power density should be asymmetrically distributed ahead of the arc and behind it. Thus, the double-ellipsoid volume is taken, as shown in Figure 2.7.

The front half of the heat source is the quadrant of one ellipsoidal source, and the rear half is the quadrant of another ellipsoid. The semiaxes of the double ellipsoid are set as (a_h, a_r, b_h, c_h). The fractions f_f and f_r of the heat deposited in the front and rear quadrants, respectively, are needed. By using Eq. (2.37), the heat density distribution inside the front and rear quadrants of the heat source can be written respectively as

$$q_f(x,y,z) = \frac{6\sqrt{3}(f_f Q)}{a_f b_h c_h \pi \sqrt{\pi}} \exp\left(-\frac{3x^2}{a_f^2} - \frac{3y^2}{b_h^2} - \frac{3z^2}{c_h^2} \right), \quad x \geq 0 \tag{2.38}$$

$$q_r(x,y,z) = \frac{6\sqrt{3}(f_r Q)}{a_r b_h c_h \pi \sqrt{\pi}} \exp\left(-\frac{3x^2}{a_r^2} - \frac{3y^2}{b_h^2} - \frac{3z^2}{c_h^2} \right), \quad x < 0 \tag{2.39}$$

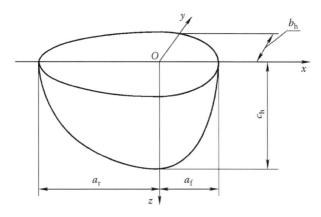

FIGURE 2.7
Schematic of double ellipsoid.

For the front part (half of semiellipsoid), the heat deposition is

$$2\int_0^\infty\int_0^\infty\int_0^\infty q_f(x,y,z)dxdydz = 2\times\frac{6\sqrt{3}f_fQ}{a_fb_hc_h\pi\sqrt{\pi}}\int_0^\infty\exp\left(-\frac{3x^2}{a_f^2}\right)dx$$

$$\times\int_0^\infty\exp\left(-\frac{3y^2}{b_h^2}\right)dy\int\exp\left(-\frac{3z^2}{c_h^2}\right)dz$$

$$= 2\times\frac{6\sqrt{3}\,f_fQ}{a_fb_hc_h\pi\sqrt{\pi}}\times\frac{a_f}{\sqrt{3}}\frac{\sqrt{\pi}}{2}\times\frac{b_h}{\sqrt{3}}\frac{\sqrt{\pi}}{2}\times\frac{c_h}{\sqrt{3}}\frac{\sqrt{\pi}}{2}=\frac{1}{2}(f_fQ)$$

Similarly, for the rear part (half of another semiellipsoid), the deposited heat is

$$2\int_0^\infty\int_0^\infty\int_0^\infty q_r(x,y,z)dxdydz=\frac{1}{2}(f_rQ)$$

Since

$$\eta IU_a = Q=\frac{1}{2}(f_fQ)+\frac{1}{2}(f_rQ)=\frac{1}{2}Q(f_f+f_r)$$

then

$$f_f + f_r = 2,\quad Q=\eta IU_a \tag{2.40}$$

2.6.3 Other Modes of Volumetric Heat Source

Beside the volumetric heat source modes mentioned, there are other modes of volumetric sources applicable to describe high-density energy welding processes (laser beam welding, electron beam welding, and plasma arc welding), such as conic body source, rotational body based on curves, and so on. These heat source modes are discussed in Chapter 9.

3

Analytical Approaches to Weld Thermal Process Calculation

3.1 Introduction

In the 1940s, Rosenthal and Rykalin developed a set of formulae to calculate the weld thermal process by means of analytical solution.[2-5] However, they did not give a detailed derivation of the basic equations for an instantaneous centralized heat source and just demonstrated indirect proof. This chapter provides direct derivation details to lay a solid foundation for analytic solutions, introduces a main improvement on classical analytic formulae and their dimensionless version, and gives the modified version as well.

3.2 Mathematical Description of Thermal Conduction

When a temperature gradient exists in a body, experience has shown that there is an energy transfer from the high-temperature region to the low-temperature region. We say that the energy is transferred by *conduction*. The heat flux q is defined as the thermal energy transferred through unit area in unit time. The unit of heat flux is watts per square meter. The heat flux is proportional to the normal temperature gradient. Fourier's law of heat conduction is used to describe the correlation between the heat flux and the temperature gradient. For the three-dimensional case, it may be expressed as

$$\vec{q} = -\lambda \operatorname{grad} T = -\lambda \frac{\partial T}{\partial n}\vec{n} = -\lambda \left(\frac{\partial T}{\partial x}\vec{i} + \frac{\partial T}{\partial y}\vec{j} + \frac{\partial T}{\partial z}\vec{k} \right) \tag{3.1}$$

where λ is the thermal conductivity of the material [W/(m K)], T is the temperature (K), grad T is the temperature gradient (K/m), $\vec{n}$ is the unit normal

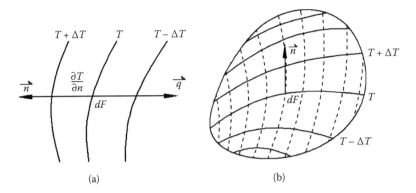

FIGURE 3.1
Schematic of temperature gradient, heat flux, and isothermal.

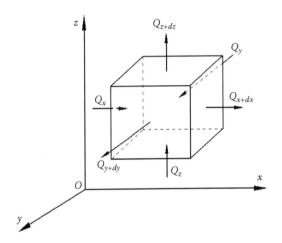

FIGURE 3.2
Elemental volume for three-dimensional heat conduction analysis.

vector, $\partial T / \partial n$ is the derivative of temperature in the direction $\vec{n}$, (x, y, z) are the coordinates, and $\vec{i}$, $\vec{j}$, and $\vec{k}$ are the unit vectors along the x-, y-, z-axes, respectively. The minus sign is inserted so that the second principle of thermodynamics will be satisfied; that is, heat must flow from the high-temperature region to the low-temperature region. Figure 3.1(a) shows the relationship between the temperature gradient and heat flux vector $\vec{q}$; Figure 3.1(b) shows the relation of isothermal and heat flux direction.

We now set ourselves the problem of determining the basic equation that governs the heat conduction transfer in a solid, using Eq. (3.1) as a starting point. Consider the elemental volume shown in Figure 3.2. Heat flux in any direction can be divided into the components along the x-, y-, z-axes. For a unit volume, the heat conducted into it at the left face during the period dt in the x-direction is

$$Q_x = q_x dy dz dt$$

The heat conducted out of the volume at the right face during the period dt in the x-direction is

$$Q_{x+dx} = (q_x + dq_x)dy dz dt$$

Thus, the heat variation in the volume in the x-direction is given by

$$dQ_x = Q_x - Q_{x+dx} = -dq_x dy dz dt = -\frac{\partial q_x}{\partial x} dx dy dz dt$$

Similarly, the heat variation in the y-direction and in the z-direction, respectively, is

$$dQ_y = -\frac{\partial q_y}{\partial y} dx dy dz dt$$

$$dQ_z = -\frac{\partial q_z}{\partial z} dx dy dz dt$$

The total heat saved in the volume is written as

$$dQ = dQ_x + dQ_y + dQ_z = -\left(\frac{\partial q_x}{\partial x} + \frac{\partial q_y}{\partial y} + \frac{\partial q_z}{\partial z}\right) dx dy dz dt$$

Substituting Fourier's law into this equation,

$$dQ = -\left[\frac{\partial}{\partial x}\left(-\lambda \frac{\partial T}{\partial x}\right) + \frac{\partial}{\partial y}\left(-\lambda \frac{\partial T}{\partial y}\right) + \frac{\partial}{\partial z}\left(-\lambda \frac{\partial T}{\partial z}\right)\right] dx dy dz dt \qquad (3.2)$$

During the time interval dt, the heat amount dQ is gathered due to heat conduction in the unit volume of $dxdydz$. If the material in the unit volume has specific heat capacity ρC_p [J/(m³ K)] and its temperature is increased by the amount $dT = (\partial T/\partial t)dt$, then we have

$$dQ = \rho C_p \frac{\partial T}{\partial t} dt \cdot dx dy dz \qquad (3.3)$$

where $\partial T/\partial t$ is the transient varying rate for a given point, ρ is the density of the material (kg/m³), and C_p is the specific heat of the material [J/(kg K)]. Equation (3.2) should be equal to Eq. (3.3). After simplifying, we obtain

$$\rho C_p \frac{\partial T}{\partial t} = \frac{\partial}{\partial x}\left(\lambda \frac{\partial T}{\partial x}\right) + \frac{\partial}{\partial y}\left(\lambda \frac{\partial T}{\partial y}\right) + \frac{\partial}{\partial z}\left(\lambda \frac{\partial T}{\partial z}\right) \tag{3.4}$$

Generally, ρC_p and λ are dependent on the coordinates (x, y, z) and the temperature T. For constant thermal properties (ρC_p and λ are constants), Eq. (3.4) is written as

$$\frac{\partial T}{\partial t} = \frac{\lambda}{\rho C_p}\left(\frac{\partial^2 T}{\partial x^2} + \frac{\partial^2 T}{\partial y^2} + \frac{\partial^2 T}{\partial z^2}\right) = a\nabla^2 T \tag{3.5}$$

where $a = (\lambda/\rho C_p)$ is called the thermal diffusivity of the material (m²/s). It is the ability of temperature values at various points in a body to be uniform when the body experiences heating or cooling. The larger the value of a, the faster heat will diffuse through the material.

Equation (3.4) is the generalized form of heat conduction, which describes the transient temperature field $T(x, y, z, t)$. If $(\partial T/\partial t) = 0$, temperature is a function of coordinates but not time; that is, if it does not change with time, then it is the steady-state temperature field.

Equation (3.5) may be transformed into either cylindrical or spherical coordinates by standard calculus techniques. The results are as follows:

Cylindrical coordinates:

$$\frac{\partial T}{\partial t} = a\left(\frac{\partial^2 T}{\partial r^2} + \frac{1}{r}\frac{\partial T}{\partial r} + \frac{1}{r^2}\frac{\partial^2 T}{\partial \theta^2} + \frac{\partial^2 T}{\partial z^2}\right) \tag{3.6}$$

Spherical coordinates:

$$\frac{\partial T}{\partial t} = a\left[\frac{1}{r^2}\frac{\partial}{\partial r}\left(r^2\frac{\partial T}{\partial r}\right) + \frac{1}{r^2\sin\theta}\frac{\partial}{\partial \theta}\left(\sin\theta\frac{\partial T}{\partial \theta}\right) + \frac{1}{r^2\sin^2\theta}\frac{\partial^2 T}{\partial \varphi^2}\right] \tag{3.7}$$

Solving a problem of heat conduction in a transient situation requires the boundary conditions and the initial condition. The former define the temperature or heat transfer situation at the boundary, while the latter gives the temperature distribution in a body at the initial moment. The general heat conduction equation must be combined with boundary conditions and initial condition to describe a specific heat conduction case completely. For a steady-state case, there are only boundary conditions without an initial condition.

There are three kinds of boundary conditions:

1. In the first kind of boundary condition, the temperature is known at the boundary:

$$T_s = T_s(x,y,z,t) \tag{3.8}$$

A special case is when the temperature is constant at the boundary, that is, the isothermal boundary condition.

2. In the second kind of boundary condition, the heat density is known at the boundary:

$$q_s = q_s(x,y,z,t) \tag{3.9}$$

A special case is the adiabatic boundary, $q_s = (\partial T/\partial n)|_s = 0$.

3. In the third kind of boundary condition, the coefficient α of heat transfer due to convection and radiation and the ambient temperature T_f are known:

$$-\lambda \frac{\partial T}{\partial n}\bigg|_s = \alpha(T_s - T_f) \tag{3.10}$$

If the ambient temperature is constant, then Eq. (3.10) can be written as

$$T_s - T_f = -\frac{1}{\alpha/\lambda} \frac{\partial T}{\partial n}\bigg|_s$$

Therefore, there are two special cases: (a) if $\alpha \to \infty$ or $\lambda \to 0$, that is, $(\alpha/\lambda) \to \infty$, then $T_s = T_f$; this is the isothermal boundary condition. It corresponds to a case with very small thermal conductivity and a very large heat loss coefficient, so the body surface temperature approaches the ambient temperature. (b) If $\alpha \to 0$ or $\lambda \to \infty$, that is, $(\alpha/\lambda) \to 0$, then $(\partial T/\partial n)|_s \to 0$; this is the adiabatic boundary. It is a case with very large thermal conductivity and a very small heat loss coefficient, so the heat flux passing through the body surface is near zero.

3.3 Heat Conduction in an Infinite Body

For an infinite body, the effect of boundary conditions can be neglected because the boundary is very far away from the heat source. Since there is only the initial condition, the problem is simplified. The heat conduction in such a case may be written as

$$
\begin{cases}
\dfrac{\partial T}{\partial t} = a\left(\dfrac{\partial^2 T}{\partial x^2} + \dfrac{\partial^2 T}{\partial y^2} + \dfrac{\partial^2 T}{\partial z^2}\right) \\[2ex]
T\big|_{t=0} = \varphi(x,y,z)
\end{cases}
\tag{3.11}
$$

3.3.1 Instantaneous and Centralized Heat Source

The instantaneous heat sources are based on the assumption that the heat is released instantaneously at time $t = 0$ in an infinite medium of initial temperature T_0.

3.3.1.1 Instantaneous Point Source

Assume that the initial temperature of an infinite body is zero ($T_0 = 0$), and there is a centralized source with energy Q (J) in an elementary volume $dxdydz$. The coordinate system (x, y, z) is set up to take the center of volume as the origin. First, consider a case in which the volume temperature is unitary at $t = 0$. The problem may be expressed as

$$
\begin{cases}
\dfrac{\partial T}{\partial t} = a\left(\dfrac{\partial^2 T}{\partial x^2} + \dfrac{\partial^2 T}{\partial y^2} + \dfrac{\partial^2 T}{\partial z^2}\right) \\[2ex]
T\big|_{t=0} = \delta(x,y,z)
\end{cases}
\tag{3.12}
$$

where $\delta(x, y, z)$ is the delta function.

By employing Fourier transformation,

$$
F[T(x,y,z,t)] = \tilde{T}(\lambda_1,\lambda_2,\lambda_3,t)
$$

Equation (3.12) becomes

$$
\begin{cases}
\dfrac{d\tilde{T}}{dt} = -a\left(\lambda_1^2 + \lambda_2^2 + \lambda_3^2\right)\tilde{T} \\[2ex]
\tilde{T}\big|_{t=0} = 1
\end{cases}
\tag{3.13}
$$

The solution of Eq. (3.13) is as follows:

$$
\tilde{T}(\lambda_1,\lambda_2,\lambda_3,t) = e^{-a\left(\lambda_1^2+\lambda_2^2+\lambda_3^2\right)t}
$$

Through inverse transformation, since

$$F^{-1}\left[e^{-b^2\lambda_4^2 t}\right] = \frac{1}{2\pi}\int_{-\infty}^{+\infty} e^{-b^2\lambda_4^2 t} \cdot e^{-i\lambda_4 x}d\lambda_4 = \frac{1}{2b\sqrt{b\pi t}}e^{-\frac{x^2}{4b^2 t}}$$

thus

$$F^{-1}\left[\tilde{T}(\lambda_1,\lambda_2,\lambda_3,t)\right] = \frac{1}{(2\pi)^3}\int_{-\infty}^{+\infty}\int_{-\infty}^{+\infty}\int_{-\infty}^{+\infty} e^{-a(\lambda_1^2+\lambda_2^2+\lambda_3^2)t} \times e^{-i(\lambda_1 x+\lambda_2 y+\lambda_3 z)}d\lambda_1 d\lambda_2 d\lambda_3$$

$$= \left[\frac{1}{2\pi}\int_{-\infty}^{+\infty} e^{-a\lambda_1^2 t} \cdot e^{-i\lambda_1 x}d\lambda_1\right] \times \left[\frac{1}{2\pi}\int_{-\infty}^{+\infty} e^{-a\lambda_2^2 t} \cdot e^{-i\lambda_2 y}d\lambda_2\right] \times \left[\frac{1}{2\pi}\int_{-\infty}^{+\infty} e^{-a\lambda_3^2 t} \cdot e^{-i\lambda_3 z}d\lambda_3\right]$$

$$= \frac{1}{\left(2\sqrt{a\pi t}\right)^3} \cdot e^{-\frac{x^2+y^2+z^2}{4at}}$$

Finally,

$$F^{-1}\left[\tilde{T}(\lambda_1,\lambda_2,\lambda_3,T)\right] = \frac{1}{\left(4a\pi t\right)^{3/2}}e^{-\frac{x^2+y^2+z^2}{4at}}$$

The temperature field can be written as

$$T(x,y,z,t) = \frac{1}{\left(4\pi at\right)^{3/2}}e^{-\frac{x^2+y^2+z^2}{4at}} \tag{3.14}$$

Equation (3.14) is based on the assumption of unitary temperature in the volume. If there is a centralized source with energy Q (J) in an elementary volume at time $t = 0$, its temperature is $Q/\rho C_p$ (°C), then the solution is

$$T(x,y,z,t) = \frac{Q}{\left(4\pi at\right)^{3/2}\rho C_p}\exp\left(-\frac{x^2+y^2+z^2}{4at}\right)$$

Let $R^2 = x^2 + y^2 + z^2$, which is the square of the distance from the origin to a point (x, y, z); the temperature field in an infinite body under action of the instantaneous point source can be calculated by

$$T(R,t) = \frac{Q}{\rho C_p\left(4\pi at\right)^{3/2}}\exp\left(-\frac{R^2}{4at}\right) \tag{3.15}$$

In this case, isothermal curves are a series of spheres with radius R.

3.3.1.2 Instantaneous Line Source

Assume that the initial temperature of an infinite plate is zero ($T_0 = 0$), and there is a centralized line source with linear intensity Q_1 (J/m) in an elementary volume (prismoid, its bottom area is $dxdy$, and its height is the plate thickness). First, consider a case in which the volume temperature is unitary at $t = 0$. The problem may be expressed as

$$\begin{cases} \dfrac{\partial T}{\partial t} = a\left(\dfrac{\partial^2 T}{\partial x^2} + \dfrac{\partial^2 T}{\partial y^2} \right) \\[2mm] T\Big|_{t=0} = \delta(x,y) \end{cases}$$

(3.16)

By employing Fourier transformation,

$$F[T(x,y,t)] = \tilde{T}(\lambda_1, \lambda_2, t)$$

We have

$$\begin{cases} \dfrac{d\tilde{T}}{dt} = -a\left(\lambda_1^2 + \lambda_3^2\right)\tilde{T} \\[2mm] \tilde{T}\Big|_{t=0} = 1 \end{cases}$$

Its solution is

$$\tilde{T}(\lambda_1, \lambda_2, t) = e^{-a\left(\lambda_1^2 + \lambda_3^2\right)t}$$

And the inverse transformation is

$$T(x,y,t) = \frac{1}{\left(2\sqrt{a\pi t}\right)^2} e^{-\frac{x^2+y^2}{4at}}$$

Let $r^2 = x^2 + y^2$, which is the square of the distance from the line source oz to the point (x, y). If the centralized linear intensity is Q_1, then the two-dimensional temperature field can be written as

$$T(r,t) = \frac{Q_1}{\rho C_p (4\pi a t)} \exp\left(-\frac{r^2}{4at}\right)$$

(3.17)

In such a case, isothermal curves are a series of cylinders.

3.3.1.3 *Instantaneous Plane Source*

Assume that the initial temperature of an infinite rod is zero ($T_0 = 0$), and there is a centralized plane source with intensity Q_2 (J/m²) in an elementary volume (slice with its height dx). First, consider a case in which the volume temperature is unitary at $t = 0$. The problem may be expressed as

$$\begin{cases} \dfrac{\partial T}{\partial t} = a\dfrac{\partial^2 T}{\partial x^2} \\[2mm] T\Big|_{t=0} = \delta(x) \end{cases} \tag{3.18}$$

By employing Fourier transformation,

$$F\big[T(x,t)\big] = \tilde{T}(\lambda_1, t)$$

because

$$\begin{cases} \tilde{T}\Big|_{t=0} = 1 \\[2mm] \dfrac{d\tilde{T}}{dt} = -a\lambda_1^2\,\tilde{T} \end{cases}$$

Its solution is

$$\tilde{T}(\lambda, t) = e^{-a\lambda_1^2 t}$$

The inverse transformation is

$$T(x,t) = \frac{1}{2\sqrt{a\pi t}}\, e^{-\frac{x^2}{4at}}$$

If the heat inside the volume is Q_2, the one-dimensional temperature field can be written as

$$T(x,t) = \frac{Q_2}{\rho C_p(4\pi a t)^{1/2}}\exp\left(-\frac{x^2}{4at}\right) \tag{3.19}$$

In such a case, isothermal curves are a series of planes.

3.3.2 Principle of Superposition

The principle of superposition states that when a series of heat sources acts on a body, the temperature on the body can be seen as the repeated addition of the temperature produced by every individual heat source. The linearity of the heat conduction equation is the prerequisite for the principle of superposition. It is invalid for two cases when the heat conduction equation is nonlinear: (1) The thermal properties of materials are temperature dependent, and (2) the absorption or releasing of latent heat is considered during phase transformation.

Based on the principle of superposition, (1) the temperature field under continuous action of heat sources can be deduced from that under instantaneous sources because continuous action of heat sources may be taken as the superposition of thermal actions from numerous instantaneous sources at different instants; (2) the temperature field under action of distributed heat sources can be deduced from that under centralized sources because the action of distributed heat sources may be taken as the superposition of thermal actions from numerous centralized sources at different positions.

3.4 Rosenthal-Rykalin's Simplified Approach

It is worth looking briefly at one of the earliest, and still widely used, simplified solutions of the generalized equation of heat conduction, which was developed by Rosenthal and Rykalin.[2–4] The key to their solution is the simplified treatment of both workpiece dimensions and heat sources.

3.4.1 Approach for Modeling of Arc Heating the Workpiece

A suitable approach has to be selected to calculate the weld thermal process in arc welding so that the main characteristics are taken into consideration, some minor factors are neglected, and the calculation approach is simplified.

3.4.1.1 Treatment of the Workpiece

According to the shape and size of the workpiece, it may be classified as follows:

1. *Semi-infinite body*: This occupies the space with positive z-coordinate, as shown in Figure 3.3. Heat flow is three dimensional in this case.
2. *Medium-thickness plate*: This workpiece has a medium thickness H but an infinite length and width, as shown in Figure 3.4. Heat flow is three dimensional in this case.

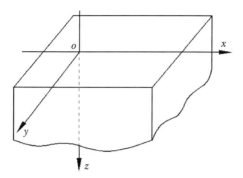

FIGURE 3.3
Schematic of semi-infinite body.

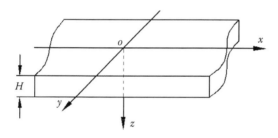

FIGURE 3.4
Schematic of medium-thickness plate.

3. *Thin plate*: This workpiece has thickness so thin that the temperature can be treated as uniform along the thickness direction. Heat flow is two dimensional in this case.

3.4.1.2 Treatment of Heat Sources

With respect to the workpiece shape and size, heat sources may be treated as (1) point source for bead-on-plate welding of thick plates, (2) line source for butt welding of thin plates, and (3) plane source for butt welding of long rods.

According to the action time, heat sources may be divided into the instantaneous source and the continuous deposition source. The former corresponds to the case of very short action of the welding arc, while the latter is for normal welding operations.

Based on the welding speed, heat sources may be categorized as (1) stationary, (2) moving, and (3) fast moving.

3.4.2 Taking Arc as Instantaneous Centralized Heat Sources

3.4.2.1 Instantaneous Point Heat Source

As shown in Figure 3.3, assume that an instantaneous point source acts on the plane *xoy* of a semi-infinite body, the plane *xoy* is adiabatic, and the initial temperature is zero. The heat conduction process when a power Q acts on such a semi-infinite body is equivalent to that when power $2Q$ acts on an infinite body. The temperature field is defined as

$$T(R,t) = \frac{2Q}{\rho C_P (4\pi at)^{3/2}} \exp\left(-\frac{R^2}{4at}\right) \tag{3.20}$$

where $R^2 = x^2 + y^2 + z^2$ (m²); Q is the thermal energy deposited on the semi-infinite body (J); t is time (s); ρC_p is the specific heat capacity [J/(m³ K)]; and a is the thermal diffusivity (m²/s).

3.4.2.2 Instantaneous Line Heat Source

As shown in Figure 3.5, assume that an instantaneous line source acts on the plate of thickness H, the initial temperature is zero, and the source power is $Q_1 = Q/H$ (W/m²). The linear heat source acted on the line segment oo'. The ambient temperature is zero, and the heat loss coefficient of the infinite plate is α. For an elementary volume $\delta dxdy$ with temperature T, the heat released from the top and bottom surfaces of the plate to the surroundings can be written as $dQ = 2\alpha\ Tdxdydt$. Due to this heat releasing, the change rate of the plate temperature is $-\partial T'/\partial t$. Then, we have

$$2\alpha T dxdydt = \rho C_p H dxdy\left(-\frac{\partial T'}{\partial t}\right)dt$$

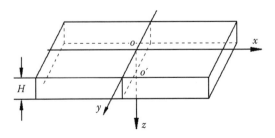

FIGURE 3.5
Schematic of a line heat source on a thin plate.

After adjustment,

$$\frac{\partial T'}{\partial t} = -\frac{2\alpha T dx dy dt}{\rho C_p H dx dy dt} = -\frac{2\alpha T}{\rho C_p H}$$

For a plate with heat releasing from its surfaces, the heat flow equation should be changed as an instantaneous line source acts on the plate:

$$\frac{\partial T}{\partial t} + \frac{\partial T'}{\partial t} = a\nabla^2 T$$

Thus, we have the following equation for the temperature profile on an infinite plate under an instantaneous line source:

$$\frac{\partial T}{\partial t} = a\left(\frac{\partial^2 T}{\partial x^2} + \frac{\partial^2 T}{\partial y^2}\right) - b_c T \tag{3.21}$$

where $b_c = (2\alpha/\rho C_p H)$, the heat loss coefficient (1/s).
   The solution of Eq. (3.21) is as follows:

$$T(r,t) = \frac{Q}{4\pi\lambda H t}\exp\left(-\frac{r^2}{4at} - b_c t\right) \tag{3.22}$$

3.4.3 Weld Thermal Process with a Moving Arc of Constant Power

3.4.3.1 Moving Point Heat Source on a Semi-Infinite Body

As shown in Figure 3.6, take the arc as a continuous action point source with constant power that moves on the surface of a semi-infinite body with a constant speed v_0. A stationary coordinate system $o_0 - x_0 y_0 z_0$ is set up, its origin o_0

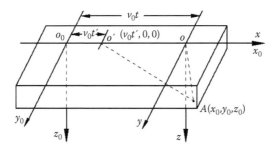

FIGURE 3.6
Moving point heat source on a semi-infinite body.

is the position of the arc starting action at time $t = 0$, o_0x_0 is along the welding direction, and o_0z_0 is along the thickness direction. A moving coordinate system $o - xyz$ is also established; its origin o coincides with the transient position of the arc, the axis ox coincides with o_0x_0, and oy and oz are parallel to o_0y_0 and o_0z_0, respectively. At instant t, the arc locates at the point $o(v_0t, 0, 0)$.

At time t', the arc locates at $o'(v_0t', 0, 0)$. During a very short time interval dt', the amount of heat released at point o' on the surface is $dQ = qdt'$. According to Eq. (3.20), this will produce an infinitesimal rise of temperature in point A (x_0, y_0, z_0) at time t:

$$dT(t') = \frac{2Qdt'}{\rho C_P \left[4\pi a(t-t')\right]^{3/2}} \exp\left[-\frac{(x_0 - v_0t')^2 + y_0^2 + z_0^2}{4a(t-t')}\right], t \geq t' > 0,$$

The total rise of temperature at point A is obtained by integrating this equation during the period $[0, t]$:

$$T(x_0, y_0, z_0, t) - T_0 = \int_0^t \frac{2Q}{\rho C_P \left[4\pi a(t-t')\right]^{3/2}} \exp\left[-\frac{(x_0 - v_0t')^2 + y_0^2 + z_0^2}{4a(t-t')}\right] dt' \quad (3.23)$$

where T_0 is the initial temperature.

In the moving coordinate system, the point A can be expressed as

$$\begin{cases} x = x_0 - v_0t \\ y = y_0 \\ z = z_0 \end{cases} \quad (3.24)$$

Introduce a time variable $t'' = t-t'$, which is the time available for conduction of heat over the distance between o' and A. Substituting Eq. (3.24) into Eq. (3.23), we get an equation for calculating temperature in a moving coordinate system:

$$T(x, y, z, t) - T_0 = \frac{2Q}{\rho C_P (4\pi a)^{3/2}} \exp\left(-\frac{v_0 x}{2a}\right) \int_0^t \frac{1}{t''^{3/2}} \exp\left(-\frac{v_0^2 t''}{4a} - \frac{R^2}{4at''}\right) dt'' \quad (3.25)$$

where $R^2 = x^2 + y^2 + z^2$.

3.4.3.2 Moving Line Heat Source on Plate

As shown in Figure 3.7, the general thin-plate model considers a line source in a wide plate of thickness H and initial temperature T_0. At time $t = 0$, the source starts to move from point o_0 at a constant speed v_0. According to

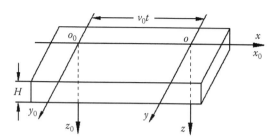

FIGURE 3.7
Moving line heat source on a thin plate.

Eq. (3.22) the elementary source $Qdt' = Q_1H$ released at position $o'(v_0t', 0, 0)$ at time t' will cause a small rise of temperature. After integrating all contributions over the time interval $[0, t]$, we obtain

$$T(x_0, y_0, t) - T_0 = \int_0^t \frac{Q}{4\pi\lambda H(t-t')} \exp\left[-\frac{(x_0 - v_0t')^2 + y_0^2}{4a(t-t')} - b_c(t-t')\right] dt'$$

Transforming it in a moving coordinate system,

$$T(x, y, t) - T_0 = \frac{Q}{4\pi\lambda H} \exp\left(-\frac{v_0 x}{2a}\right) \int_0^t \frac{1}{t''} \exp\left[-\left(\frac{v_0^2}{4a} + b_c\right)t'' - \frac{r^2}{4at''}\right] dt'' \qquad (3.26)$$

where $r^2 = x^2 + y^2$.

3.4.4 Quasi-Steady State

The temperature field related to a moving arc expands initially. After some time, it does not change any more but just moves with the arc itself. The temperature distribution around the moving arc soon becomes a constant. If standing at the arc centerline, an observer does not see any variation of temperature when the arc moves along the welding direction. This is referred to as the *quasi-steady state* of the temperature field.

Theoretically, the weld thermal process turns into a quasi-steady state if the heat source acts for a sufficient length of time, that is, $t \to \infty$.

3.4.4.1 Point Source Heating of a Semi-Infinite Body

If we let $t \to \infty$ in Eq. (3.25), we obtain the equation for calculating the quasi-steady state temperature field on a semi-infinite body under action of a point heat source:

$$T(x, y, z, \infty) - T_0 = \frac{2Q}{\rho C_P (4\pi a)^{3/2}} \exp\left(-\frac{v_0 x}{2a}\right) \int_0^\infty \frac{1}{t''^{3/2}} \exp\left(-\frac{v_0^2 t''}{4a} - \frac{R^2}{4at''}\right) dt''$$

By substituting

$$\frac{R^2}{4at''} = u^2, \quad t'' = \frac{R^2}{4a}\frac{1}{u^2}$$

$$dt'' = \frac{R^2}{4a}\cdot\frac{-2du}{u^3} = -\frac{R^2}{2a}\cdot\frac{1}{u^3}du$$

$$\frac{v_0^2 t''}{4a} = \frac{v_0^2}{4a}\cdot\frac{R^2}{4a}\cdot\frac{1}{u^2} = \left(\frac{v_0 R}{4a}\right)^2\cdot\frac{1}{u^2} = \frac{m^2}{u^2}$$

where

$$m = \frac{v_0 R}{4a}$$

we get

$$T(x,y,z,\infty) - T_0 = \frac{2Q}{\rho C_P(4\pi a)^{3/2}}\exp\left(\frac{v_0 x}{2a}\right)$$

$$\times\int_0^\infty (4a)^{3/2}\left(\frac{u}{R}\right)^3\frac{-R^2}{2a}\frac{du}{u^3}\exp\left(-u^2 - \frac{m^2}{u^2}\right)du$$

$$= \frac{Q}{\rho C_P a\pi^{3/2}R}\exp\left(-\frac{v_0 x}{2a}\right)\int_0^\infty\exp\left(-u^2 - \frac{m^2}{u^2}\right)du$$

$$= \frac{Q}{\lambda R\pi^{3/2}}\exp\left(-\frac{v_0 x}{2a}\right)\cdot\frac{\sqrt{\pi}}{2}e^{-2m}$$

$$= \frac{Q}{2\pi\lambda R}\exp\left(-\frac{v_0 x}{2\alpha}\right)\cdot\exp\left(-\frac{v_0 R}{2a}\right)$$

Hence, the quasi-steady state temperature distribution can be calculated by

$$T(R,x) - T_0 = \frac{Q}{2\pi\lambda R}\exp\left(-\frac{v_0 x}{2a} - \frac{v_0 R}{2a}\right) \tag{3.27}$$

where R is the distance from the heat source to a particular fixed point, $R = \sqrt{x^2 + y^2 + z^2}$. Equation (3.27) is often referred to as the *Rosenthal-Rykalin thick-plate solution*.

3.4.4.2 Line Source on a Thin Plate

If we let $t \to \infty$ in Eq. (3.26), we obtain the equation for calculating the quasi-steady state temperature field on an infinite plate under action of a line heat source:

$$T(x,y,t)-T_0 = \frac{Q}{4\pi\lambda H}\exp\left(\frac{v_0 x}{2a}\right)\int_0^\infty \frac{1}{t''}\exp\left[-\left(\frac{v_0^2}{4a}+b_c\right)t''-\frac{r^2}{4at''}\right]dt''$$

By substituting

$$w = \left(\frac{v_0^2}{4a}+b_c\right)t'', \quad dw = \left(\frac{v_0^2}{4a}+b_c\right)dt'', \quad \frac{dw}{w}=\frac{dt''}{t''}$$

$$\frac{r^2}{4at''}=\frac{r^2}{4a}\frac{\frac{v_0^2}{4a}+b_c}{w}=\frac{1}{4}\frac{\left(\frac{v_0^2}{4a^2}+\frac{b_c}{a}\right)r^2}{w}$$

and letting $u^2 = r^2\left(\dfrac{v_0^2}{4a^2}+\dfrac{b_c}{a}\right)$

$$\frac{r^2}{4at''}=\frac{1}{4}\frac{u^2}{w}$$

Thus, we obtain

$$T(r,x)-T_0 = \frac{Q}{4\pi\lambda H}\exp\left(-\frac{v_0 x}{2a}\right)\int_0^\infty \frac{dw}{w}\exp\left(-w-\frac{u^2}{4w}\right)$$

where

$$\int_0^\infty \frac{dw}{w}\exp\left(-w-\frac{u^2}{4w}\right)=2K_0(u)$$

$K_0(u)$ is the modified Bessel function of the second kind and zero order.

Hence, the quasi-steady state temperature field on a thin plate under action of a line source can be written as

$$T(r,x)-T_0 = \frac{Q}{2\pi\lambda H}\exp\left(-\frac{v_0 x}{2a}\right)K_0\left(r\sqrt{\frac{v_0^2}{4a^2}+\frac{b_c}{a}}\right) \qquad (3.28)$$

The value of $K_0(u)$ can be found in specific mathematical data tables and can be calculated by the following means:

$$K_0(u) = -\left[\left(\ln\frac{u}{2} + 0.57722\right)I_0(u) - \frac{2}{1}I_2(u) - \frac{2}{2}I_4(u) - \frac{2}{3}I_6(u) - \frac{2}{4}I_8(u) - \cdots\cdots\right]$$

(3.29)

where $I_n(u) = \sum_{k=0}^{\infty}((u/2)^{2k+n}/k!(k+n)!)$, modified Bessel function of the first kind and n order. The Bessel function has the following features:

$$\frac{\partial}{\partial x}K_0(u) = -K_1(u)$$

(3.30)

$K_1(u)$ is a modified Bessel function of the second kind and first order. A graphical representation of $K_0(u)$ and $K_1(u)$ is shown in Figure 3.8. $K_0(u)$ can be simplified as follows:

For a larger value of u ($u > 1$),

$$K_0(u) \approx \sqrt{\frac{\pi}{2u}}\exp(-u)$$

(3.31)

If $u \to 0$,

$$K_0(u) \to \infty$$

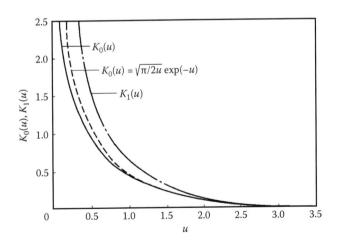

FIGURE 3.8
Graphical representation of the Bessel functions $K_0(u)$ and $K_1(u)$.

3.4.5 Solution for Medium-Thick Plate

We have discussed the solutions for heat conduction on semi-infinite body and infinite plates. The assumption of infinite dimensions of a workpiece simplifies the solution because no boundary conditions exist. However, all practical workpieces are of finite dimensions. For a medium-thickness plate, the temperature gradient in the thickness direction and the effect of its back-side surface on the heat transfer phenomenon cannot be neglected.

The model for a medium-thickness plate considers a point heat source moving at a constant speed across a wide plate of finite thickness H. With the exception of certain special cases (e.g., water cooling of the back side of the plate), it is a reasonable approximation to assume that the plate surfaces are adiabatic. Thus, to maintain the net heat flux passing through both boundaries equal to zero, it is necessary to account for mirror reflections of the source with respect to the planes of $z = 0$ and $z = H$. This can be done on the basis of the "method of images" as illustrated in Figure 3.9.

As demonstrated in Figure 3.9, the heat conduction of a medium-thickness plate with adiabatic surfaces under action of a point source can be taken as a part of the thermal process on an infinite body. Hence, we double the source power and add the imaginary sources 1, 2, 3, … of the power $2Q$ on the infinite body. The imaginary sources are successive reflections of the basic point source (source 0) on the surfaces $z = 0$ and $z = H$. For example, source 1 is a reflection of source 0 at the back-side surface ($z = H$), source 2 is a reflection of source 1 at the front-side surface ($z = 0$), source 3 is a reflection of source 2 at the back-side surface ($z = H$), and so on. An infinite number of point sources

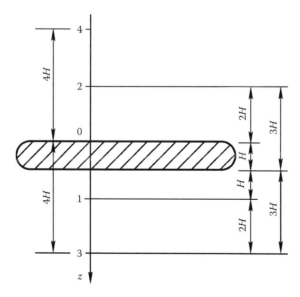

FIGURE 3.9
Multireflection method of real and imaginary point sources on a medium-thick plate.

0, 1, 2, 3, 4, ... is lined along the axis oz, and these are symmetric with respect to the planes of $z = 0$ and $z = H$. Thus, the heat fluxes passing through these two boundaries are zero to maintain their adiabatic condition.

By including all contributions from the imaging sources located symmetrically at distances $\pm 2nH$ below and above the upper surface of the plate and employing Eq. (3.25), we obtain

$$T(x,y,z,t) - T_0 = \sum_{n=-\infty}^{+\infty} \frac{2Q}{\rho C_P (4\pi a)^{3/2}} \exp\left(-\frac{v_0 x}{2a}\right) \int_0^t \frac{1}{t''^{3/2}} \exp\left(-\frac{v_0^2 t''}{4a} - \frac{R_n^2}{4at''}\right) dt''$$

(3.32)

where $R_n^2 = x^2 + y^2 + (z - 2nH)^2$, $n = \pm 1, 2, 3, \ldots$.

Let $t \to \infty$; the quasi-steady state temperature distribution is obtained for a medium-thickness plate:

$$T(x,y,z) - T_0 = \frac{Q}{2\pi\lambda} \exp\left(-\frac{v_0 x}{2a}\right) \sum_{n=-\infty}^{+\infty} \frac{1}{R_n} \exp\left(-\frac{v_0 R_n}{2a}\right)$$

(3.33)

Note that Eq. (3.33) is simply the general Rosenthal-Rykalin thick-plate solution Eq. (3.27) summed for each source.

3.4.6 Simplified Solution for a Fast-Moving High-Power Source

The heat input in welding is the quantity of thermal energy introduced per unit length of weld from a traveling heat source, which is computed as the ratio of the input power of the heat source (in watts) to its travel speed (in millimeters per second), that is, Q/v_0 (J/mm). As the arc power Q and the welding speed v_0 increase but Q/v_0 remains almost constant, it is the so-called fast-moving high-power source. If the welding speed is very high, conduction of heat will occur exclusively in directions normal to the x-axis (welding direction), and heat flow in the welding direction can be neglected. This will make the solution greatly simplified.

3.4.6.1 Fast-Moving High-Power Source on a Semi-Infinite Body

As shown in Figure 3.10, in a short time interval dt, the amount of heat released per unit length of the weld is equal to

$$\frac{Qdt}{dx} = \frac{Qdt}{v_0 dt} = \frac{Q}{v_0}$$

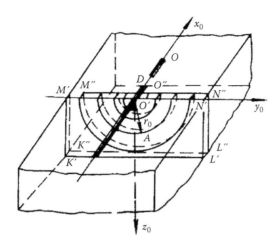

FIGURE 3.10
A fast-moving high-power source on a semi-infinite body.

This amount of heat will remain in a slice of thickness dx due to the assumption of no heat flow in the welding direction. This situation is similar to the temperature field around a linear instantaneous heat source in a thin plate, provided that the strength of the source is doubled because the line source acts on only half of the thin plate. Employing Eq. (3.17) and replacing Q_1 by $2Q/v_0$, we get the solution for the fast-moving high-power source on a semi-infinite body:

$$T - T_0 = \frac{Q}{2\pi\lambda v_0 t}\exp\left(-\frac{r_0^2}{4at}\right) \tag{3.34}$$

where $r_0^2 = y_0^2 + z_0^2$, the distance from the heat source to a point A; t is the time starting from the instant when the heat source arrives at the slice where point A locates.

Equation (3.34) can be used to estimate the weld width and thermal cycle. Taking the logarithm for both sides of Eq. (3.34),

$$\ln(T - T_0) = \ln\left(\frac{Q}{2\pi\lambda v_0}\right) - \ln t - \frac{r_0^2}{4at}$$

To differentiate it with respect to t,

$$\frac{1}{T - T_0}\frac{\partial T}{\partial t} = -\frac{1}{t} + \frac{r_0^2}{4at^2}$$

$$\frac{\partial T}{\partial t} = \frac{T - T_0}{t}\left(\frac{r_0^2}{4at} - 1\right)$$

The temperature reaches its peak value when $(\partial T/\partial t) = 0$. Meanwhile, $T = T_p$, and $t = t_p$. Thus,

$$\frac{r_p^2}{4at_p} - 1 = 0,$$

$$t_p = \frac{r_p^2}{4a} \tag{3.35}$$

Equation (3.35) indicates that, for a point with a distance r_p away from the weld centerline, its temperature rises to the peak value T_p at time t_p since the heat source passed the slice where this point locates.

Substituting r_p and t_p into Eq. (3.34),

$$T_p - T_0 = \frac{Q}{2\pi\lambda v_0 t_p} \exp\left(-\frac{r_p^2}{4at_p}\right)$$

$$= \frac{Q}{2\pi\lambda v_0} \frac{4a}{r_p^2} \exp\left(-\frac{r_p^2}{4a} \frac{4a}{r_p^2}\right) = \frac{2aQ}{\pi\lambda v_0 r_p^2} e^{-1} = \frac{2Q}{e\pi v_0 \rho C_p r_p^2}$$

After manipulations,

$$r_p^2 = \frac{2Q}{e\pi v_0 \rho C_p (T_p - T_0)} \tag{3.36}$$

When the peak temperature of a point gets the melting point ($T_p = T_m$), such a point locates at the fusion line. Its distance to the weld line is the half-weld width r_m:

$$r_m^2 = \frac{2Q}{e\pi v_0 \rho C_p (T_m - T_0)} = \frac{0.234Q}{v_0 \rho C_p (T_m - T_0)} \tag{3.37}$$

For example, gas metal arc welding (GMAW) of mild steel plates has the following parameters: $I = 300$ A, $U_a = 28$ V, $v_0 = 4$ mm/s. The thermal physical properties of the steel are as follows: $a = 5$ mm^2 s^{-1}, $\rho C_p = 0.005$ J mm^{-3} °C^{-1}, $T_m = 1520$°C. $T_0 = 20$°C, $\eta = 0.8$. Substituting these data into Eq. (3.37),

$$r_m^2 = \frac{0.234Q}{v_0 \rho C_p (T_m - T_0)} = \frac{0.234 \times 0.8 \times 300 \times 28}{4 \times 0.005 \times (1520 - 20)} = 52.416, \quad r_m = 7.2\text{(mm)}$$

The transverse cross section of weld is a circle with radius of 7.2 mm.

3.4.6.2 Fast-Moving High-Power Source on a Thin Plate

A fast-moving high-power source on a thin plate is equivalent to a case of a plane source in a rod perpendicular to the welding direction. As shown in Figure 3.11, in a short time interval dt the amount of heat released per unit length of the weld is equal to

$$\frac{Qdt}{Hdx} = \frac{Q}{Hv_0}$$

According to the assumptions, this amount of heat will remain in a rod of constant cross-sectional area due to lack of heat flow in the welding direction. Under such conditions, the mode of heat flow becomes one dimensional. By employing Eq. (3.19), we get the solution for a fast-moving high-power source on a thin plate:

$$T - T_0 = \frac{Q}{v_0 H (4\pi\lambda\rho C_p t)^{1/2}} \exp\left(-\frac{y_0^2}{4at}\right) \tag{3.38}$$

where y_0 is the distance from a point to the weld line.

Taking the logarithm for both sides of Eq. (3.38),

$$\ln(T - T_0) = \ln\left(\frac{Q}{v_0 H \sqrt{4\pi\lambda\rho C_p}}\right) - \frac{1}{2}\ln t - \frac{y_0^2}{4at}$$

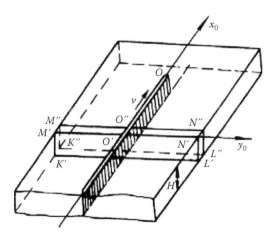

FIGURE 3.11
A fast-moving high-power source on a thin plate.

To differentiate it with respect to t,

$$\frac{1}{T-T_0}\frac{\partial T}{\partial t} = -\frac{1}{2}\frac{1}{t} + \frac{y_0^2}{4at^2}$$

When $t = t_m$, $(\partial T/\partial t) = 0$, $T = T_m$. Thus,

$$t_m = \frac{y_m^2}{2a} \qquad\qquad (3.39)$$

Substituting Eq. (3.39) into Eq. (3.38), after manipulation, we obtain

$$y_m = \sqrt{\frac{1}{2\pi e}}\frac{Q}{v_0 H \rho C_p (T_m - T_0)} \qquad\qquad (3.40)$$

This is the formula to calculate the half-weld width for butt welding of a thin plate.

Equations (3.34) and (3.38) may be used for general welding cases. The higher the welding speed is, the more accurate are the calculation results. For welding of mild steel plates, both equations can be employed if the welding speed is over 600 mm/min.

3.5 Dimensionless Forms of Rosenthal-Rykalin's Formulae

To obtain a general survey of the quasi-steady state temperature field, it is convenient to write Rosental-Ryklin's formulae in a dimensionless form. Grong transformed Rosental-Ryklin's formulae into a dimensionless version and provided some dimensionless temperature maps.[5] Next, we introduce Grong's work briefly.

3.5.1 Thick Plate

For a thick-plate solution of Eq. (3.27),

$$T(R,x) - T_0 = \frac{Q}{2\pi\lambda R}\exp\left(-\frac{v_0 x}{2a} - \frac{v_0 R}{2a}\right)$$

The following parameters are defined:

Dimensionless operating parameter:

$$n_3 = \frac{Qv_0}{4\pi a^2 \rho C_p (T_c - T_0)} = \frac{Qv_0}{4\pi a^2 (H_c - H_0)} \tag{3.41}$$

where Q is the net power, T_c is the chosen reference temperature, and $(H_c - H_0)$ is the heat content per unit volume at the reference temperature.

Dimensionless temperature:

$$\theta = \frac{T - T_0}{T_c - T_0} \tag{3.42}$$

Dimensionless coordinates:

$$\xi = \frac{v_0 x}{2a}, \; \psi = \frac{v_0 y}{2a}, \; \zeta = \frac{v_0 z}{2a} \tag{3.43}$$

Dimensionless distance:

$$\sigma_3 = \frac{v_0 R}{2a} \tag{3.44}$$

By substituting these dimensionless parameters into Eq. (3.27), we obtain

$$\frac{\theta}{n_3} = \frac{1}{\sigma_3} \exp(-\sigma_3 - \xi) \tag{3.45}$$

Equation (3.45) has been solved numerically for chosen values of σ_3 and ξ. A graphical presentation of the different solutions is shown in Figure 3.12. These maps provide a good overall indication of the thermal conditions during thick-plate welding but are not suitable for precise readings. Consequently, for quantitative analyses, the set of equations discussed next can be used.

3.5.1.1 Isothermal Zone Widths

The maximum width of an isothermal enclosure is obtained by setting $\partial \ln(\theta/n_3)/\partial\sigma_3 = 0$:

$$\frac{\partial \ln(\theta/n_3)}{\partial\sigma_3} = \frac{\partial \ln(\theta/n_3)}{\partial\xi} \cdot \frac{\partial\xi}{\partial\sigma_3} = 0$$

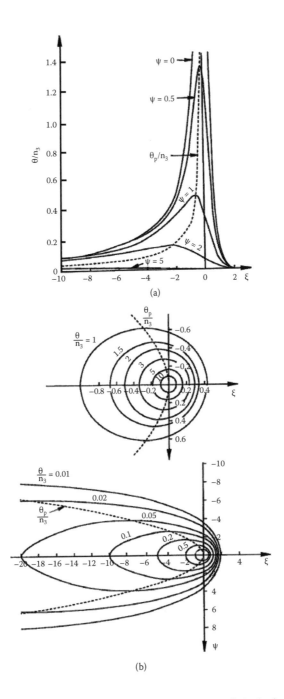

FIGURE 3.12
Dimensionless temperature maps for point sources on a semi-infinite body (from Grong Ø. *Metallurgical modeling of welding*, 2nd ed., Institute of Materials, London, 1997): (a) vertical sections parallel to the ξ–axis; (b) isothermal contours in the ξ–ψ plane for different values of θ/n_3.

From the definition of σ_3, we have

$$\frac{\partial \xi}{\partial \sigma_3} = \frac{2\sqrt{\xi^2 + \psi^2 + \zeta^2}}{2\xi} = \frac{\sigma_3}{\xi}$$

Partial differentiation of Eq. (3.45) gives

$$\frac{\partial(\ln \theta / n_3)}{\partial \sigma_3} = \frac{\partial}{\partial \sigma_3}(-\ln \sigma_3 - \sigma_3 - \xi) = -\frac{1}{\sigma_3} - 1 - \frac{\sigma_3}{\xi}$$

For a point located at the maximum width of the isothermal zone,

$$\sigma_3 = \sigma_{3m}, \quad \xi = \xi_m, \quad \frac{\partial(\ln \theta / n_3)}{\partial \sigma_3} = 0$$

and

$$-\frac{1}{\sigma_{3m}} - 1 - \frac{\sigma_{3m}}{\xi_m} = 0$$

Then,

$$\xi_m = \frac{-(\sigma_{3m})^2}{\sigma_{3m} + 1}$$

Substituting this equation into Eq. (3.45), we obtain

$$\frac{\theta_p}{n_3} = \frac{1}{\sigma_{3m}} \exp\left(-\frac{\sigma_{3m}}{\sigma_{3m} + 1}\right) \tag{3.46}$$

Equation (3.46) can be used for calculations of isothermal zone widths ψ_m and cross-sectional areas Λ_1. From Figure 3.13, we have

$$\psi_m = \zeta_m = \sqrt{(\sigma_{3m})^2 - (\xi_m)^2} = \frac{\sigma_{3m}}{\sigma_{3m} + 1}\sqrt{1 + 2\sigma_{3m}} \tag{3.47}$$

$$\Lambda_1 = \frac{\pi}{2}\psi_m^2 = \frac{\pi}{2}\left(\frac{\sigma_{3m}}{\sigma_{3m} + 1}\right)^2 (1 + 2\sigma_{3m}) \tag{3.48}$$

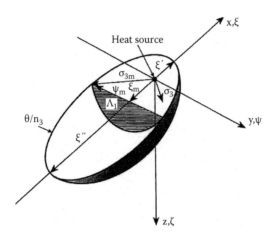

FIGURE 3.13
Three-dimensional representation of a weld pool.

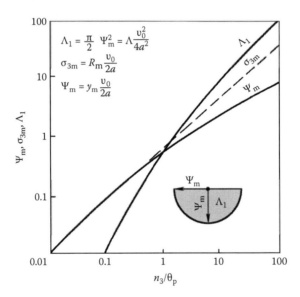

FIGURE 3.14
σ_{3m}, ψ_m, $\Lambda_1$versus n_3/θ. (From Grong Ø. *Metallurgical modeling of welding*, 2nd ed., Institute of Materials, London, 1997.)

A graphical presentation of Eqs. (3.46), (3.47), and (3.48) is shown in Figure 3.14.

3.5.1.2 Length of Isothermal Enclosures

Referring to Figure 3.13, the total length of an isothermal enclosure ξ_1 is given by

$$\xi_1 = \xi' - \xi'' \tag{3.49}$$

where ξ' and ξ'' are the distance from the heat source to the front and the rear edges of the enclosure, respectively. The coordinates $(\xi', 0, 0)$ and $(\xi'', 0, 0)$ are found by setting $\sigma_3 = \xi'$ and $\sigma_3 = \xi''$ in Eq. (3.45), respectively. This gives

$$\xi' = \frac{1}{2}\ln\left(\frac{n_3/\theta}{\xi'}\right), \quad \xi'' = -n_3/\theta \tag{3.50}$$

3.5.1.3 Volume of Isothermal Enclosures

Since the assumption of a point heat source involves semicircular isotherms in the ψ–ζ plane, the volume of an isothermal enclosure is obtained by integration over the total length from ξ' to ξ'':

$$\Gamma = \frac{\pi}{2}\int_{\xi''}^{\xi'} \psi^2 d\xi = \frac{\pi}{2}\int_{\xi''}^{\xi'} (\sigma_3)^2 d\xi - \frac{\pi}{2}\int_{\xi''}^{\xi'} \xi^2 d\xi$$

The former integral is readily evaluated by substituting

$$d\xi = -d\sigma_3\left(\frac{1+\sigma_3}{\sigma_3}\right)$$

From a differentiation of Eq. (3.45),

$$0 = -\frac{1}{\sigma_3^2}\exp(-\sigma_3 - \xi)\frac{d\sigma_3}{d\xi} + \frac{1}{\sigma_3}\exp(-\sigma_3 - \xi)\left(-\frac{d\sigma_3}{d\xi} - 1\right)$$

and

$$\frac{d\sigma_3}{d\xi}\left(\frac{1+\sigma_3}{\sigma_3}\right) = -1.$$

We obtain

$$\Gamma = -\frac{\pi}{2}\int_{\xi''}^{\xi'} \left[\sigma_3 + (\sigma_3)^2\right] d\sigma_3 - \frac{\pi}{2}\int_{\xi''}^{\xi'} \xi^2 d\xi$$

Noting that $\sigma_3 = -\xi'' = n_3/\theta$ at the lower limit of integration, we obtain

$$\Gamma = \frac{\pi}{12}\left[3(n_3/\theta)^2 - 3(\xi')^2 - 4(\xi')^3\right] \tag{3.51}$$

The dimensionless volume Γ is related to the real volume of the enclosure V_p (mm³) through the following equation:

$$V_p = \frac{8a^3}{v_0^3}\Gamma \tag{3.52}$$

Next, we give an example to demonstrate the application of a dimensionless formula.

Example 3.1

This example involves GMAW of a lower-alloy steel thick plate under the following conditions: I = 300 A, U_a = 28 V, v_0 = 4 mm/s, $U_a T_0$ = 20°C. Sketch the contours of the fusion boundary and the Ac_3 isotherm (910°C) in the $\xi - \psi(x - y)$ plane at quasi-steady state.

As shown in Figure 3.15(a), it is sufficient to calculate the coordinates in four different positions to draw a sketch of the isothermal contours. If the latent heat of melting is neglected, the θ/n_3 ratio at the melting point becomes, according to Eqs. (3.41), (3.42), and (3.45),

$$\frac{\theta_m}{n_3} = \frac{T_m - T_0}{T_c - T_0}\cdot\frac{4\pi a^2 \rho C_p(T_c - T_0)}{Qv_0} = \frac{4\pi a^2 \rho C_p(T_m - T_0)}{\eta I U_a v_0}$$

For low-alloy steel, physical properties are as follows: a = 5 (mm² s⁻¹), ρC_p = 0.005 (J mm⁻³ °C⁻¹), T_m = 1520°C. For GMAW, η = 0.8.

$$\frac{\theta_m}{n_3} = \frac{4\pi \times 5^2 \times 0.005 \times (1520 - 20)}{0.8 \times 300 \times 28 \times 4} = 0.088$$

The endpoints are readily obtained from Eq. (3.50):

$$\xi' = \frac{1}{2}\ln\left(\frac{1}{0.088\xi'}\right), \quad \xi' = 1.15 \quad \left(x = \frac{2a}{v_0}\xi = \frac{2 \times 5}{4} \times 1.15 = 2.87\text{mm}\right)$$

$$\xi'' = \frac{-n_3}{\theta} = \frac{-1}{0.088} = -11.36, (x = -28.41 \text{ mm})$$

For the maximum width of the fusion line,

$$\frac{\theta_m}{n_3} = \frac{1}{\sigma_{3m}}\exp\left(-\frac{\sigma_{3m}}{\sigma_{3m}+1}\right) = 0.088, \sigma_{3m} = 4.95$$

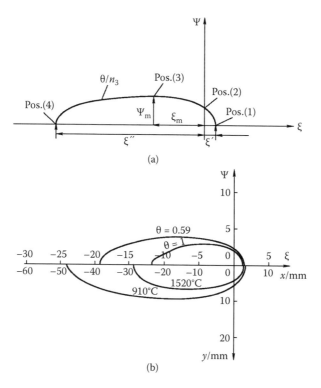

FIGURE 3.15
Calculated isothermal contours from thick-plate solution (from Grong Ø. *Metallurgical modeling of welding*, 2nd ed., Institute of Materials, London, 1997): (a) calculation procedure; (b) sketch of fusion boundary and Ac_3-isotherm at top surface (example 1).

$$\psi_m = \frac{\sigma_{3m}}{\sigma_{3m}+1}\sqrt{1+2\sigma_{3m}} = 2.75, (y_m = 6.88\text{mm})$$

and

$$\xi_m = \frac{-(\sigma_{3m})^2}{\sigma_{3m}+1} = 4.13, (x_m = 10.33\text{mm})$$

For the intersection point on the fusion line with *the* $\psi(y)$-axis,

$$\xi = \zeta = 0, \sigma_3 = \psi, \frac{\theta}{n_3} = \frac{1}{\psi}\exp(-\psi)$$

$$\psi = -\ln\left(\frac{\theta}{n}\psi\right) = -\ln(0.088\psi), \psi = 1.83, (y = 4.58\,\text{mm})$$

Similarly, the contour of the Ac_3 isotherm (910°C) can be determined by inserting $\theta_p/n_3 = 0.052$ into the same set of equations. Figure 3.15(b) shows a graphical representation of the computed isothermal contours.

3.5.2 Thin Plate

If heat losses from plate surfaces are not considered, Eq. (3.28) becomes

$$T(r,x) - T_0 = \frac{Q}{2\pi\lambda H}\exp\left(-\frac{v_0 x}{2a}\right)K_0\left(\frac{v_0 r}{2a}\right)$$

For defining the dimensionless distance and thickness,

$$\sigma_5 = \frac{v_0 r}{2a}, \; d = \frac{v_0 H}{2a}$$

Other dimensionless parameters are defined as before. By substituting these dimensionless parameters into the thin-plate equation, we obtain

$$\frac{\theta d}{n_3} = \exp(-\xi)K_0(\sigma_5) \tag{3.53}$$

Plots of this equation are shown in Figure 3.16.

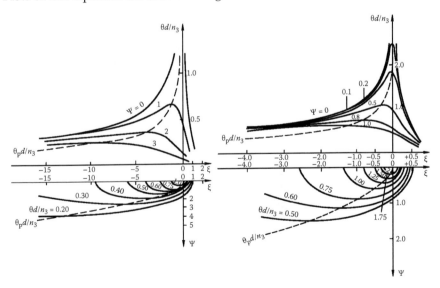

FIGURE 3.16
Dimensionless temperature maps for line sources in thin plates for different ranges of $\theta d/n_3$. (From Grong Ø. *Metallurgical modeling of welding*, 2nd ed., Institute of Materials, London, 1997.)

3.5.2.1 Isothermal Zone Widths

The maximum width of an isothermal enclosure is obtained by setting $\partial(\theta d / n_3) \,/\, \partial \xi = 0$. Noting that

$$(\partial/\partial u)K_0(u) = -K_1(u)$$

where $K_1(u)$ is the modified Bessel function of the second kind and first order, we get

$$\frac{\partial(\theta d / n_3)}{\partial \xi} = \frac{\partial(\theta d / n_3)}{\partial \sigma_5} \cdot \frac{\partial \sigma_5}{\partial \xi}$$

$$= -\exp(-\xi)\left[K_0(\sigma_5) + \frac{\xi}{\sigma_5} K_1(\sigma_5) \right]$$

For a point located at the maximum width of the isothermal enclosure,

$$\exp(-\xi_m)\left[K_0(\sigma_{5m}) + \frac{\xi_m}{\sigma_{5m}} K_1(\sigma_{5m}) \right] = 0$$

After manipulation,

$$\xi_m = -\sigma_{5m} \frac{K_0(\sigma_{5m})}{K_1(\sigma_{5m})} \tag{3.54}$$

Substituting Eq. (3.54) into (3.53), we have

$$\frac{\theta_p d}{n_3} = \exp\left[\sigma_{5m} \frac{K_0(\sigma_{5m})}{K_1(\sigma_{5m})} \right] K_0(\sigma_{5m}) \tag{3.55}$$

Equation (3.55) can be used for calculations of isothermal zone widths ψ_m and cross-sectional area Λ_2 in thin-plate welding. Referring to Figure 3.17, we have

$$\psi_m = \sqrt{(\sigma_{5m})^2 - (\xi_m)^2} = \sigma_{5m}\sqrt{1 - \left[K_0(\sigma_{5m})\right]^2 / \left[K_1(\sigma_{5m})\right]^2} \tag{3.56}$$

$$\Lambda_2/d = 2\psi_m = 2\sigma_{5m}\sqrt{1 - \left[K_0(\sigma_{5m})\right]^2 / \left[K(\sigma_{5m})\right]^2} \tag{3.57}$$

Figure 3.18 shows a graphical presentation of Eqs. (3.55), (3.56), and (3.57).

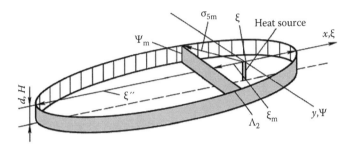

FIGURE 3.17
Graphical representation of thin-plate solution.

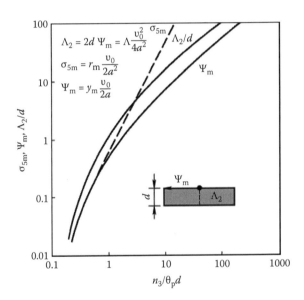

FIGURE 3.18
σ_{5m}, Ψ_m, Λ_2/δ versus $n_3/\theta_p\delta$. (From Grong Ø. *Metallurgical modeling of welding*, 2nd ed., Institute of Materials, London, 1997.)

3.5.2.2 Length of Isothermal Enclosures

The distance from the heat source to the front edge (ξ', 0) and rear edge (ξ'', 0) of an isothermal enclosure is obtained by substituting $\sigma_5 = \xi'$ and $\sigma_5 = \xi''$ into Eq. (3.53), respectively:

$$\xi' = \ln\left(\frac{n_3 K_0(\xi')}{\theta d}\right), \xi' > 0 \tag{3.58}$$

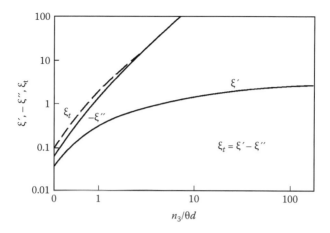

FIGURE 3.19

ξ, ξ'', ξ_t versus $n_3/\theta d$. (From Grong Ø. *Metallurgical modeling of welding*, 2nd ed., Institute of Materials, London, 1997.)

and

$$\xi'' = \ln\left(\frac{n_3 K_0(-\xi'')}{\theta d}\right), \xi'' < 0 \tag{3.59}$$

A graphical presentation of Eqs. (3.58) and (3.59) is shown in Figures 3.18 and 3.19.

Example 3.2

Consider gas tungsten arc butt welding of a 2-mm thick sheet of aluminum alloy (Al-Mg alloy) under the following conditions:

$I = 110\ A$, $U_a = 15\ V$, $v_0 = 4\ mm/s$, $\eta = 0.6$, $T_0 = 20°C$

Sketch the contours of the fusion boundary and the A_r isotherm in the $\xi - \psi$ (x-y) plane at quasi-steady state. The recrystallization temperature A_r of the base material is taken as 275°C.

For the Al-Mg alloy, $a = 55\ (mm^2\ s^{-1})$, $\rho C_p = 0.0027$ (J mm^{-3} °C^{-1}), $T_m = 650°C$. Using the definition of dimensionless parameters, we have

$$\frac{\theta_m d}{n_3} = \frac{T_m - T_0}{T_c - T_0} \cdot \frac{v_0 H}{2a} \cdot \frac{4\pi a^2 \rho C_p (T_c - T_0)}{Q v_0} = \frac{2\pi a H \rho C_p (T_m - T_0)}{\eta I U_a}$$

$$\frac{n_3}{\theta_m d} = \frac{0.6 \times 110 \times 15}{2\pi \times 55 \times 2 \times 0.0027 \times 630} = 0.84$$

The endpoints of the fusion boundary can be read from Figure 3.19:

$$\xi' = 0.25, (x = 6.88\,mm)$$

$$\xi'' = -0.29, (x = -24.75\,mm)$$

Based on Eq. (3.56), we obtain:

$$\psi_m = 0.41, (y_m = 11.41\,mm)$$

$$\xi_m = -0.29, (x_m = -7.87\,mm)$$

The intersection point of the fusion line with the $\psi(y)$-axis can be determined by setting $\xi = 0$ and $\sigma_5 = \psi$:

$$\frac{\theta_m d}{n_3} = K_0(\psi) = \frac{1}{0.84} = 1.19$$

$$\psi = 0.37, (y = 10.18\,mm)$$

Similarly, the contour of the A_r (275°C) isotherm can be determined by inserting $n_3/\theta_p d = 2.08$ into the same set of equations. Figure 3.20 shows a graphical representation of the calculated isothermal contours.

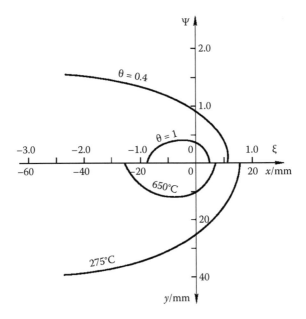

FIGURE 3.20
Calculated contours of fusion boundary and A_r-isotherm (example 2). (From Grong Ø. *Metallurgical modeling of welding*, 2nd ed., Institute of Materials, London, 1997.)

3.6 Limitations of Rosenthal-Rykalin's Formulae

Rosenthal-Rykalin's formulae were discussed in previous sections. The simplifying assumptions inherent in these analytical solutions of heat conduction can be summarized as follows:

1. All heat from the sources is concentrated into a zero-volume point or line or plane.
2. The parent material is isotropic and homogeneous at all temperatures, and no phase changes occur on heating.
3. The thermal conductivity, density, and specific heat are constant and independent of temperature.
4. The workpiece dimensions are infinite. The semi-infinite body and infinite plate correspond to point and line heat sources, respectively.
5. The workpiece is completely insulated from its surroundings; that is, there are no heat losses by convection or radiation from the boundaries.

Since there are big differences between these assumptions and practical cases of heat transfer during welding, there exist large errors for the predicted temperatures adjacent to the heat source. The positions near the heat source must be paid much attention because it is important to determine the fusion zone shape and size from the viewpoint of welding technology on one hand; on the other hand, the region heated above the phase change temperature must be emphasized from the viewpoint of metallurgy.

Because Rosenthal-Rykalin's formulae cannot give satisfactory results for temperature distributions in and near the weld pool, since the 1950s many researchers have attempted to modify the unreasonable assumptions and improve the prediction accuracy. Due to the inherent drawbacks of Rosenthal-Rykalin's formulae, it is difficult to modify all the unreasonable assumptions completely. Only one or two assumptions can be avoided, but the problem cannot be solved from the very beginning of formulae derivation.

The development of modern computers promoted the application of numerical simulation in calculating the weld thermal process. A numerical simulation technique can be used to model the welding process by a group of partial differential equations describing the physical mechanisms in welding and their boundary conditions, which are solved numerically to obtain quantitative understanding of the welding process. The numerical approach is able to deal with various kinds of complex boundary conditions, intensity distribution of heat sources, and nonlinear problems, so it has incomparable advantages that the analytic approach does not possess. Thus, numerical analysis has been widely used to investigate the weld thermal process.

3.7 Modifications to Rosenthal-Rykalin's Solutions

Since Rosenthal-Rykalin's outstanding early contribution in analysis of the weld thermal process, there have been numerous modifications of their solutions. The modifications are mainly made for two aspects: (1) a distributed heat source and (2) finite plate thickness with surface heat losses. Here, an example is given to demonstrate how a continuous-action Gaussian heat source mode is used to modify Rosenthal-Rykalin's solutions.

3.7.1 Instantaneous Gaussian Heat Source

First, consider a case of an instantaneous Gaussian heat source acting on a semi-infinite body.

The energy deposited from the welding arc can be approximated by a Gaussian function. An instantaneous Gaussian heat source releases its heat at time $t = 0$. As shown in Figure 3.21, a coordinate system is established, taking the source center as its origin. For any point $(x', y', 0)$ inside the deposition region of the heat source, there is an elementary rectangle $ds = dx'dy'$ with the heat $dQ = q(r')dx'dy'dt$. Taking this heat dQ as a point source and employing Eq. (3.15), the temperature rise at a point (x, y, z) on the workpiece caused by dQ at point $(x', y', 0)$ may be written as

$$dT(x,y,z,t) = \frac{2q(r')dx'dy'dt}{\rho C_p (4\pi at)^{3/2}} \exp\left(-\frac{R'^2}{4at}\right)$$

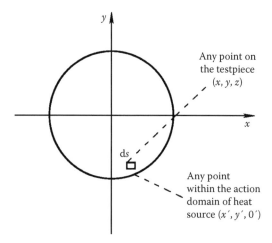

FIGURE 3.21
Schematic of instantaneous Gaussian heat source.

where $R'^2 = (x-x')^2 + (y-y')^2 + z^2$, $r' = \sqrt{x'^2 + y'^2}$。
Since

$$q(r') = \frac{Q}{2\pi\sigma_q^2} \exp\left(-\frac{r'^2}{2\sigma_q^2}\right)$$

Thus,

$$dT(x,y,z,t) = \frac{Q}{\pi\sigma_q^2} \frac{dx'dy'dt}{\rho C_p (4\pi at)^{3/2}} \exp\left(-\frac{R'^2}{4at} - \frac{r'^2}{2\sigma_q^2}\right)$$

Taking an instantaneous Gaussian heat source as a series of point heat source dQ deposited on infinite elementary rectangles ds, we have

$T(x,y,z,t) - T_0$

$$= \frac{Q}{\pi\sigma_q^2} \frac{dt}{\rho C_p (4\pi at)^{3/2}} \int_{-\infty}^{\infty}\int_{-\infty}^{\infty} \exp\left[-\frac{(x-x')^2 + (y-y')^2 + z^2}{4at} - \frac{x'^2 + y'^2}{2\sigma_q^2}\right] dx'dy'$$

$$= \frac{Q}{\pi\sigma_q^2} \frac{dt}{\rho C_p (4\pi at)^{3/2}} \exp\left(-\frac{z^2}{4at}\right) \int_{-\infty}^{\infty} \exp\left[-\frac{(x-x')^2}{4at} - \frac{x'^2}{2\sigma_q^2}\right] dx'$$

$$\times \int_{-\infty}^{\infty} \exp\left[-\frac{(y-y')^2}{4at} - \frac{y'^2}{2\sigma_q^2}\right] dy' \qquad (3.60)$$

Two integrals need to be calculated. First, deal with the first integral:

$$\int_{-\infty}^{\infty} \exp\left[-\frac{(x-x')^2}{4at} - \frac{x'^2}{2\sigma_q^2}\right] dx' \qquad (3.61)$$

Let $m = 2\sigma_q^2$, $n = 4at$. Substituting them into Eq. (3.61),

$$\int_{-\infty}^{\infty} \exp\left[-\frac{(x-x')^2}{n} - \frac{x'^2}{m}\right] dx' \qquad (3.62)$$

Let $u = \sqrt{\frac{m+n}{mn}} x' - \sqrt{\frac{mn}{m+n}} \frac{x}{n}$, $du = \sqrt{\frac{m+n}{mn}} dx'$

$$u^2 = \frac{m+n}{mn} x'^2 - 2\frac{xx'}{n} + \frac{m}{n(m+n)} x^2 = \frac{x'^2}{n} + \frac{x'^2}{m} - 2\frac{xx'}{n} + \frac{x^2}{n} - \frac{x^2}{n} + \frac{m}{n(m+n)} x^2$$

$$= \frac{x'^2}{m} + \frac{(x-x')^2}{n} - \frac{x^2}{m+n}$$

After manipulation, we obtain

$$-\frac{x'^2}{m} - \frac{(x-x')^2}{n} = -u^2 - \frac{x^2}{m+n}$$

Substituting these variables into Eq. (3.62), we have

$$\int_{-\infty}^{\infty} \exp\left[-\frac{(x-x')^2}{n} - \frac{x'^2}{m}\right] dx' = \sqrt{\frac{mn}{m+n}} \int_{-\infty}^{\infty} \exp\left(-u^2 - \frac{x^2}{m+n}\right) du$$

$$= \sqrt{\frac{mn}{m+n}} \exp\left(-\frac{x^2}{m+n}\right) \int_{-\infty}^{\infty} \exp(-u^2) du = \sqrt{\frac{mn}{m+n}} \exp\left(-\frac{x^2}{m+n}\right) \sqrt{\pi} \quad (3.63)$$

By substituting $m = 2\sigma_q^2$ and $n = 4at$ $n = 4at$ into Eq. (3.63), we get

$$\int_{-\infty}^{\infty} \exp\left[-\frac{(x-x')^2}{n} - \frac{x'^2}{m}\right] dx' = \sqrt{\frac{2\sigma_q^2 \cdot 4at}{4at + 2\sigma_q^2}} \exp\left(-\frac{x^2}{4at + 2\sigma_q^2}\right) \sqrt{\pi} \quad (3.64)$$

Similarly, we get

$$\int_{-\infty}^{\infty} \exp\left[-\frac{(y-y')^2}{n} - \frac{y'^2}{m}\right] dy' = \sqrt{\frac{2\sigma_q^2 \cdot 4at}{4at + 2\sigma_q^2}} \exp\left(-\frac{y^2}{4at + 2\sigma_q^2}\right) \sqrt{\pi} \quad (3.65)$$

Substituting Eqs. (3.64) and (3.65) into Eq. (3.60), we obtain

$$T(x,y,z,t) - T_0 = \frac{Q}{\pi\sigma_q^2} \frac{dt}{\rho C_p (4\pi at)^{3/2}} \exp\left(-\frac{z^2}{4at}\right) \frac{2\sigma_q^2 \cdot 4at \cdot \pi}{4at + 2\sigma_q^2} \exp\left(-\frac{x^2 + y^2}{4at + 2\sigma_q^2}\right)$$

After manipulations, we obtain

$$T(x,y,z,t) = \frac{2Qdt}{\pi\rho C_p} \frac{1}{(4\pi at)^{1/2}} \frac{1}{4at + 2\sigma_q^2} \exp\left(-\frac{x^2 + y^2}{4at + 2\sigma_q^2} - \frac{z^2}{4at}\right) \quad (3.66)$$

This is the temperature variation caused by an instantaneous Gaussian heat source on a semi-infinite body. For a special case, when $\sigma_q = 0$, Eq. (3.66) will become Eq. (3.20) for an instantaneous point heat source.

3.7.2 Continuous Gaussian Heat Source

As shown in Figure 3.22, a Gaussian heat source moves at the welding speed v_0 on the semi-infinite body. At time $t = 0$, the source center locates at point O' (the origin of the stationary coordinate system $x_0 y_0 z_0$). At time t, the source center locates at point O (the origin of the moving coordinate system xyz). Divide the continuous deposition time duration $[0, t]$ of a moving Gaussian heat source into infinite elementary intervals dt'. At time t', the source center locates at point B, and the deposited heat amount is Qdt'. According to Eq. (3.66), this heat amount Qdt' produces a temperature rise at point A after the time period $\tau = t - t'$. It may be written as

$$dT(x_0, y_0, z_0, \tau) = \frac{2Qdt'}{\pi \rho C_p} \frac{1}{(4\pi a\tau)^{1/2}} \frac{1}{4a\tau + 2\sigma_q^2} \exp\left(-\frac{(x_0 - v_0 t')^2 + y_0^2}{4a\tau + 2\sigma_q^2} - \frac{z_0^2}{4a\tau}\right)$$

This equation is in the stationary coordinate system. By transforming it in the moving coordinate system and integrating it over the period $[0, t]$, we obtain

$$x_0 - v_0 t' = (x + v_0 t) - v_0 t' = x + v_0(t - t') = x + v_0\tau, \quad y_0 = y, \quad z_0 = z$$

$$\tau = t - t', \quad d\tau = -dt', \quad \tau \in [t, 0], \quad -\tau \in [0, t]$$

$$T(x, y, z, t) = \frac{2Q}{\pi \rho C_p} \int_0^t \frac{1}{(4\pi a\tau)^{1/2}} \frac{1}{4a\tau + 2\sigma_q^2} \exp\left[-\frac{(x + v_0\tau)^2 + y^2}{4a\tau + 2\sigma_q^2} - \frac{z^2}{4a\tau}\right] d\tau \quad (3.67)$$

Equation (3.67) is the solution of the temperature distribution for a Gaussian distributed heat source moving on a semi-infinite body. When $\sigma_q = 0$, Eq. (3.67) becomes the version of a point heat source like Eq. (3.25).

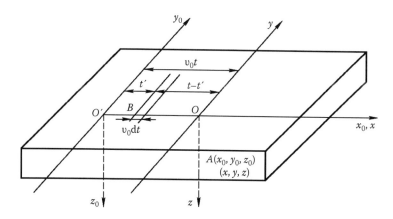

FIGURE 3.22
Schematic of continuous Gaussian heat source.

4

Finite Difference Solution of Heat Conduction in Welding

4.1 Introduction

Some problems encountered in welding engineering, such as heat conduction, thermal distortion, and hydrogen diffusion, can be attributed by the solution of specific differential equations. However, differential equations are usually solved analytically for simple cases under some simplifications. There are different practical problems with complicated boundary conditions that are difficult to solve. Hence, approximation approaches must be used to meet the requirements in production and engineering. Finite element and finite difference methods are the most widely employed numerical analysis approaches.

In mathematics, the finite difference method is a numerical method for approximating the solutions to differential equations using finite difference equations to approximate derivatives.[123] The advantages of the finite difference method are as follows: The program and algorithm are simple for the workpiece with regular geometry and homogeneous properties; the finite difference method has a good convergence characteristic; and the solution procedure is simple. The disadvantages include its limitation to a regular grid system that is only suitable for the calculation domains of square, rectangular, or regular triangular geometry.

This chapter introduces the applications of the finite difference method in analysis of heat conduction in welding.

4.2 Difference Quotients in Uniform Grids

Finite difference methods approximate the solutions to differential equations by replacing derivative expressions with approximately equivalent difference quotients. The principle of the finite difference solution of partial difference equations is to express the derivative of function $f(x)$ at a checked point based on the values of this function at adjacent points. The checked point may be on one side of the adjacent points or between the adjacent points. The distance between the adjacent points may be either identical or different.[121] In this section, we discuss the case of a uniform grid distance that is equal to Δx; the next section is about nonuniform grids.

4.2.1 Direct-Vision Method

For the function $f(x)$ shown in Figure 4.1, x_{-1}, x_0, and x_1 are the nodes that are uniformly spaced, that is, $x_1 - x_0 = x_0 - x_{-1} = \Delta x$ or $x_i = x_0 + i\Delta x$. Here, $f(x)$ and $f_i = f(x)|_{x=x_i} = f(x_i)$ represent the values of the function $f(x)$. The value of the function at a point between the nodes may be expressed by a noninteger of x, such as

$$f_{-1/2} = f\left(x_{-1/2}\right) = f\left[x_0 + \left(-\frac{1}{2}\right)\Delta x\right]$$

The direct-vision estimation of the first derivative f' at $x = x_{-1/2}$ is

$$\left.\frac{df}{dx}\right|_{x=x_{-1/2}} = \frac{f_0 - f_{-1}}{\Delta x} \tag{4.1}$$

And, at $x = x_{1/2}$,

$$\left.\frac{df}{dx}\right|_{x=x_{1/2}} = \frac{f_1 - f_0}{\Delta x} \tag{4.2}$$

If we take these two equations as the approximation of (df/dx) at $x = x_0$, they are the one-side difference. Equation (4.1) is the backward difference, and Eq. (4.2) is the forward difference. Clearly, a good estimation of the first derivative of the function $f(x)$ at $x = x_0$ can be obtained using the values of the function at two nodes on both sides of x_0, that is,

$$\left.\frac{df}{dx}\right|_{x=x_0} = \frac{f_1 - f_{-1}}{2\Delta x} \tag{4.3}$$

Equation (4.3) is the central difference based on f_1 and f_{-1}. In fact, the central difference is equal to the averaged value of two one-side differences.

The first central derivative may be expressed as

$$\left.\frac{df}{dx}\right|_{x=x_0} = \frac{f_{1/2} - f_{-1/2}}{\Delta x}$$

The second central derivative is written as

$$\frac{d^2 f}{dx^2}\Big|_{x=x_0} = \frac{d}{dx}\left(\frac{df}{dx}\right)\Big|_{x=x_0} = \frac{\frac{df}{dx}\big|_{x=x_{1/2}} - \frac{df}{dx}\big|_{x=x_{-1/2}}}{\Delta x} = \frac{\frac{f_1 - f_0}{\Delta x} - \frac{f_0 - f_{-1}}{\Delta x}}{\Delta x}$$

$$= \frac{f_1 - 2f_0 + f_{-1}}{(\Delta x)^2}$$

(4.4)

The accuracy of these expressions is improved as the distance Δx decreases.

4.2.2 Derivation from Taylor's Polynomial

Considering the case of equal node space shown in Figure 4.1 and expanding the function $f(x_1) = f(x_0 + \Delta x)$, $\Delta x = x_1 - x_0$ by Taylor's polynomial theorem, we have

$$f_1 = f_0 + \Delta x f_0' + \frac{(\Delta x)^2}{2!} f_0'' + \frac{(\Delta x)^3}{3!} f_0''' + \frac{(\Delta x)^4}{4!} f_{0,1}^{(4)}$$

where $f_{0,1}^{(4)}$ is the fourth derivative at a point belonging to the interval (x_0, x_1).

Similarly, expanding the function $f(x_{-1}) = f(x_0 - \Delta x)$, $\Delta x = x_0 - x_{-1}$ by Taylor's theorem, we have

$$f_{-1} = f_0 - \Delta x f_0' + \frac{(\Delta x)^2}{2!} f_0'' - \frac{(\Delta x)^3}{3!} f_0''' + \frac{(\Delta x)^4}{4!} f_{-1,0}^{(4)}$$

where $f_{-1,0}^{(4)}$ is the fourth derivative at a point within the interval (x_{-1}, x_0).

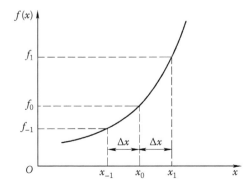

FIGURE 4.1
The values of function at adjacent points with equal node space.

When the terms after $(\Delta x)^2$ are neglected, we obtain the forward and backward differences of $f'(x_0)$ from these expansions, respectively:

$$\left.\frac{df}{dx}\right|_{x=x_0} = \frac{f_1 - f_0}{\Delta x} - \frac{\Delta x}{2}f_{0,1}''$$

$$\left.\frac{df}{dx}\right|_{x=x_0} = \frac{f_0 - f_{-1}}{\Delta x} + \frac{\Delta x}{2}f_{-1,0}''$$

These equations have a truncation error of the order Δx. On the other hand, subtracting the two Taylor's expansions, we get

$$\left.\frac{df}{dx}\right|_{x=x_0} = \frac{f_1 - f_{-1}}{2\Delta x} - \frac{(\Delta x)^2}{3!}\left(f_{-1,0}''' + f_{0,1}'''\right)$$

This equation has a truncation error of the order $(\Delta x)^2$.

To get the second derivative of the function at $x = x_0$, adding the two Taylor's expansions, we have

$$\left.\frac{d^2 f}{dx^2}\right|_{x=x_0} = \frac{f_1 - 2f_0 + f_{-1}}{(\Delta x)^2} - \frac{(\Delta x)^2}{4!}\left[f_{0,1}^{(4)} + f_{-1,0}^{(4)}\right]$$

This equation has a truncation error of the order $(\Delta x)^2$.

4.3 Derivatives of Function in Nonuniform Grids

4.3.1 Direct-Vision Method

For the points x_{-1}, x_0, and x_1 with nonequal node distance shown in Figure 4.2, we estimate the values of forward and backward difference for $f(x)$ at $x = x_0$. For the central point $x_0 + (1/2)S_E\Delta x$ between x_0 and x_1, the forward difference is

$$\left.\frac{df}{dx}\right|_{x=x_0+\frac{1}{2}S_E\Delta x} = \frac{f_1 - f_0}{S_E\Delta x} \qquad (4.5)$$

For the central point $x_0 - (1/2)S_W\Delta x$ between x_{-1} and x_0, the backward difference is

$$\left.\frac{df}{dx}\right|_{x=x_0-\frac{1}{2}S_W\Delta x} = \frac{f_0 - f_{-1}}{S_W\Delta x} \qquad (4.6)$$

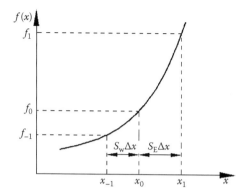

FIGURE 4.2
The values of function at adjacent points with unequal node space.

The central difference is written as:

$$\left.\frac{df}{dx}\right|_{x=x_0} = \frac{f_1 - f_{-1}}{(S_E + S_W)\Delta x} \tag{4.7}$$

Since $S_E > S_W$, the point $x = x_{-1} + (1/2)(S_E + S_W)\Delta x = x_0 + (1/2)(S_E - S_W)\Delta x$ locates at the right side of x_0 in Figure 4.2. So, the central difference is the best approximation.

It seems better to take the weighted average of two one-side derivatives. For the first derivative,

$$
\begin{aligned}
\left.\frac{df}{dx}\right|_{x=x_0} &= \frac{S_E}{S_E + S_W}\left.\frac{df}{dx}\right|_{x=x_0-\frac{1}{2}S_W\Delta x} + \frac{S_W}{S_E + S_W}\left.\frac{df}{dx}\right|_{x=x_0+\frac{1}{2}S_E\Delta X} \\
&= \frac{S_E}{S_E + S_W}\frac{f_0 - f_{-1}}{S_W\Delta x} + \frac{S_W}{S_W + S_E}\frac{f_1 - f_0}{S_E\Delta x} \\
&= \frac{1}{\Delta x}\left[\frac{S_W}{S_E(S_E + S_W)}f_1 - \frac{S_W - S_E}{S_E S_W}f_0 - \frac{S_E}{S_W(S_W + S_E)}f_{-1}\right]
\end{aligned} \tag{4.8}
$$

For the second derivative,

$$\left.\frac{d^2 f}{dx^2}\right|_{x=x_0} = \frac{\left.\dfrac{df}{dx}\right|_{x=x_0+\frac{1}{2}S_E\Delta x} - \left.\dfrac{df}{dx}\right|_{x=x_0-\frac{1}{2}S_W\Delta X}}{\dfrac{(S_E + S_W)\Delta x}{2}}$$

$$= \frac{\dfrac{f_1 - f_0}{S_E \Delta x} - \dfrac{f_0 - f_{-1}}{S_W \Delta x}}{\dfrac{(S_E + S_W)\Delta x}{2}}$$

$$= \frac{2}{(\Delta x)^2}\left[\frac{1}{S_E(S_E + S_W)}f_1 - \frac{1}{S_E S_W}f_0 + \frac{1}{S_W(S_W + S_E)}f_{-1}\right] \qquad (4.9)$$

4.3.2 Derivation from Taylor's Polynomial

For the case shown in Figure 4.2, write Taylor's expansions for $f(x_1) = f(x_0 + S_E\Delta x)$ and $f(x_{-1}) = f(x_0 - S_W\Delta x)$. After manipulations, we can also obtain Eqs. (4.8) and (4.9).

4.4 Finite Difference Equations of Steady-State Heat Conduction

For a function of multivariable $T = T(x, y, z)$, if taking y, z as constants, then its partial derivative with respect to x is its general derivative. Thus, all concepts and equations introduced in Sections 4.2 and 4.3 can be directly used to deal with functions of the multivariable. To express $(\partial T/\partial x)$ with the equations mentioned, we have to keep $y = y_0$, $z = z_0$ and vice versa.

4.4.1 Substitution Method of Partial Difference Equations

4.4.1.1 Uniform Grids

A three-dimensional domain with homogeneous thermal conductivity λ is divided into uniform grids. First, consider inner nodes. As shown in Figure 4.3, a typical node P has six adjacent nodes represented by N, S, E, W, I, O, respectively. Set $\Delta = \Delta x = \Delta y = \Delta z$, which is the node space in uniform grids. Referring to the general equation of heat conduction in the steady state,

$$\frac{\partial^2 T}{\partial x^2} + \frac{\partial^2 T}{\partial y^2} + \frac{\partial^2 T}{\partial z^2} + \frac{Q_{vd}}{\lambda} = 0$$

where Q_{vd} is the internal heat generation term.

According to Eq. (4.4), we transform the general equation of heat conduction in the steady state into its finite difference version as follows:

$$\frac{T_o - 2T_P + T_I}{\Delta^2} + \frac{T_E - 2T_P + T_W}{\Delta^2} + \frac{T_N - 2T_P + T_S}{\Delta^2} + \frac{Q_{vd}}{\lambda} = 0$$

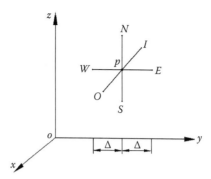

FIGURE 4.3
A node P with its six adjacent nodes.

If Q_{vd} changes with position or temperature, it takes its value at point P. This equation may be simplified:

$$T_o + T_1 + T_E + T_W + T_N + T_S - 6T_P + \frac{Q_{vd}}{\lambda}\Delta^2 = 0 \tag{4.10}$$

4.4.1.2 Nonuniform Grids

For nonuniform grids, the spaces from node P to its adjacent nodes N, S, E, W, I, O are written as $S_N\Delta$, $S_S\Delta$, $S_E\Delta$, ...$S_O\Delta$. Then, Eq. (4.9) can be used. The general equation of heat conduction in the steady state may be transformed into its finite difference version as follows:

$$\frac{2}{\Delta^2}\left[\frac{1}{S_N(S_N + S_S)}T_N - \frac{1}{S_N S_S}T_P + \frac{1}{S_S(S_N + S_S)}T_S\right]$$

$$+ \frac{2}{\Delta^2}\left[\frac{1}{S_E(S_E + S_W)}T_E - \frac{1}{S_E S_W}T_P + \frac{1}{S_W(S_W + S_E)}T_W\right] \tag{4.11}$$

$$+ \frac{2}{\Delta^2}\left[\frac{1}{S_o(S_o + S_I)}T_0 - \frac{1}{S_o S_I}T_P + \frac{1}{S_I(S_I + S_o)}T_I\right] + \frac{Q_{vd}}{\lambda} = 0$$

Similarly, if Q_{vd} changes with position or temperature, it takes its value at point P.

Note that when all S terms take a value of 1, Eq. (4.11) becomes that in uniform grids like Eq. (4.10). Generally, Eq. (4.11) is revised as

$$a_N T_N + a_S T_S + a_E T_E + a_W T_W + a_I T_I + a_o T_o + a_P T_P + a_C = 0 \tag{4.12}$$

Equation (4.12) is equivalent to Eqs. (4.11) and (4.10). For uniform grids,

$$
\begin{cases}
a_N = a_S = a_E = a_W = a_I = a_O = 1 \\[2mm]
a_P = -6 \\[2mm]
a_C = \dfrac{Q_{vd}}{\lambda}\Delta^2
\end{cases}
$$

and for nonuniform grids,

$$
\begin{cases}
a_N = \dfrac{2}{\Delta^2}\dfrac{1}{S_N\left(S_S + S_N\right)} \\[4mm]
a_P = -\dfrac{2}{\Delta^2}\left(\dfrac{1}{S_N S_S} + \dfrac{1}{S_E S_W} + \dfrac{1}{S_O S_I}\right) \\[4mm]
a_C = \dfrac{Q_{vd}}{\lambda}
\end{cases}
$$

where a_S, a_E, a_W, a_I, a_O take similar expression as a_N.

Equation (4.12) is the general form of the finite difference equation for temperature field. Divide the domain related to the workpiece into uniform or nonuniform grids. The number of nodes is n. Each node has a finite difference equation like Eq. (4.12). There are n finite difference equations in total, which are actually algebraic equations. Solving these algebraic equations simultaneously, we get the temperature values of n nodes.

4.4.2 Energy Balance Method

The finite difference equation for determining the temperature at a node may be derived by the energy balance method. There is a control volume surrounding a node. Consider the heat flux in and out of the faces of the control volume and get this node temperature equation expressed by the temperature values of the adjacent nodes. The same finite difference equations are obtained as before, but this energy balance method is more convenient and easy to use, especially for nonuniform grids and complicated boundary conditions.

4.4.2.1 Uniform Grids

We consider a control volume shown in Figure 4.4. It is a case of a uniform grid with equal node space $\Delta = \Delta x = \Delta y = \Delta z$. The control volume has six faces with area Δ^2. If the heat transfer coefficient K_{ij} between node i and its adjacent node j is calculated, we write the heat flux from j to i as

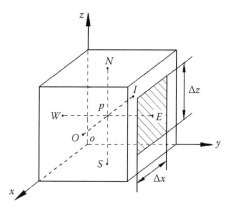

FIGURE 4.4
The control volume containing node P.

$$Q_{ij} = K_{ij}\left(T_j - T_i\right) \tag{4.13}$$

In the steady state, the energy balance equation for node i can be written as

$$\sum Q_{ij} + Q_{vd} \cdot V_i = \sum_i K_{ij}\left(T_j - T_i\right) + Q_{vd} \cdot V_i = 0 \tag{4.14}$$

where $\sum$ means to sum for all faces. This equation is based on the heat flux from six faces, adding the internal heat generation term Q_{vd}. The control volume is as follows:

$$V_i = \Delta x \Delta y \Delta z = \Delta^3$$

The heat transfer coefficient K_{ij} for inner nodes can be calculated by

$$K_{ij} = \frac{\lambda A_{ij}}{L_{ij}} \tag{4.15}$$

where λ is thermal conductivity, A_{ij} is the averaged area of the face between j and i, and L_{ij} is the space between j and i.

Considering the energy balance for node P in Figure 4.4 and using Eqs. (4.13) and (4.15), we obtain

$$\lambda\frac{\Delta^2}{\Delta}\left(T_N - T_P\right) + \lambda\frac{\Delta^2}{\Delta}\left(T_S - T_P\right) + \lambda\frac{\Delta^2}{\Delta}\left(T_E - T_P\right) +$$

$$\lambda\frac{\Delta^2}{\Delta}\left(T_W - T_P\right) + \lambda\frac{\Delta^2}{\Delta}\left(T_I - T_P\right) + \lambda\frac{\Delta^2}{\Delta}\left(T_O - T_P\right) + Q_{vd} \cdot \Delta^3 = 0$$

After manipulation,

$$\frac{\lambda}{\Delta^2}(T_N - 2T_P + T_S) + \frac{\lambda}{\Delta^2}(T_E - 2T_P + T_W) + \frac{\lambda}{\Delta^2}(T_I - 2T_P + T_O) + Q_{vd} = 0$$

$$T_N + T_S + T_W + T_I + T_O + T_E - 6T_P + \frac{Q_{vd} \cdot \Delta^2}{\lambda} = 0$$

It can be seen that this equation is identical to Eq. (4.10).

4.4.2.2 Nonuniform Grids

Here, we discuss a case like that in Figure 4.4, but the node space is varied. The node spaces from P to its adjacent nodes N, S, E, I, O are expressed as $S_N\Delta$, $S_S\Delta$, $S_E\Delta$, ... $S_O\Delta$. The control volume may be calculated as

$$V_V = \frac{1}{2}(S_N + S_S)\Delta \cdot \frac{1}{2}(S_E + S_W)\Delta \cdot \frac{1}{2}(S_I + S_O)\Delta$$

$$= \frac{\Delta^3}{8}(S_N + S_S)(S_E + S_W)(S_I + S_O)$$

For the face area for the north or south side (A_N, A_S),

$$A_N = A_S = \frac{1}{2}(S_E + S_W)\Delta \cdot \frac{1}{2}(S_I + S_O)\Delta = \frac{\Delta^2}{4}(S_E + S_W)(S_I + S_O)$$

$$= \frac{V_V}{\frac{\Delta}{2}(S_N + S_S)}$$

Other faces can be obtained in the same way.

The energy balance for this control volume can be expressed as

$$\lambda \frac{V_V}{\frac{\Delta}{2}(S_N + S_S)} \cdot \frac{1}{S_N\Delta}(T_N - T_P) + \lambda \cdot \frac{V_V}{\frac{\Delta}{2}(S_N + S_S)} \cdot \frac{1}{S_E\Delta}(T_S - T_P)$$

+ Four similar terms for nodes E, W, I, and $O + Q_{vd} \cdot V_V = 0$

This equation is equivalent to

$$\frac{2}{\Delta^2}\left[\frac{1}{S_N(S_N + S_S)}T_N - \frac{1}{S_N S_S}T_P + \frac{1}{S_S(S_N + S_S)}T_S\right] + \text{Two similar terms} + \frac{Q_{vd}}{\lambda} = 0$$

This equation is identical to Eq. (4.11).

In conclusion, two methods for getting finite difference equations of thermal conduction provide identical results. It is therefore suggested that one can use whichever is more convenient for solving the specific problem.

4.5 Finite Difference Equations of Transient Heat Conduction

The main difference between transient and steady-state heat conduction is the time variable. Thus, for transient problems, not only the geometric domain but also the time elapsed have to be discretized. In Section 4.3, we found that the finite difference equations derived by the substitution method of partial differential equations and the energy balance method are identical, even for transient cases. Therefore, this section mainly introduces how to derive finite difference equations in uniform grids for transient heat conduction by the substitution method of partial differential equations.

For a transient problem, the domain is divided into uniform grids with homogeneous thermal properties. Set $\Delta = \Delta x = \Delta y = \Delta z$ and pay attention to the inner node P shown in Figure 4.3. Without considering the internal heat generation term first, the heat conduction equation can be written as

$$\frac{\partial^2 T}{\partial x} + \frac{\partial^2 T}{\partial y} + \frac{\partial^2 T}{\partial z} = \frac{1}{a}\frac{\partial T}{\partial t} \tag{4.16}$$

where a is thermal diffusivity, $a = (\lambda/\rho C_p)$.

4.5.1 Explicit Method

For Eq. (4.16), the left-hand side takes the second-order central difference at time t, while the right-hand side uses a forward difference of temperature at node P. Then, the finite difference equation for node P is expressed as

$$\frac{T_O - 2T_P + T_I}{\Delta^2} + \frac{T_E - 2T_P + T_W}{\Delta^2} + \frac{T_N - 2T_P + T_S}{\Delta^2} = \frac{1}{a}\frac{T_P' - T_P}{\Delta t} \tag{4.17}$$

where $T_P' = T_P(t + \Delta t)$, $T_P = T_P(t)$, and other terms without (') take their values at time t. After manipulation, we have

$$T_O + T_I + T_E + T_W + T_N + T_S + \left(\frac{1}{F_0} - 6\right)T_P = \frac{1}{F_0}T_P' \tag{4.18}$$

where F_0 is Fourier number, $F_0 = (a\Delta t/\Delta^2) = (\lambda\Delta t/\rho C_p\Delta^2)$.

In Eq. (4.18), the term $T_P' = T_P(t + \Delta t)$ can be calculated from the temperature values of its own and six adjacent nodes at time t, which are all known. Thus, according to the temperature values of nodes at a former time instant, those values at the next time instant can be directly calculated by Eq. (4.18). By performing such a calculation procedure step by step, the temperature values of various nodes at each time step can be obtained.

If there is an internal heat generation term, the value of $(Q_{vd}/\lambda)\Delta^2$ at node P and time t can be supplemented in Eq. (4.18).

4.5.2 Implicit Method

For Eq. (4.16), the left-hand side takes the second-order central difference at time $t + \Delta t$, while the right-hand side uses a backward difference of temperature at node P. Then, the finite difference equation for node P is expressed as

$$\frac{T'_O - 2T'_P + T'_I}{\Delta^2} + \frac{T'_E - 2T'_P + T'_W}{\Delta^2} + \frac{T'_N - 2T'_P + T'_S}{\Delta^2} = \frac{1}{a}\left(\frac{T'_P - T_P}{\Delta t}\right) \qquad (4.19)$$

It can be simplified as

$$T'_O + T'_I + T'_E + T'_W + T'_N + T'_S - \left(6 + \frac{1}{F_0}\right)T'_P = -\frac{1}{F_0}T_P \qquad (4.20)$$

Thus, by this implicit method, it is impossible to get the temperature values at time $t + \Delta t$ just from those at time t. Rather, at each time step, a group of equations has to be solved simultaneously to get new temperature values for all nodes. The scheme is always numerically stable and convergent but usually more numerically intensive than the explicit method as it requires solving a system of numerical equations at each time step.

4.5.3 Crank-Nicolson Method

For Eq. (4.16), the left-hand side takes the second-order central difference at time $t + (1/2)\Delta t$, while the right-hand side is dealt with as before. We get

$$\frac{1}{2}\left[\frac{T'_O - 2T'_P + T'_I}{\Delta^2} + \frac{T'_E - 2T'_P + T'_W}{\Delta^2} + \frac{T'_N - 2T'_P + T'_S}{\Delta^2}\right] +$$

$$\frac{1}{2}\left[\frac{T_O - 2T_P + T_I}{\Delta^2} + \frac{T_E - 2T_P + T_W}{\Delta^2} + \frac{T_N - 2T_P + T_S}{\Delta^2}\right] = \frac{1}{a}\left(\frac{T'_P - T_P}{\Delta t}\right) \qquad (4.21)$$

After manipulations,

$$\frac{1}{2}(T'_O + T'_I + T'_W + T'_E + T'_N + T'_S) - \left(3 + \frac{1}{F_0}\right)T'_P$$

$$= -\frac{1}{2}(T_O + T_I + T_W + T_E + T_N + T_S) + \left(3 - \frac{1}{F_0}\right)T_P \qquad (4.22)$$

The scheme is always numerically stable and convergent but usually more numerically intensive as it requires solving a system of numerical equations on each time step. Usually the Crank-Nicolson scheme is the most accurate scheme for small time steps. The explicit scheme is the least accurate and can be unstable, but it is also the easiest to implement and the least numerically intensive. The implicit scheme works the best for large time steps.

4.5.4 Weighted Difference Scheme

If the node temperature values are the weighted summation of their values at time $t + \Delta t$ and t through a weighted factor $\theta_s (0 \le \theta_s \le 1)$, then we have

$$\theta_s \left(T'_O + T'_I + T'_E + T'_W + T'_N + T'_S - 6T'_P \right) + \left(1 - \theta_s \right) \left(T_O + T_I + T_E + T_W + T_N + T_S - 6T_P \right)$$

$$= \frac{1}{F_0} \left(T'_P - T_P \right) \tag{4.23}$$

After manipulation,

$$\theta_s \left(T'_O + T'_I + T'_E + T'_W + T'_N + T'_S \right) - \left(6\theta_s + \frac{1}{F_0} \right) T'_P =$$

$$- \left(1 - \theta_s \right) \left(T_O + T_I + T_W + T_E + T_N + T_S \right) + \left[6 \left(1 - \theta_s \right) - \frac{1}{F_0} \right] T_P \tag{4.24}$$

If $\theta_s = 0$, Eq. (4.24) is identical to Eq. (4.18), and it is an explicit scheme.

If $\theta_s = 1$, Eq. (4.24) is identical to Eq. (4.20), and it is an implicit scheme.

If $\theta_s = 0.5$, Eq. (4.24) is identical to Eq. (4.22), and it is a Crank-Nicolson scheme.

Therefore, different schemes may be obtained when θ_s varies from 0 to 1. Except for the case of $\theta_s = 0$, all other schemes are implicit.

Under the condition of $0 \le \theta_s \le 1$, for given Δt and Δ, the numerical error rises with increasing θ_s while it gets more stable. When $\theta_s = 2/3$, Eq. (4.24) is called Galerkin scheme, which is often applied.

4.6 Boundary, Interface, and Inhomogeneous Workpieces

4.6.1 Nodes at Boundary

Until now, we have just discussed the inner nodes. Now, the task is to provide an equation for each boundary node. For heat conduction, taking point B in Figure 4.5 as an example, there are two conditions:

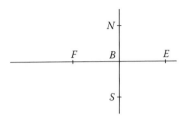

FIGURE 4.5
Boundary node B.

1. The temperature at the boundary is known, that is,

$$T_B = T_G \tag{4.25}$$

   where T_G is the known temperature for a given case. Equation (4.25) is a simple equation for node B.

2. The boundary temperature is unknown, but the temperature gradient $(\partial T/\partial n)$ is given as

$$\frac{\partial T}{\partial n} = q_0 \ (q_0 \text{ is known}) \tag{4.26}$$

$$\frac{\partial T}{\partial n} = -\frac{\alpha}{\lambda}\left(T_B - T_f\right) \tag{4.27}$$

where α is the heat loss coefficient, λ is the thermal conductivity, and T_B and T_f are the boundary and ambient temperatures, respectively.
The energy balance method is used to transform Eqs. (4.26) and (4.27) into finite difference equations at the boundary.

4.6.2 Steady-State Boundary Conditions

In Figure 4.6, node B locates at the boundary and has four adjacent nodes N, S, E, and A (ambient surroundings). For the control volume defined by the dotted line, the energy balance method is employed, and we get

$$\lambda\frac{\Delta}{2}\frac{T_N - T_B}{\Delta} + \lambda\frac{\Delta}{2}\frac{T_S - T_B}{\Delta} + \lambda\Delta\frac{T_E - T_B}{\Delta} + \alpha\Delta\left(T_f - T_B\right) = 0 \tag{4.28}$$

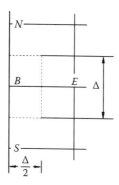

FIGURE 4.6
Node B at the boundary.

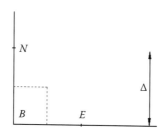

FIGURE 4.7
Node B at the corner.

After manipulation,

$$T_B = \frac{\frac{1}{2}(T_N + T_S) + T_E + B_i T_f}{2 + B_i} \tag{4.29}$$

where B_i is the Biot number, $B_i = \alpha \Delta / \lambda$.

For the corner node B shown in Figure 4.7, the energy balance method is employed, and we get

$$\lambda \frac{\Delta}{2} \frac{T_N - T_B}{\Delta} + \lambda \frac{\Delta}{2} \frac{T_E - T_B}{\Delta} + \alpha \left(\frac{\Delta}{2} + \frac{\Delta}{2} \right)(T_f - T_B) = 0 \tag{4.30}$$

After manipulation,

$$T_B = \frac{\frac{1}{2}(T_N + T_E) + B_i T_f}{1 + B_i} \tag{4.31}$$

4.6.3 Transient Boundary Condition

As shown in Figure 4.8, there are two kinds of boundary nodes: node 1 on the boundary and node 0 at the corner.

For node 0, the energy balance method is employed:

$$\rho C_p \frac{1}{4} \Delta^2 \frac{T_0' - T_0}{\Delta t} = \lambda \frac{\Delta}{2} \frac{T_3 - T_0}{\Delta} + \lambda \frac{\Delta}{2} \frac{T_1 - T_0}{\Delta} + \alpha \left(\frac{\Delta}{2} + \frac{\Delta}{2} \right) (T_f - T_0)$$

After manipulation, we get the explicit equation

$$T_0' = 2F_0 \left[T_1 + T_3 + 2B_i T_f + T_0 \left(\frac{1}{2F_0} - 2 - 2B_i \right) \right] \qquad (4.32)$$

It may be expressed by an implicit form:

$$T_0 = 2F_0 \left[-T_1' - T_3' - 2B_i T_f' + T_0' \left(\frac{1}{2F_0} + 2 + 2B_i \right) \right] \qquad (4.33)$$

where $F_0 = (\lambda \Delta t / \rho C_p \Delta^2)$, $B_i = (\alpha \Delta / \lambda)$.

For node 1, its explicit form is

$$T_1' = F_0 \left[T_0 + 2T_2 + T_4 + 2B_i T_f + T_1 \left(\frac{1}{F_0} - 4 - 2B_i \right) \right] \qquad (4.34)$$

and its implicit form is

$$T_1 = F_0 \left[-T_0' - 2T_2' - T_4' - 2B_i T_f' + T_1' \left(\frac{1}{F_0} + 4 + 2B_i \right) \right] \qquad (4.35)$$

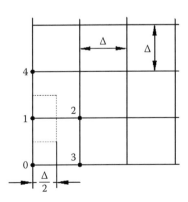

FIGURE 4.8
Boundary nodes 1 and 0.

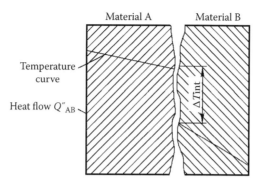

FIGURE 4.9
Schematic of material interface.

This discussion takes the two-dimensional case as an example. For the three-dimensional situation, the same method may be used to obtain finite difference equations for the boundary.

4.6.4 Interfaces

For a composite structure consisting of different materials, there are interfaces between these different materials. As shown in Figure 4.9, when heat flow conducts through the interface, there is a temperature gap between material A and material B. Assuming that the temperature gap ΔT_{int} can be determined by extrapolation of local temperatures of each material at the interface, the heat flux in the steady state is as follows:

$$Q''_{AB} = -\lambda_A \frac{\partial T}{\partial x}\bigg|_A = -\lambda_B \frac{\partial T}{\partial x}\bigg|_B \qquad (4.36)$$

The heat conduction coefficient in unit area at the interface may be written as

$$a_i = \frac{Q''_{AB}}{\Delta T_{int}} \qquad (4.37)$$

If the contact condition between two materials is perfect (e.g., melted together), there is no temperature gap ΔT_{int}. This is the case of $a_i \rightarrow \infty$, so that $T_A = T_B$. But, for usual cases, $\lambda_A \neq \lambda_B$, and the temperature gradient is not continuous. We can use the energy balance method to derive finite difference equations for these cases.

4.6.5 Inhomogeneous Material Properties

If the temperature does not change very much, then thermal conductivity may be taken as constant. But, if the temperature is expected to vary greatly, then the temperature dependence of thermal conductivity has to be considered. For example, thermal conductivity may be written as

$$\lambda = \lambda(T) \tag{4.38}$$

If the relation between λ and T is linear, the following equation may be employed:

$$\lambda\left(\frac{T_i + T_j}{2}\right) = \frac{\lambda(T_i) + \lambda(T_j)}{2} \tag{4.39}$$

In practice, the averaged value of thermal conductivity λ over the temperature range may be used. Based on the calculated temperature field, both λ and T are continuously iterated until satisfactory convergence is achieved.

4.7 Accuracy, Stability, and Convergence of Finite Difference Solutions

4.7.1 Error

The *error* in the solution of a method is defined as the difference between its approximation and the exact analytical solution. The two sources of error in finite difference methods are *round-off error*, the loss of precision due to computer rounding of decimal quantities, and *truncation error* or *discretization error*, the difference between the exact solution of the finite difference equation and the exact quantity assuming perfect arithmetic (that is, assuming no round-off).

To use a finite difference method to attempt to solve (or, more generally, approximate the solution to) a problem, one must first discretize the domain of the problem. This is usually done by dividing the domain into a uniform grid. Note that this means that finite difference methods produce sets of discrete numerical approximations to the derivative, often in a "time-stepping" manner.

An expression of general interest is the local truncation error of a method. *Local truncation error* refers to the error from a single application of a method. That is, it is the quantity $f'(x_i) - f'_i$ if $f'(x_i)$ refers to the exact value and f'_i to the numerical approximation. The remainder term of a Taylor polynomial is convenient for analyzing the local truncation error. For example, according to Taylor's expansion in Section 4.2.2, (1) forward and backward differences have a local truncation error of the order Δx, that is, the error is proportional to the space step; (2) the local truncation error of central difference is proportional to the square of the space step, that is, the order $(\Delta x)^2$; and (3) a second-order central difference for the space derivative has a local truncation error of the order $(\Delta x)^2$.

Convergence means that by making the time and space step finer, the difference between the approximation and the exact solution is narrowed. Not all finite difference solutions of partial differential equations can achieve convergence.

4.7.2 Stability

The *stability* of a finite difference scheme means that the numerical error must be controlled within an allowable range as calculation steps increase. Ensuring the stability of the finite difference scheme is essential. The initial and boundary conditions are usually based on the measured data, which contain errors inevitably, and there is also round-off error during numerical calculation. If these errors are continuously amplified during the calculation process, the solution will be unstable, and the calculated results are meaningless.

4.7.2.1 Explicit Scheme

Now, check out a general explicit scheme of transient heat conduction:

$$T_O + T_I + T_E + T_W + T_N + T_S - \left(6 - \frac{1}{F_O}\right)T_P = \frac{1}{F_O}T_P' \tag{4.40}$$

where $F_O = (\alpha\Delta t / \Delta^2) = (\lambda\Delta t / \rho C_p\Delta^2)$. In fact, for a uniform grid ($\Delta = \Delta x = \Delta y = \Delta z$), the coefficients of $T_j = (j = O, I, N, S, E, W)$ are all positive. If the coefficient of T_P is negative, then the lower is the temperature T_P at time t, the higher is the temperature T_P' (the temperature at next time step $t + \Delta t$). This will violate the principle of thermodynamics. Thus, the time and space step must be so selected that the coefficient of T_P is positive, which is the sufficient condition for keeping the explicit scheme stable. To apply this simple rule to various difference schemes, we obtain the following limitations for the Fourier number $F_0 = (a\Delta t/\Delta^2)$ and the Biot number Bi $= (\Delta\alpha/\lambda)$:

For inner nodes,
One-dimensional coordinate:

$$F_0 \leq \frac{1}{2} \tag{4.41}$$

Two-dimensional coordinate:

$$F_0 \leq \frac{1}{4} \tag{4.42}$$

Three-dimensional coordinate:

$$F_0 \leq \frac{1}{6}$$

(4.43)

For boundary nodes (heat losses),

One-dimensional coordinate:

$$F_0 \leq \frac{1}{2(1+B_i)}$$

(4.44)

Two-dimensional coordinate:

$$F_0 \leq \frac{1}{2(2+B_i)}$$

(4.45)

Three-dimensional coordinate:

$$F_0 \leq \frac{1}{2(3+B_i)}$$

(4.46)

Two-dimensional coordinate (corner node):

$$F_0 \leq \frac{1}{4(1+B_i)}$$

(4.47)

According to this simple rule, insertion of a steady inner heat source term in finite difference schemes does not affect the stability.

Clearly, the limitation of Eqs. (4.44)–(4.47) is unfavorable for the heat conduction case with a higher Biot number. For example, for given values of α, λ, Δ in a transient three-dimensional situation, the time step is limited as follows:

$$\Delta t = \frac{\Delta^2}{a} F_0 \leq \frac{\Delta^2}{2a(3+B_i)}$$

(4.48)

For larger B_i, Δt must be less to retain stability. This causes an enormous increase in workload but without marked improvement of accuracy. If the grid nodes are rearranged to increase the space step Δ, the limitation to Δt is widened, but accuracy becomes worse.

4.7.2.2 Implicit Scheme

If the Fourier number satisfies the following condition:

$$F_0 \leq \frac{1}{2(1-2\theta_s)}$$

(4.49)

then the weighted difference equation (4.24) is stable for inner nodes. For $(1/2) \le \theta_s \le 1$, Eq. (4.49) is clearly satisfied. The scheme is always stable for any value of F_0. The implicit scheme ($\theta_s = 1$), the Crank-Nicolson scheme ($\theta_s = 1/2$), and the Galerkin scheme ($\theta_s = 2/3$) are always stable.

4.7.3 Accuracy

If the difference scheme is numerically stable, then the round-off error is much less or is of no effect on the accuracy; the truncation error has marked influence on accuracy. The error for inner nodes is as follows:

Explicit scheme:

$$|E_t| \le A_1 \Delta t + B_1 (\Delta x)^2 \tag{4.50}$$

Implicit scheme:

$$|E_t| \le A_2 \Delta t + B_2 (\Delta x)^2 \tag{4.51}$$

Crank-Nicolson scheme:

$$|E_t| \le A_3 (\Delta t)^2 + B_3 (\Delta x)^2 \tag{4.52}$$

Usually, the Crank-Nicolson scheme is the most accurate scheme for small time steps. The explicit scheme is the least accurate and can be unstable, but it is also the easiest to implement and the least numerically intensive. The implicit scheme works the best for large time steps. For any scheme, the calculation accuracy with a constant time step is improved as the number of steps increases. It is better to employ the varying time step. A smaller time step is selected at the initial stage of calculation, while the time step is gradually increased after some time elapses. In such a way, the accuracy is ensured while computer time is saved.

4.8 Finite Difference Equations in Nonrectangular Coordinates

It is easier to describe geometry of some bodies by a nonrectangular coordinate system. For example, a cylindrical coordinate system can describe axis-symmetric problems more conveniently. The general equation of heat conduction in a cylindrical coordinate system can be transformed to finite difference equations in the same way as in a rectangular coordinate system.

The general equation of heat conduction in a cylindrical coordinate system is as follows:

$$\frac{\partial^2 T}{\partial r^2} + \frac{1}{r}\frac{\partial T}{\partial r} + \frac{1}{r^2}\frac{\partial^2 T}{\partial \theta^2} + \frac{\partial^2 T}{\partial z^2} + \frac{Q_{vd}}{\lambda} = \frac{1}{a}\frac{\partial T}{\partial t} \tag{4.53}$$

Figures 4.10 and 4.11 demonstrate the geometrical drawings of a cylindrical coordinate system, and z, r, and θ have equal increments, respectively.

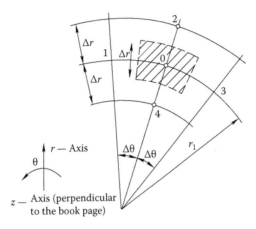

FIGURE 4.10
Schematic of cylindrical coordinate system. (From Croft, D.R. *Finite difference calculation of heat transfer* [translated by Zhang F L], Metallurgy Industry Press, Beijing, 1982.)

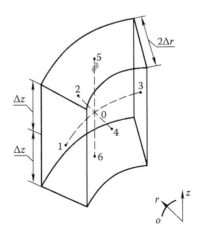

FIGURE 4.11
Control volume in cylindrical coordinate system. (From Croft D.R. *Finite difference calculation of heat transfer* [translated by Zhang F L], Metallurgy Industry Press, Beijing, 1982.)

Employing the substitution method of partial differential equations in Eq. (4.53), we get the finite difference equation for node 0:

$$\left[\frac{T_2 - 2T_0 + T_4}{(\Delta r)^2} + \frac{T_2 - T_4}{2r_0 \Delta r} + \frac{T_1 - 2T_0 + T_3}{r_0^2 (\Delta\theta)^2} + \frac{T_5 - 2T_0 + T_6}{(\Delta z)^2} + \frac{Q_{vd}}{\lambda}\right] = \frac{T_0' - T_0}{a\Delta t} \qquad (4.54)$$

This equation is an explicit scheme. The energy balance method can also be used to derive the finite difference equation in a cylindrical coordinate system. Referring to Figures. 4.10 and 4.11, according to the heat conduction rate at node 0 from its adjacent nodes, the internal heat generation rate in the control volume around node 0, and the temperature increase rate in the same control volume, we obtain the energy balance of node 0.

Heat conduction rate from node 1 to node $0 = \lambda(\Delta z \Delta r)\dfrac{T_1 - T_0}{r_0 \Delta\theta}$

Heat conduction rate from node 2 to node $0 = \lambda(\Delta\theta \Delta z)\left(r_0 + \dfrac{\Delta r}{2}\right)\dfrac{T_2 - T_0}{\Delta r}$

Heat conduction rate from node 3 to node $0 = \lambda(\Delta z \Delta r)\dfrac{T_3 - T_0}{r_0 \Delta\theta}$

Heat conduction rate from node 4 to node $0 = \lambda(\Delta\theta \Delta z)\left(r_0 - \dfrac{\Delta r}{2}\right)\dfrac{T_4 - T_0}{\Delta r}$

Heat conduction rate from node 5 to node $0 = \lambda(r_0 \Delta\theta \Delta r)\dfrac{T_5 - T_0}{\Delta z}$

Heat conduction rate from node 6 to node $0 = \lambda\left(r_0 \Delta\theta \Delta r\dfrac{T_6 - T_0}{\Delta z}\right)$

Energy variation of control volume $= \rho C_p (r_0 \Delta\theta \Delta z \Delta r)\dfrac{T_0' - T_0}{\Delta t}$

Internal heat generation rate in the control volume $= Q_{vd}\,(r_0 \Delta\theta \Delta r \Delta z)$

Thus, the energy balance at node 0 can be expressed as

$$\left[\frac{T_2 - 2T_0 + T_4}{(\Delta r)^2} + \frac{T_2 - T_4}{2r_0 \Delta r} + \frac{T_1 - 2T_0 + T_3}{r_0^2 (\Delta\theta)^2} + \frac{T_5 - 2T_0 + T_6}{(\Delta z)^2} + \frac{Q_{vd}}{\lambda}\right] = \frac{T_0' - T_0}{a\Delta t} \qquad (4.55)$$

Eq. (4.54) is identical to Eq. (4.55). Except a node at the symmetric axis, Eq. (4.55) is applicable to any node. At the symmetric axis, $(\partial T/\partial r) = 0$, so that $1/r(\partial T/\partial r)$ is under range $(= 0/0)$. However, if $r \to 0$, then

$$\frac{1}{r}\frac{\partial T}{\partial r} \to \frac{\partial^2 T}{\partial r^2} \tag{4.56}$$

For a case of cylindrical axis symmetry and $Q_{vd} = 0$, the temperature of the nearest adjacent node may be taken as that at $r = 0$. For the nonsymmetric steady-state case $(\partial T/\partial \theta \neq 0)$, taking the averaged temperature value of adjacent nodes as the temperature at $r = 0$.

For Eq. (4.55), if the terms at the left-hand side are calculated at the previous time step, it is an explicit scheme. If the terms at the left-hand side are calculated at a new time step, it is an implicit scheme:

$$\frac{T_2' - 2T_0' + T_4'}{(\Delta r)^2} + \frac{T_3' - T_4'}{2r_0\Delta r} + \frac{T_1' - 2T_0' + T_2'}{r_0^2(\Delta\theta)^2} + \frac{T_5' - 2T_0' + T_6'}{(\Delta z)^2} = \frac{T_0' - T_0}{a\Delta t} \tag{4.57}$$

4.9 Solution Method for Finite Difference Equations

The finite difference scheme of heat conduction in steady and transient states transforms the partial differential equation of heat flow into a set of algebraic equations of the nodes; these equations have to be solved simultaneously to get the temperature values at the nodes. First, we discuss the equation sets with n unknowns and n equations (corresponding to the heat conduction case of n nodes):

$$\begin{cases} a_{11}T_1 + a_{12}T_2 + \cdots\cdots + a_{1n}T_n = b_1 \\ a_{21}T_1 + a_{22}T_2 + \cdots\cdots + a_{2n}T_n = b_2 \\ a_{n1}T_1 + a_{n2}T_2 + \cdots\cdots + a_{nn}T_n = b_n \end{cases} \tag{4.58}$$

Because each unknown node temperature is just related to the temperature values of its adjacent nodes, the coefficient matrix of linear equations set Eq. (4.58) is a sparse matrix, which is more suitable to solve by the iteration method.

Write Eq. (4.58) in a compact way:

$$\sum_{j=1}^{n} a_{ij}T_j = b_i \qquad (i = 1, 2, \cdots, n) \tag{4.59}$$

The principle of iteration is as follows: To constitute a vector sequence $\{T_1, T_2, ..., T_n\}$, make it be convergent to an aim vector sequence $\{T_1^*, T_2^*, \cdots, T_n^*\}$, while $\{T_1^*, T_2^*, \cdots, T_n^*\}$ is the exact solution of the set of Eq. (4.58). Based on the constituting approaches of the vector sequence, the iteration methods may be classified as simple iteration, Gauss-Seidel iteration, superrelaxation iteration, and so on.

4.9.1 Simple Iteration Method

The final aim of iteration is to get the solution of the set of Eq. (4.58), that is, T_1, T_2, ..., T_n.

If the elements at the diagonal line of the coefficient matrix $[A]$ are not zero, that is, $a_{ii} \neq 0$ ($i = 1, 2, ..., n$), then Eq. (4.58) may be written as

$$\begin{cases} T_1 = a_{11}^{-1}\left(b_1 - a_{12}T_2 - \cdots - a_{1n}T_n\right) \\ T_2 = a_{22}^{-1}\left(b_2 - a_{21}T_1 - a_{23}T_3 - \cdots - a_{2n}T_n\right) \\ \qquad \vdots \qquad \vdots \\ T_n = a_{nn}^{-1}\left(b_n - a_{n1}T_1 - a_{n2}T_2 - \cdots - a_{nn-1}T_{n-1}\right) \end{cases} \tag{4.60}$$

Any equation in this set may be written as

$$T_i = a_{ii}^{-1}\left(b_i - \sum_{\substack{j=1 \\ j \neq i}}^{n} a_{ij}T_j\right) \qquad (i = 1,2,\cdots,n) \tag{4.61}$$

By arbitrarily selecting $T_i^{(0)}$ $(i = 1,2,\cdots,n)$ as the 0th approximation of the solution and substituting it into the right-hand side of Eq. (4.61), we obtain

$$T_i^{(1)} = a_{ii}^{-1}\left(b_i - \sum_{\substack{j=1 \\ j \neq i}}^{n} a_{ij}T_j^{(0)}\right) \qquad (i = 1,2,\cdots,n)$$

This is the first approximation of the solution. Substituting the first approximation into the right-hand side of Eq. (4.61) again, we obtain the second approximation of the solution. Generally, by substituting the Kth approximation $T_i^{(K)}$ into the right-hand side of Eq. (4.61), we get the ($K + 1$)th approximation:

$$T_i^{(K+1)} = a_{ii}^{-1}\left(b_i - \sum_{\substack{j=1 \\ j \neq i}}^{n} a_{ij}T_j^{(K)}\right) \qquad (i = 1,2,\cdots,n) \tag{4.62}$$

The sequence $\{T_1^{(K)}, T_2^{(K)}, \cdots, T_n^{(K)}\}$ $(K = 0,1,2,\cdots)$ obtained in such a way is the approximation solution of Eq. (4.61). As long as Eq. (4.61) has a unique solution, the sequence $\{T_1^{(K)}, T_2^{(K)}, \cdots, T_n^{(K)}\}$ $(K = 0,1,2,\cdots)$ surely converges to the exact solution $\{T_1^*, T_2^*, \cdots, T_n^*\}$ if $K \to \infty$, whatever 0th approximation is selected. In fact, as long as K is sufficiently large, $\{T_1, T_2, ..., T_n\}$ approximates the exact solution with enough accuracy. Usually, if the difference between two successive iterations $T_i^{(K+1)}$ and $T_i^{(K)}$ $(i = 1, 2, ... , n)$ is less than a suitable infinitesimal ε $(\varepsilon > 0)$,

$$\left| T_i^{(K+1)} - T_i^{(K)} \right| < \varepsilon \qquad (i = 1,2,\cdots,n) \qquad (4.63)$$

then iteration stops. $T_i^{(K+1)}$ $(i = 1,2,\cdots,n)$ is taken as the final approximation of Eq. (4.58).

4.9.2　Gauss-Seidel Method

The simple iteration method needs duplicate storage of both new and old component values $T_i^{(K+1)}$, $T_i^{(K)}$ $(i = 1,2,\cdots,n)$ for the vector $\{T_1, T_2, ..., T_n\}$. To save computer storage and speed convergence, the Gauss-Seidel iteration method remedies this drawback by using each new component of the solution as soon as it has been computed, not like simple iteration, for which new component values are used only after the entire sweep has been completed.

Write the first equation of Eq. (4.60) as

$$T_1^{(K+1)} = a_{11}^{-1}\left(b_1 - a_{12}T_2^{(K)} - \cdots - a_{1n}T_n^{(K)}\right)$$

$$T_1^{(K+1)} = a_{11}^{-1}\left(b_1 - \sum_{j=2}^{n} a_{1j}T_j^{(K)}\right)$$

In the second equation in Eq. (4.60), substitute $T_1^{(K)}$ by the computed $T_1^{(K+1)}$, that is,

$$T_2^{(K+1)} = a_{22}^{-1}\left(b_2 - a_{21}T_1^{(K+1)} - \sum_{j=3}^{n} a_{2j}T_j^{(K)}\right)$$

Similarly, we get the general expression for $T_i^{(K+1)}$:

$$T_i^{(K+1)} = a_{ii}^{-1}\left(b_i - \sum_{j=1}^{i-1} a_{ij}T_j^{(K+1)} - \sum_{j=i+1}^{n} a_{ij}T_j^{(K)}\right) \qquad (4.64)$$

This is the Gauss-Seidel iteration method. Use of this method requires arranging the unknown variables in order and to iterate one by one in sequence.

4.9.3 Successive Overrelaxation

The convergence rate of the Gauss-Seidel method can be accelerated by a technique called *successive overrelaxation*.

For the set of Eq. (4.58), first assume the 0th approximated vector $\{T_1^{(0)}, T_2^{(0)}, \cdots, T_n^{(0)}\}$, then compute the first approximation $\{T_1^{(1)}, T_2^{(1)}, \cdots, T_n^{(1)}\}$ by two steps. For the first step, we use the Gauss-Seidel method

$$Z_i^{(1)} = a_{ii}^{-1}\left(b_i - \sum_{j=1}^{i-1} a_{ij}T_j^{(1)} - \sum_{j=i+1}^{n} a_{ij}T_j^{(0)}\right) \qquad (i = 1, 2, \cdots, n) \tag{4.65}$$

to get the first approximation $Z_i^{(1)}$. For the second step, we use the following equation to improve $Z_i^{(1)}$ and get the first approximation $T_i^{(1)}$:

$$T_i^{(1)} = (1 - \omega)T_i^{(0)} + \omega Z_i^{(1)} \qquad (i = 1, 2, \cdots n) \tag{4.66}$$

where ω is a fixed parameter. We use this $T_i^{(1)}$ to replace $Z_i^{(1)}$.

To eliminate $T_i^{(1)}$ from Eqs. (4.65) and (4.66), we get the equation to determine $T_i^{(1)}$ by $T_i^{(0)}$. Generally, based on $T_i^{(K)}$, compute $T_i^{(K+1)}(K = 0, 1, 2, \cdots, n)$ as follows:

$$T_i^{(K+1)} = (1 - \omega)T_i^{(K)} + \omega a_{ii}^{-1}\left(b_i - \sum_{j=i}^{i-1} a_{ij}T_j^{(K+1)} - \sum_{j=i+1}^{n} a_{ij}T_j^{(K)}\right) \tag{4.67}$$

This is the successive overrelaxation scheme; ω is the relaxation factor.

When $\omega = 1$, Eq. (4.67) is simplified to the equation of the Gauss-Seidel method. For different values of ω, the convergence rate is different. It is possible to find an optimum value ω^* to obtain the fastest convergence. Usually, $1 < \omega^* < 2$. For cases of $1 < \omega < 2$, it is underrelaxation, while it is overrelaxation for $\omega > 1$.

4.10 Case Study of Finite Difference Analysis for a Transient Welding Temperature Field

The general equation of transient heat conduction in welding has the following form:

$$\rho C_p \frac{\partial T}{\partial t} = \frac{\partial}{\partial \xi}\left(\lambda \frac{\partial T}{\partial \xi}\right) + \frac{\partial}{\partial y}\left(\lambda \frac{\partial T}{\partial y}\right) + \frac{\partial}{\partial z}\left(\lambda \frac{\partial T}{\partial z}\right) \tag{4.68}$$

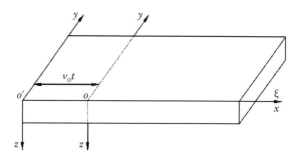

FIGURE 4.12
Schematic of stationary coordinates (ξ, y, z) and moving coordinates (x, y, z).

where (ξ, y, z) is the stationary coordinates with the origin o'; (x, y, z) is the moving coordinates with origin o; the welding arc is moving along the $\xi(x)$-axis at speed v_0, as shown in Figure 4.12.

The relation of stationary and moving coordinates is

$$\xi = x + v_0 t, \ x = \xi - v_0 t$$

Thus,

$$\frac{\partial T}{\partial \xi} = \frac{\partial T}{\partial x}, \qquad \frac{\partial T}{\partial t} = \frac{\partial T}{\partial x}\frac{\partial x}{\partial t} + \frac{\partial T}{\partial t'}\frac{\partial t'}{\partial t}$$

where t' is the time in moving coordinates. In fact, the stationary or moving coordinate just corresponds to space. There is no difference for time at two coordinate systems, that is, ($\partial t'/\partial t$) = 1 . Since $x = \xi - v_0 t$, then ($\partial x/\partial t$) = $-v_0$. Thus, we have

$$\frac{\partial T}{\partial t} = -v_0 \frac{\partial T}{\partial x} + \frac{\partial T}{\partial t'} \tag{4.69}$$

Substituting Eq. (4.69) into Eq. (4.68), we obtain

$$\rho C_p \left(\frac{\partial T}{\partial t'} - v_0 \frac{\partial T}{\partial x} \right) = \frac{\partial}{\partial x}\left(\lambda \frac{\partial T}{\partial x} \right) + \frac{\partial}{\partial y}\left(\lambda \frac{\partial T}{\partial y} \right) + \frac{\partial}{\partial z}\left(\lambda \frac{\partial T}{\partial z} \right) \tag{4.70}$$

For isotropic materials, λ is independent of space coordinates, and Eq. (4.70) becomes

$$\frac{\partial T}{\partial t'} - v_0 \frac{\partial T}{\partial x} = a \left(\frac{\partial^2 T}{\partial x^2} + \frac{\partial^2 T}{\partial y^2} + \frac{\partial^2 T}{\partial z^2} \right) \tag{4.71}$$

When the weld thermal process reaches the quasi-steady state, $(\partial T'/\partial t) = 0$. Thus, the welding thermal conduction equation in the quasi-steady state is written as

$$-\rho C_p v_0 \frac{\partial T}{\partial x} = \frac{\partial}{\partial x}\left(\lambda \frac{\partial T}{\partial x}\right) + \frac{\partial}{\partial y}\left(\lambda \frac{\partial T}{\partial y}\right) + \frac{\partial}{\partial z}\left(\lambda \frac{\partial T}{\partial z}\right) \tag{4.72}$$

$$-v_0 \frac{\partial T}{\partial x} = a\left(\frac{\partial^2 T}{\partial x^2} + \frac{\partial^2 T}{\partial y^2} + \frac{\partial^2 T}{\partial z^2}\right) \tag{4.73}$$

4.10.1 Model of Transient Welding Heat Conduction

The governing equation

$$\frac{\partial T}{\partial t} - v_0 \frac{\partial T}{\partial x} = a\left(\frac{\partial^2 T}{\partial x^2} + \frac{\partial^2 T}{\partial y^2} + \frac{\partial^2 T}{\partial z^2}\right) \tag{4.74}$$

The initial condition:

$$T = T_0 \tag{4.75}$$

Since the symmetry of temperature with respect to the y-axis (weld line),

$$\frac{\partial T}{\partial y}\Big|_{y=0} = 0$$

only half the workpiece is considered in the computation.
On the top surface of the workpiece is considered:

$$\lambda \frac{\partial T}{\partial z} = q(r) - \alpha_{cr}(T - T_\infty) \tag{4.76}$$

where $q(r)$ is the arc heat density, α_{cr} is the surface coefficient of heat losses, and T_∞ is the ambient temperature.
On the bottom surface and side surfaces of the workpiece,

$$-\lambda \frac{\partial T}{\partial z} = \alpha_{cr}(T - T_\infty) \tag{4.77}$$

The arc heat density is assumed to distribute in the double-elliptic mode:

$$q_f(x,y) = \frac{6Q_f}{\pi a_f b_h} \exp\left(-\frac{3x^2}{a_f^2} - \frac{3y^2}{b_h^2}\right), \quad x > 0 \tag{4.78}$$

$$q_r(x,y) = \frac{6Q_r}{\pi a_r b_h} \exp\left(-\frac{3x^2}{a_r^2} - \frac{3y^2}{b_h^2}\right), \quad x \leq 0 \tag{4.79}$$

where

$$Q = \eta IU_a = Q_f + Q_r, \quad Q_f = \frac{a_f}{a_f + a_r} Q, \quad Q_r = \frac{a_r}{a_f + a_r} Q \tag{4.80}$$

4.10.2 Varying Nonuniform Grid

In numerical analysis of weld thermal conduction, generating a suitable grid system (i.e., discretization of the computation domain) is important and directly affects both calculation accuracy and cost. As mentioned, there is discretization error when a partial differential equation is transformed into finite difference equations, which is related to space step in the grid system. The finer is the grid (or the smaller the space step), the lower is the error, but the finer is the grid, the larger is the number of grid points, so that the longer is the computation time. Thus, improvement of accuracy is at the cost of computation time. It is critical to solve the contradiction between the calculation accuracy and time.

During welding, the temperature gradient around the weld pool is steeper and changing fast. It is not suitable to employ a uniform grid for the whole workpiece. For quasi-steady state cases, if a finer uniform grid is used near the weld pool and a coarser nonuniform grid is used for regions far away from the heat source, it will give attention to both calculation accuracy and time to a great degree.

However, for transient problems, the position of the welding heat source varies with time. Then, the finer grid region near the heat source also has to change with time. If the weld pool locates in the finer grid zone during the entire welding process, the number of grid points (nodes) is enormous, which results in lower computation speed and higher expenses.

Therefore, by employing a varying nonuniform grid system (i.e., finer grids are selected for the designated zone while coarser grids are taken for other zones), the designated zone with finer grids moves together with the welding heat source so that the grid automatically becomes finer in the zone under the heat source action while the grid gets coarser when the heat source moves out of this zone. This means that, at any time step, the grid system is regenerated to make the zone around the weld pool take a finer grid and other zones take a coarser grid. In this way, the number of nodes is greatly decreased, so the computation time and workload can be lowered. The numerical analysis efficiency is improved, and the accuracy is ensured. Figure 4.13 is a schematic of a three-dimensional varying nonuniform grid, and 13(a), (b), and (c) are the distribution situations of finer and coarser grids on the workpiece at different time steps.

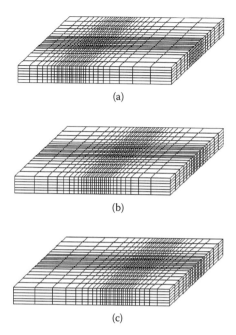

FIGURE 4.13
Schematic of three-dimensional varying nonuniform grid.

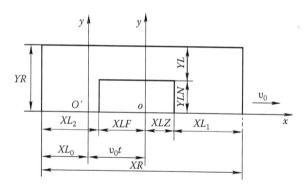

FIGURE 4.14
Schematic of domain division on the top surface of a workpiece.

4.10.2.1 Generation of Varying Nonuniform Grids

The generating method for a three-dimensional varying nonuniform grid is similar to that for the two-dimensional case. For convenience, we take a two-dimensional varying nonuniform grid as an example.

As shown in Figure 4.14, the grid geometry parameters on the top surface of a workpiece are defined according to the relation between stationary and moving coordinates. Here, o' is the starting point of welding, and o is the point where the heat source locates at the moment. XL_0 is the distance from o' to the workpiece edge. XLZ and YLN are the length and width of the zone ahead of the heat source, which should be discretized in a finer grid, respectively. XLF is the length of the zone behind the heat source, which should be discretized in a finer grid. XL1 and XL2 are the length of coarser grid zones ahead and behind the heat source, respectively. YL is the distance from the finer grid zone to the workpiece edge. The length and half-width of the workpiece are represented by XR and YR, respectively.

Define NXZ, NXF, NY, NX1, NX2, NY2 as the number of nodes at segments of XLZ, XLF, YLN, XL1, XL2, YL, respectively. They all take integral values. XL0 takes its value according to specific welding conditions.

4.10.2.1.1 Determining the Region of the Finer Grid

The zone of the finer grid is defined by XLZ, XLF, and YLN. By changing the values of XLZ, XLF, and YLN, the zone of the finer grid can be adjusted. These values are determined based on the rule that ensures that the heat density of the welding heat source is distributed inside the zone of the finer grid. If a double-elliptic heat source mode is used and the distribution parameters are expressed by a_f, a_r, b_h, then XLZ $= a_0 a_f$, XLF $= b_0 a_r$, YLN $= c_0 b_h$, and a_0, b_0, c_0 are adjusting parameters. The range of the finer grid is determined by the weld pool size computed at the last time step. If the half-width of the weld pool at the last time step is W and the distances from the arc centerline to the front edge and rear edge of the pool are L_1 and L_2, respectively, then XLZ $= a_3 L_1$, XLF $= b_3 L_2$, YLN $= c_3 W$. Here, a_3, b_3, and c_3 are adjusting parameters. All six adjusting parameters should have a value larger than 1.

4.10.2.1.2 Determining the Space Step and Node Number in the Finer Grid Zone

Usually, the number of grid points inside the weld pool should be greater than 200. Define the space step in the x and y directions as dx and dy, respectively, then NXZ = XLZ/dx, NXF = XLF/dx, NY = YLN/dy because of the uniform grid system in the finer grid zone.

4.10.2.1.3 Determining the Space Step and Node Number in Coarser Grid Zone

The space step in the coarser zone gradually expands. Referring to Figure 4.14, we get

$$XL_1 = XR - XL_0 - XLZ - v_0 t$$

$$XL_2 = XL_0 - XLF + v_0 t$$

$$YL = YR - YLN$$

where v_0 is the welding speed, and t is the time elapsed from the start of welding. The number of nodes is determined based on the values of XL_1, XL_2, and YL. After checking computations, the following formula may be used to determine the values of NX_1, NX_2, and NY_2:

$$N = \begin{cases} 1 & XL < 2mm \\ XL/2 & 2mm \leq XL < 8mm \\ 4 & 8mm \leq XL < 40mm \\ XL/10 & XL \geq 40mm \end{cases} \qquad (4.81)$$

where XL is the length of zones (NX_1, NX_2, NY), N is the node number (NX_1, NX_2, NY_2).

For the space step in the coarser grid zone,

$$D = \frac{XL}{N} \qquad (4.82)$$

$$dx(i) = \frac{E^{i-1}}{E-1} D \qquad (4.83)$$

where E is the nonuniform factor, and $dx(i)$ is the space step.

Substituting XL_1, XL_2, YL, and NX_1, NX_2, NY_2 into Eq. (4.82) to replace XL and N, respectively, the space step in the segments of XL_1, XL_2, YL can be determined.

If the weld start point is so near to the workpiece edge that XL_2 is less than the space step in the finer grid zone, no coarser grid zone is set behind the arc. In this region,

$$XLF = XL_0 + v_0 t$$
$$NXF = XLF/dx$$
$$dx(i) = XLF/NXF$$

If the weld endpoint is so near to the workpiece edge that XL_1 is less than the space step in the finer grid zone, no coarser grid zone is set ahead of the arc. In this region,

$$XLZ = XR - XL_0 - v_0 t$$
$$NXZ = XLNZ/dx$$
$$dx(i) = XLZ/NXZ$$

The case mentioned is to generate a nonuniform grid system just in the x-direction for a thin plate. If the workpiece is of larger shape and size, a varying nonuniform grid system should also be generated in the y- and z-directions.

By employing the method introduced, a three-dimensional varying nonuniform grid system may be generated on the workpiece. Based on the specific situation, a uniform grid may be used in one direction while a nonuniform grid may be used for other directions.

The study case in this section deals with the workpiece dimension of 150 × 80 × 3 mm. Because of the thinner plate (3 mm thick), nonuniform grids are produced along the length (x-axis) and width (y-axis) directions, while a uniform grid is used along the thickness (z-axis) direction. Other parameters take the following values: $XL_0 = 60$ mm, $a_0 = a_3 = 1.5$, $b_0 = c_0 = b_3 = c_3 = 2$, $dx = 0.5$ mm, $dy = 0.4$ mm, $dz = 0.15$ mm.

Figure 4.15 is the block diagram of the subroutine in generating the varying nonuniform grid system.

4.10.2.2 Data Transfer between New and Old Grids

As the time step increases, the generated varying nonuniform grid moves at the same speed as the arc. At each time step, the grid needs to regenerate, so the former coarser grid zone may become a finer one and vice versa. Thus, the data transfer between any two successive time steps must be done in an appropriate way.

For the variables at the nodes in a new grid system (at the current time step), the relevant data must be obtained from the nodes in the former grid system (at the last time step). Figure 4.16 shows the schematic of the position relation between the old and new grid nodes. The variable value at the new node may be determined by interpolation of its values at former nodes.

4.10.3 Discretization of the Governing Equation

As shown in Figure 4.17, for any inner node (i, j, k), there are six adjacent nodes. Use T(i, j, k, m) to represent the temperature of node (i, j, k) at the mth time-step (t = mΔt). According to the implicit scheme, discretize Eq. (4.74):

$$\frac{T(i,j,k,m+1)-T(i,j,k,m)}{\Delta t} - v_0\frac{T(i+1,j,k,m+1)-T(i-1,j,k,m+1)}{dx(i)+dx(i-1)}$$

$$= a\frac{2}{dx(i)+dx(i-1)}\left\{\frac{T(i+1,j,k,m+1)}{dx(i)}-\left[\frac{1}{dx(i)}+\frac{1}{dx(i-1)}\right]T(i,j,k,m+1)+\frac{T(i-1,j,k,m+1)}{dx(i-1)}\right\}$$

$$+a\frac{2}{dy(j)+dy(j-1)}\left\{\frac{T(i,j+1,k,m+1)}{dy(j)}-\left[\frac{1}{dy(j)}+\frac{1}{dy(j-1)}\right]T(i,j,k,m+1)+\frac{T(i,j-1,k,m+1)}{dy(j-1)}\right\}$$

$$+a\left[\frac{T(i,j,k+1,m+1)-2T(i,j,k,m+1)+T(i,j,k-1,m+1)}{(dz)^2}\right]$$

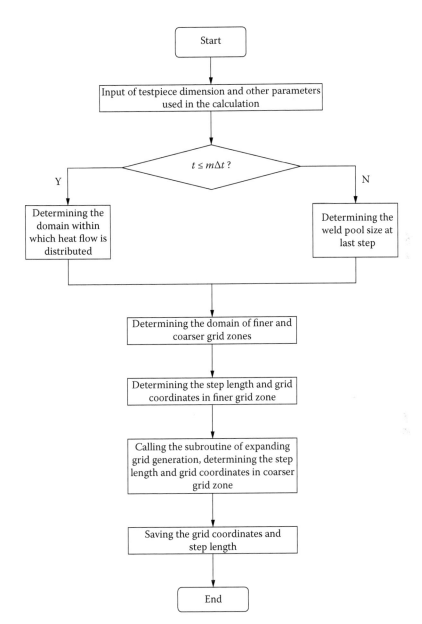

FIGURE 4.15
Block diagram of the subroutine in generating the varying nonuniform grid system.

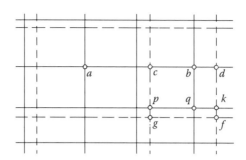

FIGURE 4.16
The schematic of position relation between old and new grid nodes. Solid line, new grid; dotted line, former grid.

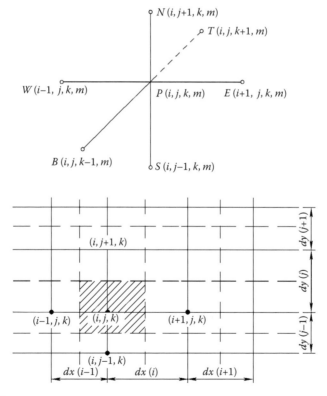

FIGURE 4.17
Schematic of the inner node and its adjacent nodes.

After manipulation, we get

$$g(i,j,k)T(i,j,k,m+1) = g(i+1,j,k)T(i+1,j,k,m+1)$$
$$+ g(i-1,j,k)T(i-1,j,k,m+1) + g(i,j+1,k)T(i,j+1,k,m+1)$$
$$+ g(i,j-1,k)T(i,j-1,k,m+1) + g(i,j,k+1)T(i,j,k+1,m+1)$$
$$+ g(i,j,k-1)T(i,j,k-1,m+1) + g_0(i,j,k)T(i,j,k,m) \quad (4.84)$$

The coefficients of Eq. (4.84) are as follows:

$$\bar{x} = dx(i) + dx(i-1), \quad \bar{y} = dy(j) + dy(j-1)$$

$$g(i+1,j,k) = \frac{2a}{\bar{x}dx(i)} + \frac{v_0}{\bar{x}}, \quad g(i-1,j,k) = \frac{2a}{\bar{x}dx(i-1)} - \frac{v_0}{\bar{x}}$$

$$g(i,j+1,k) = \frac{2a}{\bar{y}dy(j)}, \quad g(i,j-1,k) = \frac{2a}{\bar{y}dy(j-1)}$$

$$g(i,j,k+1) = \frac{a}{(dz)^2}, \quad g(i,j,k-1) = \frac{a}{(dz)^2}$$

$$g(i,j,k) = \frac{1}{\Delta t} + \frac{2a}{\bar{x}}\left(\frac{1}{dx(i)} + \frac{1}{dx(i-1)}\right) + \frac{2a}{\bar{y}}\left[\frac{1}{dy(j)} + \frac{1}{dy(j-1)}\right] + \frac{2a}{(dz)^2}$$

$$g_0(i,j,k) = \frac{1}{\Delta t}$$

For boundary nodes, similar finite difference equations may be derived.

4.10.4 Computed Results

Employing the developed model and algorithm, we computed the transient temperature field in tungsten inert gas (TIG) welding. The workpiece material was a 304L stainless steel plate of dimensions 150 × 80 × 3 mm. The thermophysical properties were as follows:

$$\rho = 7200 \ (\text{kg} \cdot \text{m}^{-3})$$

$$\alpha_{cr} = \begin{cases} (1.0 + 0.0119T) \times 10 & T < 1073\text{K} \\ [10.5 + 0.0363 \times (T - 1073)] \times 10 & T \geq 1073\text{K} \end{cases} \quad (\text{W} \cdot \text{m}^{-2} \cdot \text{K}^{-1}) \quad (4.85)$$

$$\lambda = \begin{cases} 10.717 + 0.014955 \times T & T < 780\text{K} \\ 12.076 + 0.013213 \times T & 780\text{K} \leq T < 1672\text{K} \\ 217.12 - 0.1094 \times T & 1672\text{K} \leq T < 1727\text{K} \\ 8.278 + 0.0115 \times T & T \geq 1727\text{K} \end{cases} \quad (\text{W} \cdot \text{m}^{-1} \cdot \text{K}^{-1}) \ (4.86)$$

$$C_p = \begin{cases} 438.95 + 0.198 \times T & T < 773\text{K} \\ 137.93 + 0.59 \times T & 773\text{K} \leq T < 873\text{K} \\ 871.25 - 0.25 \times T & 873\text{K} \leq T < 973\text{K} \\ 555.2 + 0.0775 \times T & T \geq 973\text{K} \end{cases} \quad (\text{J} \cdot \text{kg}^{-1} \cdot \text{K}^{-1}) \ (4.87)$$

The study case corresponded to the following welding process parameters: I = 140 A welding current, U_a = 12 V arc voltage, v_0 = 120 mm/min welding speed.

4.10.4.1 Transient Development of Welding Temperature Field

Figures 4.18, 4.19, and 4.20 illustrate transient development of the temperature field at different cross sections of the workpiece. The zero point in the x-axis is the welding start point. The temperature difference of isotherms in these figures is 400 K. The dotted line represents the fusion line with the temperature of the melting point. Clearly, as time goes on, the temperature on the workpiece gradually rises, the higher-temperature range expands, and the temperature field around the arc moves along the x-axis. Both length and width of the temperature field on the top surface of the workpiece increase with time, but its length rises more quickly, so that the shape of the temperature field becomes longer but narrower. At the longitudinal cross section, the higher-temperature range behind the arc is larger than that ahead of the arc due to the arc moving along the welding direction. Under the condition of this study case, the temperature field does not change markedly after 4.5 s, which indicates that quasi-steady state is reached at 4.5 s.

4.10.4.2 Transient Evolution of Weld Pool Shape and Size

Figure 4.21 shows the transient evolution of the weld pool shape and size on the top surface of the workpiece after arc ignition. The zero point in the x-axis is the welding start point. As the welding operation continues, the weld pool expands and moves from left to right. Due to the effect of welding speed, the weld pool length increases more quickly than its width. The weld

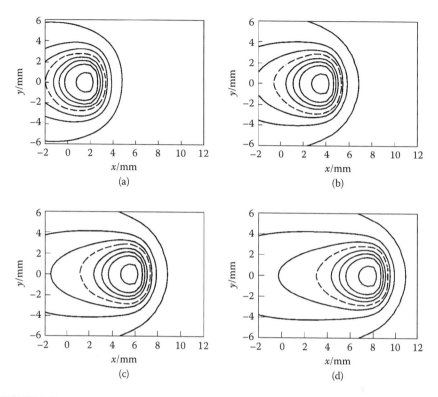

FIGURE 4.18
Transient evolution of temperature field on the top surface of a workpiece. (a) 1 s; (b) 2 s; (c) 3 s; (d) 4 s.

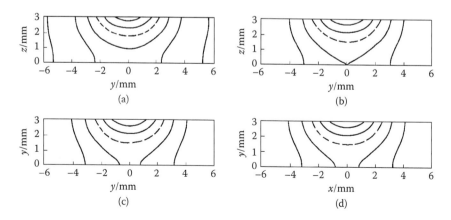

FIGURE 4.19
Temperature field at a transverse cross section ($x = 0$): (a) 1 s; (b) 2 s; (c) 3 s; (d) 4 s. Dashed line, $T = T_m$; temperature difference between isotherms is 400 K.

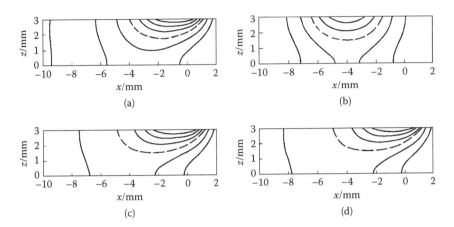

FIGURE 4.20
Temperature field at a longitudinal cross section ($y = 0$): (a) 1 s; (b) 2 s; (c) 3 s; (d) 4 s. Dashed line, $T = T_m$; temperature difference between isotherms is 400 K.

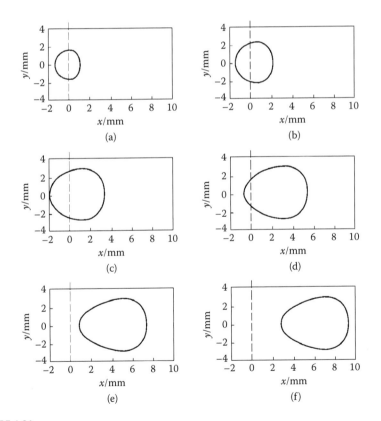

FIGURE 4.21
Transient development of weld pool shape and size on the top surface: (a) 0.2 s; (b) 0.5 s; (c) 1 s; (d) 2 s; (e) 3 s; (f) 4 s. $x = 0$ mm is the welding start point.

pool surface shape changes gradually from the initial circle-like shape to a nonstandard elliptic shape.

Figure 4.22 demonstrates the dynamic variation of the weld pool at a longitudinal cross section. The moving of the weld pool and rising of its length and depth can be clearly observed from this figure.

Figures 4.23, 4.24, and 4.25 show the relation of weld pool width, depth, and length versus time. In the figure captions, t_w is the time that it takes

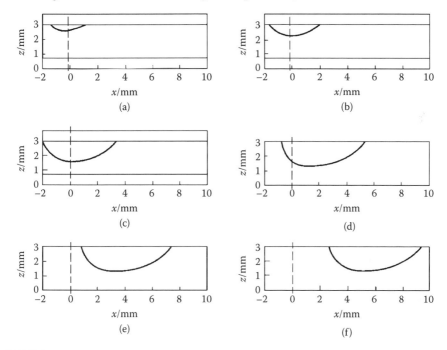

FIGURE 4.22
Transient variation of weld pool at a longitudinal cross section: (a) 0.2 s; (b) 0.5 s; (c) 1 s; (d) 2 s; (e) 3 s; (f) 4 s. $x = 0$ mm is the welding start point.

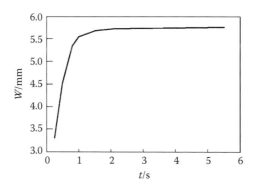

FIGURE 4.23
Weld pool width versus time ($t_w = 3$ s).

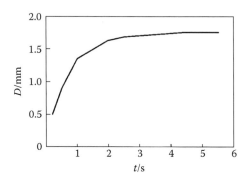

FIGURE 4.24
Weld pool depth versus time ($t_w = 4.5$ s).

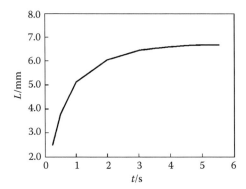

FIGURE 4.25
Weld pool length versus time ($t_w = 4.5$ s).

for each geometric parameter to reach the quasi-steady state. At the initial period of welding, each pool geometry parameter increases fast with time and keeps almost constant at 4.5 s when the quasi-steady state is established. But, the pool width reaches its quasi-steady state at about 3.0 s, sooner than the pool depth.

5

Finite Element Analysis of Weld Thermal Conduction

5.1 Introduction

The finite difference method (FDM) and finite element method (FEM) are the two most important numerical approaches in engineering computations. In Chapter 4, we introduced finite difference analysis of weld thermal conduction. This chapter discusses the principles and application of the FEM.

The FEM (sometimes referred to as finite element analysis, FEA) is a numerical technique for finding approximate solutions of partial differential equations. The FEM originated from the need to solve complex elasticity and structural analysis problems in civil and aeronautical engineering and then was popularized to fluid dynamics, heat conduction, and so on. The procedure of FEA of heat conduction is as follows:

1. To convert the problem of solving partial differential equation for heat conduction into its variational equivalents;
2. To discretize the domain and approximate the variational equivalents by a linear equations set;
3. To solve the algebraic equations set and get the approximation solution for heat conduction.

The differences between FEM and FDM are as follows:

- The most attractive feature of the FEM is its ability to handle complicated geometries (and boundaries) with relative ease. While FDM in its basic form is restricted to handling rectangular shapes and simple alterations thereof, the handling of geometries in FEM is theoretically straightforward.

- The most attractive feature of FDM is that it can be easy to implement.

- There are reasons to consider the mathematical foundation of the finite element approximation more sound, for instance, because the quality of the approximation between grid points is poor in FDM.
- The quality of an FEM approximation is often higher than that in the corresponding FDM approach, but this is extremely problem dependent, and several examples to the contrary can be provided.

Generally, FEM is the method of choice in all types of analysis in structural mechanics (i.e., solving for deformation and stresses in solid bodies or dynamics of structures), while computational fluid dynamics (CFD) tends to use FDM or other methods, like the finite volume method (FVM). CFD problems usually require discretization of the problem into a large number of cells or grid points (millions and more); therefore, the cost of the solution favors simpler, lower-order approximation within each cell. This is especially true for "external flow" problems, like airflow around a car or airplane or weather simulation in a large area.

In this chapter, first the element analysis and synthetic population in FEM are discussed for analysis of two-dimensional heat conduction, then three-dimensional FEM is introduced briefly, and computational examples of welding thermal conduction are provided.

5.2 Variation of Weld Heat Conduction

The following is a mathematical description of two-dimensional transient heat conduction:

$$\frac{\partial T}{\partial t} = \frac{\lambda}{\rho C_p}\left(\frac{\partial^2 T}{\partial x^2} + \frac{\partial^2 T}{\partial y^2}\right)$$

$$T\big|_{t=0} = f(x,y) \tag{5.1}$$

$$-\lambda\frac{\partial T}{\partial n}\bigg|_{\Gamma} = \alpha(T - T_f)\big|_{\Gamma}$$

The initial temperature distribution function $f(x, y)$ is known. Temporally, keep the time variable constant (i.e., $\partial T/\partial t$ is only a function of space) and consider the functional variation at a specific instant. Then, deal with time t and expand $\partial T/\partial t$ by means of the finite difference. In such way, we get the functional corresponding to Eq. (5.1):

$$J[T(x,y,t)] = \iint_D \left\{ \frac{\lambda}{2}\left[\left(\frac{\partial T}{\partial x}\right)^2 + \left(\frac{\partial T}{\partial y}\right)^2\right] + \rho C_p \frac{\partial T}{\partial t}T \right\} dxdy + \oint_\Gamma \alpha\left(\frac{1}{2}T^2 - T_f T\right) ds$$

(5.2)

Equation (5.2) does not include the initial condition, which is taken into account elsewhere in this chapter.

According to the variation principle, solution of the partial difference equation (5.1) can be replaced by variation calculation of the functional (5.2) and vice versa. Since the function $T = T(x,y,t)$ makes the functional achieve its extremum, it meets the corresponding Euler equation naturally. Thus, it can transform the solution of the heat conduction partial differential equation into the variation calculus that computes the functional extremum.

Solving the differential equation by variation calculus appeared first in elastic mechanics. Because the balance condition of an elastic structure is of minimum potential energy, solving the differential equation including boundary conditions of elastic mechanic naturally turns a problem of variation. There exists no concept of minimum energy in solving the heat conduction equation, so its physical meaning is not as clear as that in elastic mechanics. Here, we do not care about the maximum or minimum value of the functional. What we care about is that the required necessary and sufficient condition when a functional achieves its extremum is mathematically equivalent to a corresponding differential equation with boundary conditions. As long as there exists such a unique equivalent relation in mathematics, we can use the variation computation of the functional extremum to replace solving for differential equation and boundary conditions.

The classical variation calculus (Ritz method) used the variation computation of the functional extremum to replace solving of the differential equation. To have deep insight into the origin and development of the FEM, we give a simple example for a brief introduction of the Ritz method.[124]

For a differential equation,

$$y'' + y + 1 = 0$$

(5.3)

Its boundary conditions are

$$y(0) = 0, y(1) = 0$$

(5.4)

For the exact solution,

$$y = \cos x + \frac{1 - \cos 1}{\sin 1}\sin x - 1$$

(5.5)

According to Eq. (5.5), we have

$$y\left(\frac{1}{2}\right) = 0.13949$$

Now, we employ variation computation to get an approximation solution of the differential equation. Establish a functional

$$J[y(x)] = \int_0^1 \left[\frac{1}{2}(y')^2 - \frac{1}{2}y^2 - y\right]dx \qquad (5.6)$$

The variation principle may be used to prove that the curve $y(x)$, which meets Eq. (5.4) and achieves its extremum, is the solution of differential equation Eq. (5.3) with boundary conditions of Eq. (5.4).

Because we do not know function $y(x)$, we can select an approximate trying function and substitute it into Eq. (5.6). Trying functions are some polynomes with good differential properties. Such a polynome should first meet the boundary condition of Eq. (5.4):

$$y = a_1(x - x^2) + a_2(x - x^3) + a_3(x - x^4) + \dots \qquad (5.7)$$

Clearly, the greater the item number of the polynome is, the more accurate is the result. We select n items so that there are n unknowns $a_1, a_2, \dots, a_n$. Substituting Eq. (5.7) into Eq. (5.6), $J[y(x)]$ actually becomes a multivariable function $J(a_1, a_2, \dots, a_n)$. Then, the variation problem transforms into an extremum problem of multivariable function. We take

$$\frac{\partial J}{\partial a_k} = 0 \qquad k = 1, 2, \dots n \qquad (5.8)$$

Equation (5.8) is an equation set with n algebraic equations. After solving, we obtain n unknowns $a_1, a_2, \dots, a_n$. Then, Eq. (5.7) is a definite polynome $y(x)$ that is the solution of Eq. (5.3) satisfying with Eq. (5.4).

For simplification, just select one item for the trying function:

$$y = a_1(x - x^2) \qquad (5.9)$$

Substituting Eq. (5.9) into Eq. (5.6), we get

$$J[y(x)] = \int_0^1 \left[\frac{1}{2}a_1^2(1 - 2x)^2 - \frac{1}{2}a_1^2(x - x^2)^2 - a_1(x - x^2)\right]dx$$

$$\frac{\partial J}{\partial a_1} = \int_0^1 [a_1(1-2x)^2 - a_1(x-x^2)^2 - (x-x^2)]dx = 0$$

After manipulation,

$$-\frac{1}{6} + \frac{3a_1}{10} = 0$$

$$a_1 = \frac{5}{9}$$

Finally,

$$y = \frac{5}{9}(x - x^2) \tag{5.10}$$

This is the approximate solution of the extremum curve. Based on Eq. (5.10), we get

$$y\left(\frac{1}{2}\right) = 0.1389$$

It can be seen that for the simplest trying function of Eq. (5.9), the error between the exact solution 0.13949 and the approximation 0.1389 is just 3 millionths.

The advantage of the Ritz method is the ability to obtain approximate solutions for some differential equations that are hard to solve by selecting trying functions. Its drawback is that variation computation in the entire domain must be performed. There is nothing it can do when boundary conditions or domain structures are complicated.

Enlightened by the FDM, if we divide the domain into many small pieces, conduct variation computation for each piece, and integrate the results for all pieces together, then any complicated problem seems simple in small pieces. We do not even need to pay much attention to selecting the trying function carefully. Some of the simplest interpolation functions may be used for sufficient accuracy. This is the so-called FEM.

Now, we imagine that the domain for Eq. (5.1) is discretized into finite elements. If variation computation is conducted for each element, Eq. (5.2) becomes

$$J^e = \iint_\Delta F(x,y,T,T_x,T_y)dxdy \tag{5.11}$$

where J^e is the functional defined in an element, and Δ is the element area.

If domain D is divided into E elements with n nodes, the temperature field is discretized as n nodes of temperature $T_1, T_2, ..., T_n$. Substitute the discretized temperature functions into J, then the functional $J[T(x,y)]$ actually becomes a multivariable function $J(T_1, T_2, ..., T_n)$. In this way, the variation problem in Eq. (5.2) transforms into an extremum problem of a multivariable function. We get

$$\frac{\partial J}{\partial T_k} = 0 \qquad k = 1, 2 \cdots, n \tag{5.12}$$

Equation (5.12) has n algebraic equations. After solving, we obtain n unknowns $T_1, T_2, ..., T_n$.

Since

$$J = \sum_{e=1}^{E} J^e$$

after substituting it into Eq. (5.12), we obtain

$$\frac{\partial \sum_{e=1}^{E} J^e}{\partial T_k} = \sum_{e=1}^{E} \frac{\partial J^e}{\partial T_k} = 0 \qquad k = 1, 2 \cdots, n \tag{5.13}$$

The element analysis in the FEM is to calculate the value of $\partial J^e / \partial T_k$ for each element. The synthetic population in the FEM is to sum the values of $\partial J^e / \partial T_k$ for all elements and establish n linear algebraic equations like Eq. (5.12).

5.3 Mesh Generation and Discretization of the Temperature Field

The first step is to divide a solution region into finite elements. The finite element mesh is typically generated by a preprocessor program. The description of the mesh consists of several arrays; the main ones are nodal coordinates and element connectivities. For a planar region, divide it into a series of triangular or quadrangle elements, while for a three-dimensional body, divide it into hexahedron, tetrahedron, or pentahedron elements.

As shown in Figure 5.1, the region D with boundary Γ may be divided into triangle elements of any shape. Each node has its sequence number 1, 2, 3, ..., and each element has its series number ①, ②, ③, ... One element is related to its adjacent elements through its vertexes. For each element itself, its three vertexes are numbered as i, j, m in the counterclockwise direction.

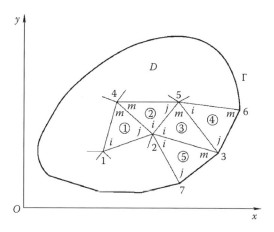

FIGURE 5.1
Mesh generation.

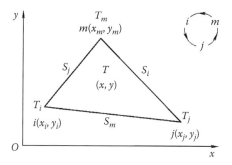

FIGURE 5.2
Triangle element.

The elements without a region boundary, such as ①, ②, ③ and so on, are called the *inner elements*. The elements with a region boundary, such as ④, ⑤, and so on, are called *boundary elements*. Usually, inner elements are numbered first. Then, the elements of the first kind of boundary condition are numbered. Finally, the elements of the third kind of boundary condition (including the adiabatic boundary) are numbered. For simplification, only one side of the boundary elements locates at the region boundary, and this side is arranged as *jm*, while node *i* faces the boundary side. For inner elements, node number *i,j,m* can be arranged arbitrarily in a counterclockwise direction. But, node number *i* is always arranged at a node with a minimum sequence number so that it is convenient to look for it.

Any one element from region D in Figure 5.1 is taken and shown in Figure 5.2. After discretization of the region, the coordinates of three vertexes are already known, and the length S_i, S_j, S_m of three sides corresponding to the vertexes *i,j,m* and the triangle area Δ are also known. The temperature

T at any point (x,y) in the triangle is discretized at three nodes of the element; that is, three nodal temperature values T_i, T_j, T_m are used to represent the temperature field in the element.

$$T = f(T_i, T_j, T_m)$$

The FEM only computes the nodal temperature values.

Discretization is an approximation approach. Generally, the lower the element size is, the higher is the accuracy. We can change the size and shape of the triangle flexibly, such as by generating smaller elements in the zone where temperature changes severely or geometry is complicated, while arranging larger elements at other positions. In this way, it is possible to improve computation accuracy, but the numbers of both elements and nodes are not increased.

Try to avoid obtuse triangle elements if possible because such a triangle has a side with a comparatively long length, and the computation accuracy is controlled by the ratio of the length of the longest side to the length of the shortest side of the element.[122]

In the two-dimensional case, except for triangle elements, rectangle elements may also be used. Because rectangle elements are not suitable for curved boundaries, arbitrary quadrangle elements are selected, which are referred to as *isoparametric elements*. In three-dimensional situations, hexahedron, tetrahedron, or pentahedron elements are used.

5.4 Selection of Interpolation Functions for Temperature

The interpolation function for temperature describes specifically how to represent the temperature at any point in the element by the nodal temperature values. Selection of a suitable interpolation function is a key problem in FEA. Because it is comparatively convenient to conduct the mathematical operation (differential and integral) for a polynomial that can fit a smooth function locally, the polynomial is usually selected as the interpolation function for temperature. To select the number of polynomial terms and orders, the degree of freedom of the element and the convergence requirement of the solution must be considered. Generally, the number of polynomial terms is equal to the number of degrees of freedom, while its orders should contain a constant term and a linear term.

For a triangle element, the linear function is usually employed:

$$T^e = a_1 + a_2 x + a_3 y \tag{5.14}$$

where a_1, a_2, a_3 are undetermined constants. They are determined by the nodal temperature values. Substituting the coordinates and temperature values of three nodes into Eq. (5.14), we get one simple equation set with three unknowns. After solving, we obtain the values of (a_1, a_2, a_3), so that the interpolation function is specified as follows:

$$T^e = N_i T_i + N_j T_j + N_m T_m \tag{5.15}$$

Write it in a compact way:

$$T^e = [N]^e \{T\}^e \tag{5.16}$$

where

$$[N]^e = [N_i, N_j, N_m] \tag{5.17}$$

$$\{T\}^e = \begin{Bmatrix} T_i \\ T_j \\ T_m \end{Bmatrix} \tag{5.18}$$

where $[N]^e$ is called a shape function and is determined by the element shape and size. Based on the shape function, we obtain the equation to represent the temperature at any point in the element by the nodal temperature values.

After mesh generation, the coordinates $(x_i, y_i), (x_j, y_j), (x_m, y_m)$ of three vertexes of the triangle element are known, so the shape function is known. Substituting the coordinates and temperature values of three nodes into Eq. (5.14), after a series of manipulations, we get the expression for the shape function:

$$
\left. \begin{aligned}
N_i &= \frac{1}{2\Delta}(a_i + b_i x + c_i y) \\
N_j &= \frac{1}{2\Delta}(a_j + b_j x + c_j y) \\
N_m &= \frac{1}{2\Delta}(a_m + b_m x + c_m y)
\end{aligned} \right\} \tag{5.19}
$$

where

$$
\left. \begin{aligned}
a_i &= x_j y_m - x_m y_j, & b_i &= y_j - y_m, & c_i &= x_m - x_i \\
a_j &= x_m y_i - x_i y_m, & b_j &= y_m - y_i, & c_j &= x_i - x_m \\
a_m &= x_i y_j - x_j y_i, & b_m &= y_i - y_j, & c_m &= x_j - x_i \\
\Delta &= (b_i c_j - b_j c_i)/2
\end{aligned} \right\} \tag{5.20}
$$

where Δ is the area of the triangle element.

5.5 Element Analysis in FEM

To conduct variational computation for one element in the FEM, just define Eq. (5.2) in the domain of the element, that is,

$$J_B^e = \iint_e \left[\frac{\lambda}{2} \left(\frac{\partial T}{\partial x} \right)^2 + \frac{\lambda}{2} \left(\frac{\partial T}{\partial y} \right)^2 + \rho C_p \frac{\partial T}{\partial t} T \right] dxdy$$

$$+ \int_{jm} \left(\frac{1}{2} \alpha T^2 - \alpha T_f T \right) ds$$

(5.21)

where e means the element. Here, only the side jm of the boundary element is one part of the domain boundary, while the inner element has no boundary. The side jm of the boundary element is not closed, so we write the contour integral $\oint_\Gamma$ as $\int_{jm}$ in Eq. (5.21). Clearly, Eq. (5.21) is suitable for boundary elements. For inner elements, Eq. (5.21) can be simplified as

$$J_{In}^e = \iint_e \frac{\lambda}{2} \left[\left(\frac{\partial T}{\partial x} \right)^2 + \left(\frac{\partial T}{\partial y} \right)^2 + \rho C_p \frac{\partial T}{\partial t} T \right] dxdy$$

(5.22)

The temperature field on the element e has been discretized as interpolation functions related to the nodal temperatures (T_i, T_j, T_m). According to Eq. (5.13), the variational computation of the element is to obtain the values of the following terms:

$$\frac{\partial J^e}{\partial T_i}, \frac{\partial J^e}{\partial T_j} \text{ and } \frac{\partial J^e}{\partial T_m}$$

Equation (5.15) is usable for the entire element, including the element boundary. Since the element boundary is a line segment jm, the linear interpolation method can be used to construct a simpler interpolation function:

$$T = (1 - \tau)T_j + \tau T_m$$

(5.23)

where $0 \leq \tau \leq 1$, τ is a factor, $\tau = 0$ at node j, and $\tau = 1$ at node m. Clearly,

$$S_i = \sqrt{(x_j - x_m)^2 + (y_j - y_m)^2} = \sqrt{b_i^2 + c_i^2}$$

(5.24)

S_i is the length of side jm, and it is known after mesh generation. The integral variable s in Eq. (5.21) is related to S_i through τ:

$$s = S_i \tau, \quad ds = S_i d\tau$$

(5.25)

5.5.1 Variational Computation of the Boundary Element

For variational computation of the boundary element, according to Eq. (5.21),

$$\frac{\partial J^e}{\partial T_i} = \iint_e \left[\lambda \frac{\partial T}{\partial x} \frac{\partial}{\partial T_i} \left(\frac{\partial T}{\partial x} \right) + \lambda \frac{\partial T}{\partial y} \frac{\partial}{\partial T_i} \left(\frac{\partial T}{\partial y} \right) + \rho C_P \frac{\partial T}{\partial t} \frac{\partial T}{\partial T_i} \right] dxdy$$

$$+ \int_0^1 (\alpha T - \alpha T_f) \frac{\partial T}{\partial T_i} S_i d\tau \tag{5.26}$$

Based on Eq. (5.15),

$$\frac{\partial T}{\partial x} = \frac{\partial N_i}{\partial x} T_i + \frac{\partial N_j}{\partial x} T_j + \frac{\partial N_m}{\partial x} T_m \tag{5.27}$$

$$\frac{\partial}{\partial T_i} \left(\frac{\partial T}{\partial x} \right) = \frac{\partial N_i}{\partial x} \tag{5.28}$$

$$\frac{\partial T}{\partial y} = \frac{\partial N_i}{\partial y} T_i + \frac{\partial N_j}{\partial y} T_j + \frac{\partial N_m}{\partial y} T_m \tag{5.29}$$

$$\frac{\partial}{\partial T_i} \left(\frac{\partial T}{\partial y} \right) = \frac{\partial N_i}{\partial y} \tag{5.30}$$

According to Eq. (5.15), inside the element,

$$\frac{\partial T}{\partial T_i} = N_i \tag{5.31}$$

and at boundary side *jm*,

$$\frac{\partial T}{\partial T_i} = 0 \tag{5.32}$$

Because T_i, T_j, T_m are functions of time in the transient temperature field, then

$$\frac{\partial T}{\partial t} = N_i \frac{\partial T_i}{\partial t} + N_j \frac{\partial T_j}{\partial t} + N_m \frac{\partial T_m}{\partial t} \tag{5.33}$$

Substituting Eqs. (5.27) to (5.33) into Eq. (5.26), we obtain

$$
\begin{aligned}
\frac{\partial J^e}{\partial T_i} = \iint_e \lambda & \left\{ \begin{array}{l} \left[\left(\frac{\partial N_i}{\partial x} \right)^2 + \left(\frac{\partial N_i}{\partial y} \right)^2 \right] T_i + \left(\frac{\partial N_i}{\partial x} \frac{\partial N_j}{\partial x} + \frac{\partial N_i}{\partial y} \frac{\partial N_j}{\partial y} \right) T_j \\[2mm] + \left(\frac{\partial N_i}{\partial x} \frac{\partial N_m}{\partial x} + \frac{\partial N_i}{\partial y} \frac{\partial N_m}{\partial y} \right) T_m \end{array} \right\} dxdy \\[4mm]
& + \iint_e \rho C_p \left(\frac{\partial T_i}{\partial t} N_i^2 + \frac{\partial T_j}{\partial t} N_i N_j + \frac{\partial T_m}{\partial t} N_i N_m \right) dxdy
\end{aligned}
\tag{5.34}
$$

To compute the following equation:

$$
\frac{\partial J^e}{\partial T_j} = \iint_e \left[\lambda \frac{\partial T}{\partial x} \cdot \frac{\partial}{\partial T_j} \left(\frac{\partial T}{\partial x} \right) + \lambda \frac{\partial T}{\partial y} \cdot \frac{\partial}{\partial T_j} \left(\frac{\partial T}{\partial y} \right) + \rho C_p \frac{\partial T}{\partial t} \frac{\partial T}{\partial T_j} \right] dxdy
$$

$$
+ \int_0^1 \alpha (T - T_f) \frac{\partial T}{\partial T_j} S_i d\tau
\tag{5.35}
$$

its first double integration is similar to that introduced, and its second integral is calculated as follows:

$$
\int_0^1 \alpha (T - T_f) \frac{\partial T}{\partial T_j} S_i d\tau
$$

$$
= \int_0^1 \alpha [(1 - \tau) T_j + \tau T_m - T_f](1 - \tau) S_i d\tau
$$

$$
= -\alpha S_i \int_0^1 [(1 - \tau) T_j - (1 - \tau) T_f] d(1 - \tau) + \alpha S_i T_m \int_0^1 (\tau - \tau^2) d\tau
$$

$$
= \frac{\alpha S_i}{3} T_j - \frac{\alpha S_i T_f}{2} + \frac{\alpha S_i}{6} T_m
\tag{5.36}
$$

Then, we obtain

$$
\frac{\partial J^e}{\partial T_j} = \iint_e \lambda \left\{ \begin{array}{l} \left[\left(\dfrac{\partial N_i}{\partial x} \dfrac{\partial N_j}{\partial x} + \dfrac{\partial N_i}{\partial y} \dfrac{\partial N_j}{\partial y} \right) T_i + \left[\left(\dfrac{\partial N_j}{\partial x} \right)^2 + \left(\dfrac{\partial N_j}{\partial y} \right)^2 \right] T_j \right. \\[3mm] \left. + \left(\dfrac{\partial N_j}{\partial x} \dfrac{\partial N_m}{\partial x} + \dfrac{\partial N_j}{\partial y} \dfrac{\partial N_m}{\partial y} \right) T_m \end{array} \right\} dxdy
$$

$$
+ \iint_e \rho C_p \left(\frac{\partial T_i}{\partial t} N_i N_j + \frac{\partial T_j}{\partial t} N_j^2 + \frac{\partial T_m}{\partial t} N_j N_m \right) dxdy + \frac{\alpha S_i}{3} T_j + \frac{\alpha S_i}{6} T_m - \frac{\alpha S_i T_f}{2}
$$

$$(5.37)$$

Similarly, we obtain

$$
\frac{\partial J^e}{\partial T_m} = \iint_e \lambda \left\{ \begin{array}{l} \left(\dfrac{\partial N_i}{\partial x} \dfrac{\partial N_m}{\partial x} + \dfrac{\partial N_i}{\partial y} \dfrac{\partial N_m}{\partial y} \right) T_i + \left(\dfrac{\partial N_j}{\partial x} \dfrac{\partial N_m}{\partial x} + \dfrac{\partial N_j}{\partial y} \dfrac{\partial N_m}{\partial y} \right) T_j \\[3mm] + \left[\left(\dfrac{\partial N_m}{\partial x} \right)^2 + \left(\dfrac{\partial N_m}{\partial y} \right)^2 \right] T_m \end{array} \right\} dxdy
$$

$$
+ \iint_e \rho C_p \left(\frac{\partial T_i}{\partial t} N_i N_m + \frac{\partial T_j}{\partial t} N_j N_m + \frac{\partial T_m}{\partial t} N_m^2 \right) dxdy + \frac{\alpha S_i}{6} T_j + \frac{\alpha S_i}{3} T_m - \frac{\alpha S_i T_f}{2}
$$

$$(5.38)$$

Usually, Eqs. (5.34), (5.37), and (5.38) are written in matrix form:

$$
\left\{ \begin{array}{c} \dfrac{\partial J^e}{\partial T_i} \\[3mm] \dfrac{\partial J^e}{\partial T_j} \\[3mm] \dfrac{\partial J^e}{\partial T_m} \end{array} \right\} = \begin{bmatrix} k_{ii} & k_{ij} & k_{im} \\ k_{ji} & k_{jj} & k_{jm} \\ k_{mi} & k_{mj} & k_{mm} \end{bmatrix} \left\{ \begin{array}{c} T_i \\ T_j \\ T_m \end{array} \right\} + \begin{bmatrix} h_{ii} & h_{ij} & h_{im} \\ h_{ji} & h_{jj} & h_{jm} \\ h_{mi} & h_{mj} & h_{mm} \end{bmatrix} \left\{ \begin{array}{c} \dfrac{\partial T_i}{\partial t} \\[3mm] \dfrac{\partial T_j}{\partial t} \\[3mm] \dfrac{\partial T_m}{\partial t} \end{array} \right\} - \left\{ \begin{array}{c} p_i \\ p_j \\ p_m \end{array} \right\} \qquad (5.39)
$$

$$
= [k]^e \{T\}^e + [h] \left\{ \frac{\partial T}{\partial t} \right\}^e - \{p\}^e
$$

where

$$k_{ii} = \iint_e \lambda \left[\left(\frac{\partial N_i}{\partial x} \right)^2 + \left(\frac{\partial N_i}{\partial y} \right)^2 \right] dxdy,$$

$$k_{jj} = \iint_e \lambda \left[\left(\frac{\partial N_j}{\partial x} \right)^2 + \left(\frac{\partial N_j}{\partial y} \right)^2 \right] dxdy + \frac{\alpha S_i}{3}$$

$$k_{mm} = \iint_e \lambda \left[\left(\frac{\partial N_m}{\partial x} \right)^2 + \left(\frac{\partial N_m}{\partial y} \right)^2 \right] dxdy + \frac{\alpha S_i}{3}$$

$$k_{ij} = k_{ji} = \iint_e \lambda \left(\frac{\partial N_i}{\partial x} \frac{\partial N_j}{\partial x} + \frac{\partial N_i}{\partial y} \frac{\partial N_j}{\partial y} \right) dxdy \qquad (5.40)$$

$$k_{im} = k_{mi} = \iint_e \lambda \left(\frac{\partial N_i}{\partial x} \frac{\partial N_m}{\partial x} + \frac{\partial N_i}{\partial y} \frac{\partial N_m}{\partial y} \right) dxdy$$

$$k_{jm} = k_{mj} = \iint_e \lambda \left(\frac{\partial N_j}{\partial x} \frac{\partial N_m}{\partial x} + \frac{\partial N_j}{\partial y} \frac{\partial N_m}{\partial y} \right) dxdy + \frac{\alpha S_i}{6}$$

$$h_{ii} = h_{jj} = h_{mm} = \rho C_p \iint_e N_i^2 dxdy = \rho C_p \iint_e N_j^2 dxdy = \rho C_p \iint_e N_m^2 dxdy$$

$$h_{ij} = h_{ji} = \rho C_p \iint_e N_i N_j dxdy$$

$$h_{im} = h_{mi} = \rho C_p \iint_e N_i N_m dxdy \qquad (5.41)$$

$$h_{jm} = h_{mj} = \rho C_p \iint_e N_j N_m dxdy$$

$$p_i = 0, \ p_j = p_m = \frac{\alpha S_i T_f}{2}$$

Substituting the shape function Eq. (5.19) into Eqs. (5.40) and (5.41) and employing the following relations given in Ref. 122:

$$\iint_e N_i^2 dxdy = \iint_e N_j^2 dxdy = \iint_e N_m^2 dxdy = \frac{\Delta}{6}$$

$$\iint_e N_i N_j dxdy = \iint_e N_i N_m dxdy = \iint_e N_j N_m dxdy = \frac{\Delta}{12}$$

We easily obtain the coefficients in Eqs. (5.40) and (5.41):

$$k_{ii} = \frac{\lambda}{4\Delta}(b_i^2 + c_i^2), \quad k_{jj} = \frac{\lambda}{4\Delta}(b_j^2 + c_j^2) + \frac{\alpha S_i}{3}, \quad k_{mm} = \frac{\lambda}{4\Delta}(b_m^2 + c_m^2) + \frac{\alpha S_i}{3}$$

$$k_{ij} = k_{ji} = \frac{\lambda}{4\Delta}(b_i b_j + c_i c_j), \quad k_{im} = k_{mi} = \frac{\lambda}{4\Delta}(b_i b_m + c_i c_m)$$

$$k_{jm} = k_{mj} = \frac{\lambda}{4\Delta}(b_j b_m + c_j c_m) + \frac{\alpha S_i}{6}, \quad h_{ii} = h_{jj} = h_{mm} = \frac{\rho C_p \Delta}{6}$$

$$h_{ij} = h_{ji} = h_{im} = h_{mi} = h_{jm} = h_{mj} = \frac{\rho C_p \Delta}{12}$$

$$p_i = 0, \quad p_j = p_m = \frac{\alpha S_i T_j}{2}$$

5.5.2 Variational Computation of Inner Elements

Based on the functional expressed by Eq. (5.22), variational computation of inner elements can be conducted. After similar derivation, we get one equation set like Eq. (5.39), for which the coefficient matrix is also like Eqs. (5.40) and (5.41) but $S_i = 0$.

5.6 Assembly Element Equations in FEM

The final results of FEA are the computed temperature distributions in the region D. As shown in Figure 5.1, the region D is divided into finite triangle elements with n nodes, and the temperature field is discretized as nodal values of temperature. The task now is to calculate the nodal values of temperature $T_1, T_2, \ldots, T_n$.

Define J as the functional in the whole region D, and J^e as the functional in the triangle element, then we have

$$J = \sum_e J^e \tag{5.42}$$

where Σ_e means the summation for all elements.

Because the temperature field has been discretized to all nodes, the functional actually becomes a multivariable function describing these nodal values of temperature. Thus, the variational problem of the functional transforms into an extremum problem of multivariable function.

If n nodal values of temperature in the region D are all unknowns, the multivariable function has the form like $J(T_1, T_2, \ldots, T_n)$. The condition for J to obtain its extremum is as follows:

$$\frac{\partial J}{\partial T_k} = \sum_{e=1}^{E} \frac{\partial J^e}{\partial T_k} = 0, \quad k = 1, 2, 3, \cdots, n \tag{5.43}$$

Equation (5.43) has n algebraic equations. After solving, we get n nodal values of temperature $(T_1, T_2, \ldots, T_n)$.

Equation (5.43) is the basis of assembly of the element equations. Substituting Eq. (5.39) that resulted from element analysis for all elements into Eq. (5.43), we complete the assembly task for the element equations. Finally, we obtain a linear equation set with n algebraic equations:

$$\begin{bmatrix} k_{11} & k_{12} & \cdots & k_{1n} \\ k_{21} & k_{22} & \cdots & k_{2n} \\ \vdots & \vdots & \ddots & \vdots \\ k_{n1} & k_{n2} & \cdots & k_{nn} \end{bmatrix} \begin{Bmatrix} T_1 \\ T_2 \\ \vdots \\ T_N \end{Bmatrix} + \begin{bmatrix} h_{11} & h_{12} & \cdots & h_{1n} \\ h_{21} & h_{22} & \cdots & h_{2n} \\ \vdots & \vdots & \ddots & \vdots \\ h_{n1} & h_{n2} & \cdots & h_{nn} \end{bmatrix} \begin{Bmatrix} \dfrac{\partial T_1}{\partial t} \\ \dfrac{\partial T_2}{\partial t} \\ \vdots \\ \dfrac{\partial T_n}{\partial t} \end{Bmatrix} = \begin{Bmatrix} p_1 \\ p_2 \\ \vdots \\ p_n \end{Bmatrix} \tag{5.44}$$

It may be written in matrix form:

$$[K]\{T\} + [H]\left\{\frac{\partial T}{\partial t}\right\} = \{P\} \tag{5.45}$$

where $[K]$ is the temperature stiffness matrix; $[H]$ is the temperature variation matrix, which is a coefficient matrix accounting for the time dependence of temperature in transient heat conduction; $\{T\}$ is the column vector of unknown nodal values of temperature; and $\{P\}$ is the column vector consisting of other right-hand-side terms in Eq. (5.45).

At any time instant t, Eq. (5.45) may be written as

$$[K]\{T\}_t + [H]\left\{\frac{\partial T}{\partial t}\right\}_t = \{P\}_t \tag{5.46}$$

where the subscript t indicates that the corresponding terms are time dependent. If the boundary conditions are not changed with time, then

$$\{P\}_{t-\Delta t} = \{P\}_t = \{P\}_{t+\Delta t} = \cdots \tag{5.47}$$

For transient heat conduction, the initial temperature distribution and boundary conditions are known, but $\{\partial T/\partial t\}_t$ is unknown. It is inconvenient to use Eq. (5.46) to get $\{T\}_t$. Thus, expand $\{\partial T/\partial t\}_t$ by the FDM. The following explicit schemes are usually employed: two-point backward scheme, Crank-Nicolson scheme, Galerkin's scheme, and three-point backward scheme.

5.6.1 Two-Point Backward Scheme

For the two-point backward scheme, assume that during the period of Δt there exists

$$\left. \begin{array}{l} \left\{\dfrac{\partial T}{\partial t}\right\}_t = \dfrac{\{T\}_t - \{T\}_{t-\Delta t}}{\Delta t} \\[3mm] \{T\}_t = \{T\}_t \end{array} \right\} \tag{5.48}$$

Substituting Eq. (5.48) into Eq. (5.46), we get

$$\left([K] + \frac{1}{\Delta t}[H]\right)\{T\}_t = \frac{1}{\Delta t}[H]\{T\}_{t-\Delta t} + \{P\}_t \tag{5.49}$$

Equation (5.49) is stable numerically for a large time step, but it is less accurate.

5.6.2 Crank-Nicolson Scheme

For the Crank-Nicolson scheme, assume that during the period of Δt there exists

$$\left. \begin{array}{l} \left\{\dfrac{\partial T}{\partial t}\right\}_t = \dfrac{\{T\}_t - \{T\}_{t-\Delta t}}{\Delta t} \\[3mm] \{T\}_t = \dfrac{\{T\}_t + \{T\}_{t-\Delta t}}{2} \end{array} \right\} \tag{5.50}$$

Substituting Eq. (5.50) into Eq. (5.46), we obtain

$$\left(\frac{1}{2}[K]+\frac{1}{\Delta t}[H]\right)\{T\}_t = \left(\frac{1}{\Delta t}[H]-\frac{1}{2}[K]\right)\{T\}_{t-\Delta t}$$

$$+\frac{1}{2}\{P\}_t+\frac{1}{2}\{P\}_{t-\Delta t} \tag{5.51}$$

Although the Crank-Nicolson scheme is stable numerically, it requires smaller time step Δt. According to Eq. (5.51), the stable condition is as follows:

$$\frac{2[H]}{\Delta t}-[K]\geq 0 \tag{5.52}$$

For a regular triangle element in plane and constant boundary conditions, the limitation to Δt may be estimated as

$$\Delta t \leq \frac{\rho C_P(\Delta x)^2}{6\lambda} \tag{5.53}$$

where Δx is the side length of the regular triangle element. If it is not a regular triangle, the averaged side length is used.

5.6.3 Galerkin's Scheme

For Galerkin's scheme, assume that during the period of Δt there exists

$$\left.\begin{array}{c}\left\{\dfrac{\partial T}{\partial t}\right\}_t = \dfrac{\{T\}_t-\{T\}_{t-\Delta t}}{\Delta t} \\[3mm] \{T\}_t = \dfrac{2}{3}\{T\}_t+\dfrac{1}{3}\{T\}_{t-\Delta t}\end{array}\right\} \tag{5.54}$$

Then, we obtain

$$\left(2[K]+\frac{3[H]}{\Delta t}\right)\{T\}_t = \left(\frac{3[H]}{\Delta t}-[K]\right)\{T\}_{t-\Delta t}$$

$$+2\{P\}_t+\{P\}_{t-\Delta t} \tag{5.55}$$

Equation (5.55) is stable numerically if the following condition is satisfied:

$$\frac{3[H]}{\Delta t}-[K]\geq 0 \tag{5.56}$$

For a regular triangle element in a plane and first kind of constant boundary condition, the limitation to Δt may be estimated as

$$\Delta t \leq \frac{\rho C_P (\Delta x)^2}{4\lambda} \tag{5.57}$$

This scheme oscillates only slightly if Δt exceeds the limit given by Eq. (5.57). This is the advantage of Galerkin's scheme.

5.6.4 Three-Point Backward Scheme

For the three-point backward scheme, express the temperature at t, $t - \Delta t$, and $t - 2\Delta t$ as T_t, T_{t-1}, and T_{t-2}, respectively. Employ Taylor's series to describe T_{t-1}, T_{t-2}:

$$T_{t-1} = T_t - T_t'(\chi\Delta t) + \frac{1}{2!}T_t''(\chi\Delta t)^2 - \cdots$$

$$T_{t-2} = T_t - T_t'[(\chi + 1)\Delta t] + \frac{1}{2!}T_t''[(\chi + 1)\Delta t]^2 - \cdots$$

where $0 < \chi < 1$. Solve these two equations simultaneously, and eliminating T_t'', we obtain

$$T_t' = \frac{1}{\Delta t}\left[\left(\frac{\chi}{\chi + 1}\right)T_{t-2} - \left(\frac{\chi + 1}{\chi}\right)T_{t-1} + \left(\frac{2\chi + 1}{\chi^2 + \chi}\right)T_t\right] + 0(\Delta t)^2 \tag{5.58}$$

Substituting Eq. (5.58) into Eq. (5.46), we get

$$\left([K] + \frac{2\chi + 1}{\chi^2 + \chi}\frac{[H]}{\Delta t}\right)\{T\}_t = \{P\}_t + \frac{[H]}{\Delta t}\left(\frac{\chi + 1}{\chi}\{T\}_{t-1} - \frac{\chi}{\chi + 1}\{T\}_{t-2}\right) \tag{5.59}$$

Equation (5.59) is stable numerically if the following condition is met:

$$\frac{\{\theta\}_{t-1}}{\{\theta\}_{t-2}} \geq \left(\frac{\chi}{\chi + 1}\right)^2 \tag{5.60}$$

where $\theta = (T - T_\infty)/(T_0 - T_\infty)$, and T_0 and T_∞ are the initial and final temperatures, respectively. The limit defined by Eq. (5.60) is less at the starting stage, which is the advantage of this scheme.

After discretizing the time domain by these schemes, the problem finally becomes the solution of the following linear algebraic equation set:

$$[A]\{T\} = \{B\} \tag{5.61}$$

where the coefficient matrix $[A]$ is typically symmetric, positive definite, and sparse.

5.7 Solving for the Nodal Values of Temperature

Through derivation and manipulation mentioned, the solution of the heat conduction partial differential equation has been transformed to a linear algebraic equation set like Eq. (5.61). By solving this equation set, we obtain the nodal values of the temperature field. Various methods, such as Gauss-Seidel iteration, overrelaxation iteration, and so on, can be used to solve the algebraic equation set.

For a transient temperature field, we need to start from the initial temperature distribution and obtain the temperature field at each time step gradually, so the workload of computation is heavy. For steady-state problems, by only conducting the solution of the algebraic equation set once, we obtain the results because the temperature does not change with time.

5.8 Three-Dimensional FEM

In previous sections, we discussed FEA of a two-dimensional temperature field. This section introduces finite element computation of a three-dimensional temperature field. Generally, the main steps of the finite element solution procedure in the three-dimensional case, such as mesh generation, element analysis, assembly of the element equations, and solution of the algebraic equation set, are similar to those for the two-dimensional situation. However, it is more complicated for three-dimensional FEA.

5.8.1 Variational Problem of Three-Dimensional Heat Conduction

Three-dimensional heat conduction is described by the following equation:

$$\frac{\partial}{\partial x}\left(\lambda \frac{\partial T}{\partial x}\right) + \frac{\partial}{\partial y}\left(\lambda \frac{\partial T}{\partial y}\right) + \frac{\partial}{\partial z}\left(\lambda \frac{\partial T}{\partial z}\right) = \rho C_P \frac{\partial T}{\partial t} - Q_V \tag{5.62}$$

where Q_v is the inner heat source term.

Boundary conditions:

$$-\lambda \frac{\partial T}{\partial n} = \alpha(T - T_f), \text{ at convection boundary } sc \tag{5.63}$$

$$T = \varphi, \text{ at specified temperature boundary } s_1 \tag{5.64}$$

Initial condition:

$$T\big|_{t=0} = T(x,y,z) \tag{5.65}$$

The functional equivalent to Eqs. (5.62)–(5.65) is written as

$$
I = \int_V \frac{\lambda}{2} \left[\left(\frac{\partial T}{\partial x} \right)^2 + \left(\frac{\partial T}{\partial y} \right)^2 + \left(\frac{\partial T}{\partial z} \right)^2 \right] dV
$$
$$
- \int_{sc} \alpha \left(TT_f - \frac{1}{2}T^2 \right) ds - \int_V T \left(\rho C_p \frac{\partial T}{\partial t} - Q_V \right) dV \tag{5.66}
$$

where dV is the volume integral over the three-dimensional region V, and ds is the area integral over the boundary zone.

Taking the first-order variational of I and setting it equal to zero, we get

$$
\delta I = \int_V \lambda \left[\frac{\partial T}{\partial x} \cdot \delta \left(\frac{\partial T}{\partial x} \right) + \frac{\partial T}{\partial y} \cdot \delta \left(\frac{\partial T}{\partial y} \right) + \frac{\partial T}{\partial z} \cdot \delta \left(\frac{\partial T}{\partial z} \right) \right] dV
$$
$$
- \int_{sc} \alpha(T_f - T)\delta T ds - \int_v \left(\rho C_p \frac{\partial T}{\partial t} - Q_V \right) \delta T dV = 0 \tag{5.67}
$$

Equation (5.67) may be written as

$$
\delta I = \int_V [\delta(T')^T] \cdot [\lambda] \cdot (T') dV
$$
$$
- \int_{sc} \alpha(T_f - T)\delta T ds - \int_V \left(\rho C_p \frac{\partial T}{\partial t} - Q_V \right) \delta T dV = 0 \tag{5.68}
$$

where

$$[T']^T = \left[\frac{\partial T}{\partial x}, \frac{\partial T}{\partial y}, \frac{\partial T}{\partial z}\right]$$
(5.69)

$$[\lambda] = \begin{bmatrix} \lambda, & 0, & 0 \\ 0, & \lambda, & 0 \\ 0, & 0, & \lambda \end{bmatrix}$$

$[\lambda]$ is the thermal conductivity matrix.

5.8.2 Eight-Node Hexahedron Isoparametric Element

In the three-dimensional case, divide the body into an arbitrary hexahedron element. For any arbitrary hexahedron element, take its vertexes as nodes. The node coordinates are expressed by $(x_i, y_i, z_i)(i = 1,2,3,4,5,6,7,8)$. Through a coordinate transformation from independent variable (x,y,z) to a new independent variable (ξ, η, s), turn the element into a regular hexahedron with a center that is coincident with the origin of coordinate (ξ, η, s), and the length is equal to 2. The nodes 1, 2, 3, 4, 5, 6, 7, 8 in space (x,y,z) correspond to the nodes ①, ②, ③, ④, ⑤, ⑥, ⑦, ⑧ in space, as shown in Figure 5.3. This coordinate transformation is not conducted to the whole body but is conducted to each element. Thus, the coordinate (x,y,z) is called the *global coordinate system* that is applied to all elements (i.e., the whole body). The coordinate (ξ, η, s) is the local coordinate system, which is just applied to each element. This kind of hexahedron element is called an isoparametric element.

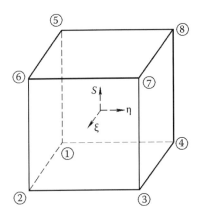

FIGURE 5.3
Schematic of hexahedron element.

Under the local coordinate system (ξ, η, s), select the interpolation function as follows:

$$T = a_1 + a_2\xi + a_3\eta + a_4 s + a_5\xi\eta + a_6\eta s + a_7\xi s + a_8\xi\eta s \qquad (5.70)$$

The constants $a_1 \sim a_8$ are well determined according to the nodal temperature $T_i (i = 1 \sim 8)$.

Write the interpolation function Eq. (5.70) as

$$T = \sum_{i=1}^{8} N_i(\xi, \eta, s) T_i \qquad (5.71)$$

where $N_i(\xi, \eta, s)(i = 1 \sim 8)$ are shape functions, which are well determined by the following two conditions:

1. $N_i(\xi, \eta, s)$ is a polynomial function like Eq. (5.70).
2. $N_i(\xi, \eta, s)$ takes the value of 1 at node i and 0 at other nodes j ($j \neq i$).

(ξ_i, η_i, s_i) is the local coordinates at node i. Referring to Figure 5.3, we have

$$\left.\begin{array}{ll}
(\xi_1, \eta_1, s_1) = (-1, -1, -1), & (\xi_2, \eta_2, s_2) = (1, -1, -1) \\
(\xi_3, \eta_3, s_3) = (1, 1, -1), & (\xi_4, \eta_4, s_4) = (-1, 1, -1) \\
(\xi_5, \eta_5, s_5) = (-1, -1, 1), & (\xi_6, \eta_6, s_6) = (1, -1, 1) \\
(\xi_7, \eta_7, s_7) = (1, 1, 1), & (\xi_8, \eta_8, s_8) = (-1, 1, 1)
\end{array}\right\} \qquad (5.72)$$

Take $N_1(\xi, \eta, s)$ as an example; its values at nodes ② – ⑧ are zero. Note that the planes 2376, 3487, and 5678 pass through these nodes, respectively, and their equations are written as

$$\xi - 1 = 0, \quad \eta - 1 = 0, \quad s - 1 = 0$$

We obtain

$$N_1(\xi, \eta, s) = \frac{(\xi - 1)(\eta - 1)(s - 1)}{[(\xi - 1)(\eta - 1)(s - 1)]_{(-1, -1, -1)}} = \frac{1}{8}(1 - \xi)(1 - \eta)(1 - s)$$

Similarly, we get the expressions of N_2 to N_8. Note that for Eq. (5.72), we write these expressions of shape functions in a unifying version:

$$N_i(\xi, \eta, s) = \frac{1}{8}(1 + \xi_i\xi)(1 + \eta_i\eta)(1 + s_i s), \ (i = 1, 2, \cdots, 8)$$
$$\qquad (5.73)$$

When shape functions $N_i(\xi,\eta,s)$ are known, the interpolation equations like Eq. (5.71) are completely determined.

Equation (5.71) is just the expression of interpolation function T in the local coordinate system. The interpolation function needed in actual computation must be expressions in the global coordinate system (x,y,z). Thus, we must write the transformation relation between the global and local coordinate systems (x,y,z) and (ξ,η,s):

$$
\begin{cases}
x = \sum_{i=1}^{8} N_i(\xi,\eta,s)x_i \\[2ex]
y = \sum_{i=1}^{8} N_i(\xi,\eta,s)y_i \\[2ex]
z = \sum_{i=1}^{8} N_i(\xi,\eta,s)z_i
\end{cases}
\tag{5.74}
$$

where $(x_i,y_i,z_i)(i = 1,2,\ldots,8)$ are the nodal global coordinates.

To form the FEM scheme, we must calculate the transformation matrix:

$$
J = \begin{bmatrix}
\dfrac{\partial x}{\partial \xi} & \dfrac{\partial y}{\partial \xi} & \dfrac{\partial z}{\partial \xi} \\[2ex]
\dfrac{\partial x}{\partial \eta} & \dfrac{\partial y}{\partial \eta} & \dfrac{\partial z}{\partial \eta} \\[2ex]
\dfrac{\partial x}{\partial s} & \dfrac{\partial y}{\partial s} & \dfrac{\partial z}{\partial s}
\end{bmatrix}
\tag{5.75}
$$

Based on Eqs. (5.73) and (5.74), we obtain

$$
J = \begin{bmatrix}
\sum_{i=1}^{8} \dfrac{\partial N_i(\xi,\eta,s)}{\partial \xi}x_i, & \sum_{i=1}^{8} \dfrac{\partial N_i(\xi,\eta,s)}{\partial \xi}y_i, & \sum_{i=1}^{8} \dfrac{\partial N_i(\xi,\eta,s)}{\partial \xi}z_i \\[2ex]
\sum_{i=1}^{8} \dfrac{\partial N_i(\xi,\eta,s)}{\partial \eta}x_i, & \sum_{i=1}^{8} \dfrac{\partial N_i(\xi,\eta,s)}{\partial \eta}y_i, & \sum_{i=1}^{8} \dfrac{\partial N_i(\xi,\eta,s)}{\partial \eta}z_i \\[2ex]
\sum_{i=1}^{8} \dfrac{\partial N_i(\xi,\eta,s)}{\partial s}x_i, & \sum_{i=1}^{8} \dfrac{\partial N_i(\xi,\eta,s)}{\partial s}y_i, & \sum_{i=1}^{8} \dfrac{\partial N_i(\xi,\eta,s)}{\partial s}z_i
\end{bmatrix}
$$

We have

$$
\begin{bmatrix} \dfrac{\partial}{\partial \xi} \\[2mm] \dfrac{\partial}{\partial \eta} \\[2mm] \dfrac{\partial}{\partial s} \end{bmatrix} = J \begin{bmatrix} \dfrac{\partial}{\partial x} \\[2mm] \dfrac{\partial}{\partial y} \\[2mm] \dfrac{\partial}{\partial z} \end{bmatrix}, \quad \begin{bmatrix} \dfrac{\partial}{\partial x} \\[2mm] \dfrac{\partial}{\partial y} \\[2mm] \dfrac{\partial}{\partial z} \end{bmatrix} = J^{-1} \begin{bmatrix} \dfrac{\partial}{\partial \xi} \\[2mm] \dfrac{\partial}{\partial \eta} \\[2mm] \dfrac{\partial}{\partial s} \end{bmatrix} \tag{5.76}
$$

The volume integrals in the global and local coordinate systems have the following relation:

$$
dxdydz = |J| \, d\xi d\eta ds \tag{5.77}
$$

where $|J| = \det J$ is the transformation determinant.

When the element stiffness matrix and load vector are computed, the following integral must be calculated:

$$
\frac{1}{2} \iiint_e \{T'\}^T [\lambda] \{T'\} dxdydz \tag{5.78}
$$

Employing Eq. (5.77), we transform the integral into one in the local coordinate system:

$$
\frac{1}{2} \int_{-1}^{1} \int_{-1}^{1} \int_{-1}^{1} \{T'\}^T [\lambda] \{T'\} |J| \, d\xi d\eta ds \tag{5.79}
$$

For this case, the integral domain is very simple, but the integrand becomes complicated, so that the integral has to be approximately computed by Gaussian quadrature.[125] According to computation like Eq. (5.79), the stiffness matrix and load vector for each element are determined. Then, through assembly, the algebraic equation set of nodal values of temperature is obtained.

The discussion here provides the principles of three-dimensional FEM. Based on these principles, we may develop an FEM program by ourselves for specific problems. On the other hand, there are many commercial FEM software packages now. People can use these commercial software packages to conduct FEM analysis. However, deep understanding of FEM principles will be beneficial for the appropriate application of these commercial software packages.

5.9 FEM Analysis of the Temperature Field in Metal Inert Gas Welding

With consideration of the process characteristics of metal inert gas (MIG) welding, a FEA model is developed to numerically analyze the weld thermal process. To this end, the functions of SYSWELD, a standard software package for numerical analysis of welding processes,[126] were extended; the method for processing the weld reinforcement was simplified, and an appropriate mode for the welding heat source was proposed. Based on the FEA results of the temperature profile in MIG welding, the weld dimensions at the transverse cross section and temperature distribution on the workpiece were obtained. The predicted and experimentally measured results matched satisfactorily.

During the MIG welding process, the weld pool surface is severely deformed under the combined action of arc pressure, droplet impact, surface tension, gravity, and so on. At the front of the weld pool, the pool surface is depressed due to droplet impingement and arc pressure. At the rear of the pool, the surface is humped to form weld reinforcement due to the depressed front part and filler metal addition. The software package was improved to deal with the front and rear parts of the weld pool.

5.9.1 Description of Weld Reinforcement

In the FEM analysis of the temperature field in MIG welding, the reinforcement is usually added to the workpiece in advance, and the "dead-activation elements" technique is used. Thus, prescribing the reinforcement shape and size in mesh generation has a direct effect on the computation results.

Here, the following assumptions were used to take advantage of the SYSWELD software package: (1) The curvature of pool surface along the welding direction was negligible; (2) the surface tension was homogeneous on the pool surface; (3) the effect of the front pool surface depression on the reinforcement was neglected; (4) there was no arc pressure action on the rear of the pool surface. For simplification, assume that the transverse cross-sectional contour of reinforcement is parabolic, that is, $y = a_w x^2$, $(a_w < 0)$, as shown in Figure 5.4.

According to the surface tension balance at the three-phase contact point, we have

$$\cos\theta = \frac{\sigma_{sg} - \sigma_{sl}}{\sigma_{lg}}$$

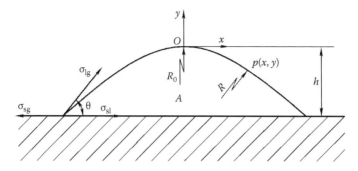

FIGURE 5.4
Schematic of weld reinforcement.

where θ is the contact angle; σ_{sg} is the surface tension of the solid-gas interface; σ_{sl} is the surface tension of the solid-liquid interface; σ_{lg} is the surface tension of the liquid-gas interface.

Under certain conditions, $\sigma_{sg}, \sigma_{sl}, \sigma_{lg}$ may only be dependent on the material properties, so that the contact angle θ may be taken as a constant. From Figure 5.4, we get

$$\left.\frac{dy}{dx}\right|_{y=-h} = \tan\theta \tag{5.80}$$

Based on the parabolic geometry of reinforcement cross section,

$$\left.\frac{dy}{dx}\right|_{y=-h} = 2a_w x\Big|_{x=\sqrt{-\frac{h}{a_w}}} = 2a_w\sqrt{-\frac{h}{a_w}} = \tan\theta$$

Thus,

$$h = -\frac{\tan^2\theta}{4a_w} \tag{5.81}$$

Since the amount of deposited filler metal in unit time is known, the area of reinforcement cross section A_R is expressed as

$$A_R = \frac{\pi d_w^2 S_m}{4v_0}$$

where d_w is the wire diameter, S_m is the wire feed rate, and v_0 is the welding speed.

By deriving

$$A_R = 2h\sqrt{\frac{h}{-a_w}} - 2\int_0^{\sqrt{\frac{h}{a_w}}} a_w x^2 dx = \frac{4h}{3}\sqrt{\frac{h}{-a_w}} = \frac{1}{6}\frac{\tan^3\theta}{a_w^2}$$

we get

$$a_w^2 = \frac{2v_0 \tan^3\theta}{3\pi d_w^2 S_m} \tag{5.82}$$

By solving Eqs. (5.81) and (5.82) simultaneously, the value of a_w can be obtained, then the height of reinforcement h is known.

5.9.2 Mesh Generation and Heat Source Mode

Bead-on-plate MIG welding was performed. The mild steel plate was 150 mm long, 80 mm wide, and 6 mm thick. The thermo properties used in computation are given in Section 6.5. For a quasi-steady state temperature field, a smaller calculation domain of dimensions 80 × 80 × 6 (mm) was used to speed up the computation process.

To deal with filler metal deposition, define the filler metal in the weld bead as FMVD. Employ the dead-activation elements technique; that is, the elements of FMVD ahead of the arc are taken as "dead" so that they are not considered in the computation, while the elements of FMVD behind the arc and below the arc are activated so that they are taken into consideration in the FEM analysis. Therefore, the weld pool is divided into two parts (front and rear) in analysis.

A nonuniform mesh system of eight-node hexahedron elements was used to balance between computation accuracy and speed. Figure 5.5 demonstrates the mesh generated. Figures 5.6 and 5.7 show the mesh for the front and rear parts of the weld pool, respectively.

Due to filler metal deposition, the heat transfer mechanism in the weld pool takes some changes. The simple mode of the heat source, such as Gaussian or double-ellipsoid distribution, cannot produce satisfactory results. In this study case, a combined heat source mode was employed: (1) a Gaussian planar heat source acted on the deformed pool surface to account for the arc heat input, and (2) a uniformly distributed heat flux of constant value acted within a domain related to the weld pool dimensions to account for heat content from droplets. The sum of both parts is equal to the effective arc power $\eta I U_a$.

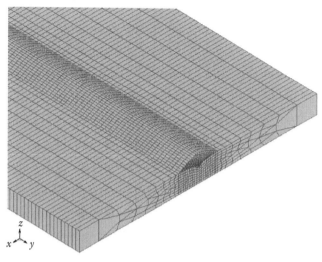

FIGURE 5.5
Mesh system.

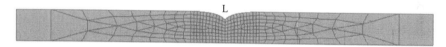

FIGURE 5.6
Mesh ahead of the arc centerline.

FIGURE 5.7
Mesh behind the arc centerline.

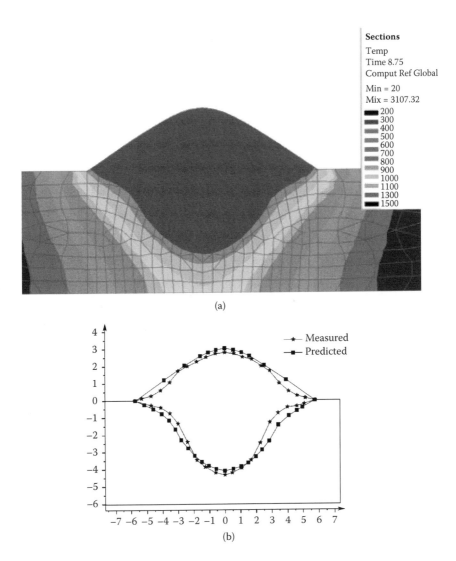

(a)

(b)

FIGURE 5.8
The computed temperature field and weld dimension: (a) The computed temperature profile;
(b) comparison between the predicted and measured weld shape and size (270-A welding current, 430-mm/min welding speed, 20-mm wire extension).

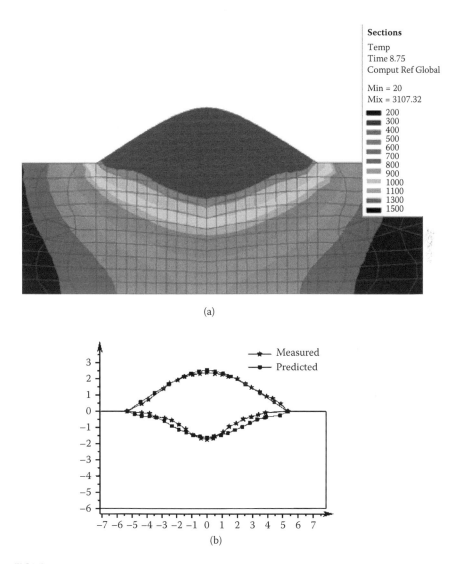

(a)

(b)

FIGURE 5.9
The computed temperature field and weld dimension: (a) The computed temperature profile; (b) comparison between the predicted and measured weld shape and size (200-A welding current, 430-mm/min welding speed, 20-mm wire extension).

5.9.3 Computational Results

For this study case, the temperature field on the workpiece was calculated by FEM analysis. The weld shape and dimension were determined. Figures 5.8(a) and 5.9(a) show the computed temperature profiles at a transverse cross section of the workpiece, while Figures 5.8(b) and 5.9(b) compare the predicted and measured weld shape and size.

6

Numerical Simulation of Fluid Flow and Heat Transfer in Transient TIG Weld Pools

6.1 Introduction

Many physical and chemical processes are involved in welding technology, such as thermal conduction, convection, evaporation, radiation, fusion, and solidification. To simulate the welding process numerically, every physical phenomenon taking place during welding needs to be analyzed. However, some simplification and assumptions have to be made first, and then mathematical methods (differential equations and their definite conditions) are used to describe the regularities behind these phenomena. So, quantitative understanding of the welding processes can be acquired. In this chapter, numerical simulation of heat transfer and fluid flow in transient tungsten inert gas (TIG) weld pools is addressed.

6.2 Mathematical Descriptions of TIG Weld Pool Behaviors

Figure 6.1 illustrates the TIG welding process with a moving arc. To describe the dynamic behaviors of a transient weld pool, the heat transfer and fluid flow in the weld pool at every time instant need to be predicted. This means that a transient process should be considered. After the welding starts, heat is deposited from the arc to the base metal. The temperature of base metal rises quickly, and then part of the metal melts to form a weld pool. Because of the continuing heat input from the welding arc, the weld pool grows with a rather rapid speed. Subjected by some forces acting on it, the top surface of the weld pool begins to deform. When the weld pool becomes fully penetrated, the bottom surface of the weld pool also deforms. Within the weld pool, a violent flow of molten metal is generated by some forces, and convection is the main heat transfer mechanism. But, in the solid area out of the weld pool, heat conduction predominates. As the arc moves forward, the

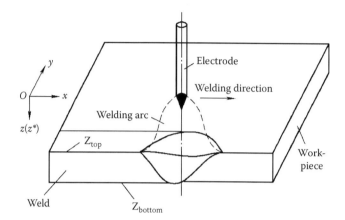

FIGURE 6.1
Schematic of TIG welding in Cartesian coordinate: (left) before the step change; (right) after the step change.

temperature distribution in base metal under the arc tends to be constant, so the shape of the weld pool remains stable and no longer expands. The weld pool obtains a quasi-steady state and moves along the welding direction with the same welding speed. Thus, it is a three-dimensional (3-D) transient question involving fluid flow and heat transfer.

For simplification, some assumptions are made in the establishment of a 3-D transient model for TIG welding:

1. The molten metal in the weld pool is viscous incompressible fluid, and its flow style is laminar.
2. The change of density is only considered in the gravity term of the momentum equation, and it follows the Boussinesq hypothesis.[39,127]
3. Except for the specific heat, thermal conductivity, viscosity, and surface tension coefficient, other thermophysical parameters such as density and heat transfer coefficient are independent of temperature.

Although one article pointed out that the fluid flow in the weld pool has turbulent characteristics,[128] almost all researchers used laminar assumption to simplify the model and formulation. It has been found that such an assumption would not cause large error.[129] Laminar assumption is also adopted here.

6.2.1 Governing Equations in the Cartesian Coordinate

The Cartesian coordinate is shown in Figure 6.1, and it is fixed in the base metal. The governing equations that describe every physical variable subjected to the conservation laws include the continuity, energy, and momentum equations, which are written as follows:

Energy conservation equation:

$$\rho C_P \left(\frac{\partial T}{\partial t} + U \frac{\partial T}{\partial x} + V \frac{\partial T}{\partial y} + W \frac{\partial T}{\partial z} \right) = \lambda \left(\frac{\partial^2 T}{\partial x^2} + \frac{\partial^2 T}{\partial y^2} + \frac{\partial^2 T}{\partial z^2} \right) \tag{6.1}$$

Momentum conservation equation:

x-axis:

$$\rho \left(\frac{\partial U}{\partial t} + U \frac{\partial U}{\partial x} + V \frac{\partial U}{\partial y} + W \frac{\partial U}{\partial z} \right) = F_x - \frac{\partial P}{\partial x} + \mu \left(\frac{\partial^2 U}{\partial x^2} + \frac{\partial^2 U}{\partial y^2} + \frac{\partial^2 U}{\partial z^2} \right) \tag{6.2}$$

y-axis:

$$\rho \left(\frac{\partial V}{\partial t} + U \frac{\partial V}{\partial x} + V \frac{\partial V}{\partial y} + W \frac{\partial V}{\partial z} \right) = F_y - \frac{\partial P}{\partial y} + \mu \left(\frac{\partial^2 V}{\partial x^2} + \frac{\partial^2 V}{\partial y^2} + \frac{\partial^2 V}{\partial z^2} \right) \tag{6.3}$$

z-axis:

$$\rho \left(\frac{\partial W}{\partial t} + U \frac{\partial W}{\partial x} + V \frac{\partial W}{\partial y} + W \frac{\partial W}{\partial z} \right) = F_z - \frac{\partial P}{\partial z} + \mu \left(\frac{\partial^2 W}{\partial x^2} + \frac{\partial^2 W}{\partial y^2} + \frac{\partial^2 W}{\partial z^2} \right) \tag{6.4}$$

Continuity equation:

$$\frac{\partial U}{\partial x} + \frac{\partial V}{\partial y} + \frac{\partial W}{\partial z} = 0 \tag{6.5}$$

where T is the temperature; U, V, and W are the three components of the fluid velocity in the x-, y-, and z-directions, respectively; t is the time; ρ is the density; C_p is the specific heat; λ is the thermal conductivity; P is the pressure in the liquid; μ is the dynamic viscosity of the liquid metal; and F_x, F_y, and F_z are the components of body forces in the x-, y-, and z-directions, respectively.

For these governing equations, the solution domain of the continuity equation and the momentum equation is within the weld pool. As the liquid velocity in the solid region becomes zero, the energy equation in the solid region will degrade into a pure heat conduction equation. Therefore, its solution region contains the weld pool and other regions out of the weld pool.

6.2.2 The Free Surface Deformation of the Weld Pool

During the welding process, some forces acting on the surface of the weld pool include the arc pressure, surface tension, gravity of weld pool, and more. Also, droplet impact is involved in gas metal arc welding (GMAW). Acted on by these forces, the weld pool surface deforms in three dimensions. When the workpiece is fully penetrated, obvious deformations will generate on both the top and the bottom surfaces of the weld pool. These deformations alter the heat transfer conditions in the weld pool, which in turn changes the 3-D shape of the weld pool correspondingly. Therefore, the surface deformation of the weld pool has a great deal of influence on the weld quality and welding efficiency.

First, the partially penetrated case is considered. Take the top surface of the weld pool as the study object and regard it as a curved surface that can deform freely under the effect of external forces. The deformable surface is a gas-liquid interface, as shown in Figure 6.2. The lengths of an infinitesimal element in the x- and y-directions are written as ds_1, ds_2, respectively; curvature radii are R_1, R_2, respectively; and pressures on the two sides of the interface are p_1, p_2, respectively. So, the pressure difference acting on the interface is expressed as $(p_1 - p_2) \, ds_1 ds_2$. There are two surface tension terms γds_1 on the opposite edges ds_1 and another two terms γds_2 on ds_2. The angles between these two surface tension forces are $d\alpha = (ds_1/R_1)$ and $d\beta = (ds_2/R_2)$, respectively. When a dynamic balance is achieved, we have

$$(p_1 - p_2)ds_1 ds_2 = 2\gamma ds_1 \frac{1}{2}\frac{ds_2}{R_2} + 2\gamma ds_2 \frac{1}{2}\frac{ds_1}{R_1} = \gamma \left(\frac{1}{R_1} + \frac{1}{R_2}\right) ds_1 ds_2$$

$$p_1 - p_2 = \gamma \left(\frac{1}{R_1} + \frac{1}{R_2}\right) \tag{6.6}$$

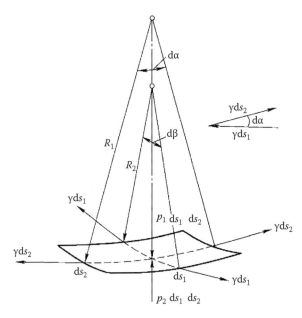

FIGURE 6.2
Force analysis of an element on the gas-liquid interface.

Equation (6.6) is the Young-Laplace equation.[130] For a curved surface $F(x, y, z) = 0$, its curvature radii can be written as

$$\frac{1}{R_1} + \frac{1}{R_2} = -\nabla \cdot \left(\frac{\nabla F}{|\nabla F|} \right) \tag{6.7}$$

For the top surface of a weld pool, its shape function is $z = \varphi(x, y)$; then

$$F(x, y, z) = \varphi(x, y) - z = 0.$$

$$\frac{\nabla F}{|\nabla F|} = \frac{(\varphi_x, \varphi_y, -1)}{\sqrt{1 + \varphi_x^2 + \varphi_y^2}}$$

where $\varphi_x = \dfrac{\partial \varphi}{\partial x}$, $\varphi_y = \dfrac{\partial \varphi}{\partial y}$. Substituting these equations into Eq. (6.7), we obtain

$$\frac{1}{R_1} + \frac{1}{R_2} = -\left(\frac{\partial}{\partial x} \vec{i} + \frac{\partial}{\partial y} \vec{j} + \frac{\partial}{\partial z} \vec{k} \right)$$

$$\times \left(\frac{\varphi_x}{\sqrt{1 + \varphi_x^2 + \varphi_y^2}} \vec{i} + \frac{\varphi_y}{\sqrt{1 + \varphi_x^2 + \varphi_y^2}} \vec{j} + \frac{-1}{\sqrt{1 + \varphi_x^2 + \varphi_y^2}} \vec{k} \right)$$

$$= -\frac{\partial}{\partial x} \left(\frac{\varphi_x}{\sqrt{1 + \varphi_x^2 + \varphi_y^2}} \right) - \frac{\partial}{\partial y} \left(\frac{\varphi_y}{\sqrt{1 + \varphi_x^2 + \varphi_y^2}} \right)$$

$$= -\frac{(1 + \varphi_y^2)\varphi_{xx} - 2\varphi_x \varphi_y \varphi_{xy} + (1 + \varphi_x^2)\varphi_{yy}}{(1 + \varphi_x^2 + \varphi_y^2)^{3/2}} \tag{6.8}$$

where

$$\varphi_{xx} = \frac{\partial^2 \varphi}{\partial x^2}, \quad \varphi_{yy} = \frac{\partial^2 \varphi}{\partial y^2}, \quad \varphi_{xy} = \frac{\partial^2 \varphi}{\partial x \partial y}$$

On the top surface of the weld pool, the arc pressure is P_{arc} (downward direction, equivalent to p_1), and the gravity of liquid metal is $\rho g \varphi$ (for a concave surface, it serves to restore to the original shape; upward direction). The shape of the top surface of the weld pool is defined as $z = \varphi(x, y)$. According to Eqs. (6.6)–(6.8), the top curved surface of the weld pool should satisfy the following equation:

$$P_a - \rho g \varphi + C_1 = -\gamma \frac{(1 + \varphi_y^2)\varphi_{xx} - 2\varphi_x \varphi_y \varphi_{xy} + (1 + \varphi_x^2)\varphi_{yy}}{(1 + \varphi_x^2 + \varphi_y^2)^{3/2}} \tag{6.9}$$

where P_a is the arc pressure, ρ is the density of liquid metal, g is the gravitational acceleration, γ is the surface tension, and C_1 is a constant.

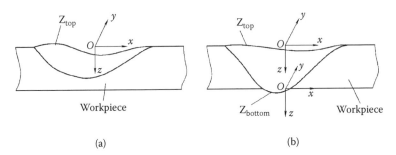

FIGURE 6.3
Schematic of surface deformation in a TIG weld pool: (a) partial penetration; (b) full penetration.

Outside the weld pool, $\varphi(x, y) = 0$. Because the total volume of the weld pool is not changed before or after the surface deformation if no filler metal is added, there is the following constraint:

$$\iint_{\Omega_1} \varphi(x,y)dxdy = 0$$

(6.10)

where Ω_1 is the surface area of the weld pool at the top surface.

As shown in Figure 6.3(b), both top and bottom surfaces of the weld pool will deform when the workpiece is fully penetrated. Suppose the shape functions of the top and bottom surfaces are $z = \varphi(x, y)$ and $z = \psi(x, y)$ and the relative coordinate origins are located on the top and bottom surfaces of the workpiece, respectively. For a fully penetrated weld pool, its shape function of the top surface should satisfy

$$P_a - \rho g\varphi + C_2 = -\gamma\frac{(1+\varphi_y^2)\varphi_{xx} - 2\varphi_x\varphi_y\varphi_{xy} + (1+\varphi_x^2)\varphi_{yy}}{(1+\varphi_x^2+\varphi_y^2)^{3/2}}$$

(6.11)

and its shape function of the bottom surface should satisfy

$$\rho g(\psi + H - \varphi) + C_2 = -\gamma\frac{(1+\psi_y^2)\psi_{xx} - 2\psi_x\psi_y\psi_{xy} + (1+\psi_x^2)\psi_{yy}}{(1+\psi_x^2+\psi_y^2)^{3/2}}$$

(6.12)

where H is the thickness of the workpiece, and C_2 is a constant. The subscripts of function ψ are its first and second partial derivatives. Although both the top and the bottom surfaces of the pool undergo deformation, the total volume of the weld pool does not vary at a specific time instant. Therefore,

$$\iint_{\Omega_1} \varphi(x,y)dxdy = \iint_{\Omega_2} \psi(x,y)dxdy$$

(6.13)

where Ω_1 and Ω_2 are the surface areas of the weld pool at the top and the bottom surface, respectively. If point (x, y) locates out of the weld pool, then $\varphi(x, y) = 0$ and $\psi(x, y) = 0$.

In Eqs. (6.9), (6.11), and (6.12), C_1 and C_2 are determinant constants. They are the total sum of other forces that act on the weld pool surface except the arc pressure, gravity, and surface tension. In the calculation, C_1 is derived as follows:

Integrate both sides of Eq. (6.9) within Ω_1 and then substitute it into Eq. (6.10); we obtain

$$C_1 \iint_{\Omega_1} dxdy = \iint_{\Omega_1} (-P_a)dxdy - \iint_{\Omega_1} \frac{(1+\varphi_y^2)\varphi_{xx} - 2\varphi_x\varphi_y\varphi_{xy} + (1+\varphi_x^2)\varphi_{yy}}{(1+\varphi_x^2+\varphi_y^2)^{3/2}/\gamma}dxdy \quad (6.14)$$

Similarly, by integrating both sides of Eqs. (6.11) and (6.12) within Ω_1 and Ω_2, respectively, and then substituting them into Eq. (6.13), we get the calculating equation for C_2:

$$C_2\left(\iint_{\Omega_1} dxdy + \iint_{\Omega_2} dxdy\right)$$

$$= \iint_{\Omega_1} (-P_a)dxdy - \iint_{\Omega_1} \frac{(1+\varphi_y^2)\varphi_{xx} - 2\varphi_x\varphi_y\varphi_{xy} + (1+\varphi_x^2)\varphi_{yy}}{(1+\varphi_x^2+\varphi_y^2)^{3/2}/\gamma}dxdy$$

$$-\rho g \iint_{\Omega_2} (H-\varphi)dxdy - \iint_{\Omega_2} \frac{(1+\psi_y^2)\psi_{xx} - 2\psi_x\psi_y\psi_{xy} + (1+\psi_x^2)\psi_{yy}}{(1+\psi_x^2+\psi_y^2)^{3/2}/\gamma}dxdy \quad (6.15)$$

The arc pressure can be expressed as

$$P_a = \frac{\mu_m I^2}{8\pi^2\sigma_j^2}\exp(-\frac{r^2}{2\sigma_j^2}) \quad (6.16)$$

where $r = \sqrt{(x-v_0t)^2+y^2}$, μ_m is the magnetic permeability in free space, I is the welding current, σ_j is the current distribution parameter, v_0 is the welding speed, and t is the time.

6.2.3 Governing Equations in Body-Fitted Coordinates

During the welding process, the shape of the calculating domain changes with the movement of the arc when the weld pool surfaces deform. Governing Eqs. (6.1) to (6.5) are derived in the Cartesian coordinate system. The finite difference method is suitable for objects with regular surfaces, but when the free surface of the weld pool deforms, regular surfaces no longer exist, and 3-D curved surfaces appear. If the curved surfaces are still dealt with in a Cartesian coordinate system, a very complex zigzag boundary for the energy equation will be generated, which causes increased difficulties in calculation. As shown in Figure 6.4, the body-fitted coordinate system is employed to treat the deformation of the weld pool surfaces. It could describe the surface deformation much better and deal with the top and bottom surface boundaries of the weld pool more easily. In this section, the relationship between the body-fitted coordinate system (x^*, y^*, z^*) and the original Cartesian coordinate system (x, y, z) is written as

$$x^* = x \tag{6.17}$$

$$y^* = y \tag{6.18}$$

$$z^* = \frac{z - \varphi(x,y)}{H + \psi(x,y) - \varphi(x,y)} \tag{6.19}$$

Obviously, on the top surface of the weld pool, $z = \varphi$, $z^* = 0$; on the bottom surface of the weld pool, $z = H + \psi$, $z^* = 1$. By this algebraic transformation, the irregular physical space in the Cartesian coordinate system is changed into the regular computing domain.

In the body-fitted coordinate system as shown in Figure 6.4, every physical variable and governing equation will alter their forms. Because $x^* = x$ and $y^* = y$, the variable $T(x^*, y^*, z^*, t)$ can be expressed as $T(x, y, z^*, t)$. If no particular explanation is given, the following physical quantities related to the body-fitted coordinate system will be expressed by (x, y, z^*): Take the variable $T(x, y, z, t)$ in physical space as an example; by using the chain rule for

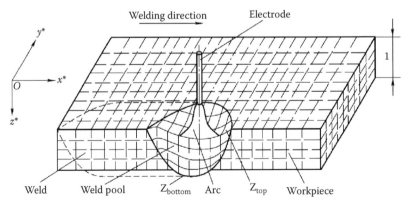

FIGURE 6.4
Schematic of grid system in the body-fitted coordinate system.

differentiation, the first- and second-order partial derivatives of variable $T(x, y, z^*, t)$ in the computing domain can be written as

$$\frac{\partial T}{\partial x^*} = \frac{\partial T}{\partial x} + \frac{\partial T}{\partial z^*}\frac{\partial z^*}{\partial x}, \quad \frac{\partial T}{\partial y^*} = \frac{\partial T}{\partial y} + \frac{\partial T}{\partial z^*}\frac{\partial z^*}{\partial y}, \quad \frac{\partial T}{\partial z^*} = \frac{\partial T}{\partial z^*}\frac{\partial z^*}{\partial z}$$

$$\frac{\partial^2 T}{\partial x^{*2}} = \frac{\partial^2 T}{\partial x^2} + 2\frac{\partial z^*}{\partial x}\frac{\partial^2 T}{\partial x \partial z^*} + \left(\frac{\partial z^*}{\partial x}\right)^2 \frac{\partial^2 T}{\partial z^{*2}} + \frac{\partial^2 z^*}{\partial x^2}\frac{\partial T}{\partial z^*}$$

$$\frac{\partial^2 T}{\partial y^{*2}} = \frac{\partial^2 T}{\partial y^2} + 2\frac{\partial z^*}{\partial y}\frac{\partial^2 T}{\partial y \partial z^*} + \left(\frac{\partial z^*}{\partial y}\right)^2 \frac{\partial^2 T}{\partial z^{*2}} + \frac{\partial^2 z^*}{\partial y^2}\frac{\partial T}{\partial z^*}$$

$$\frac{\partial^2 T}{\partial z^{*2}} = \left(\frac{\partial z^*}{\partial z}\right)^2 \frac{\partial^2 T}{\partial z^{*2}} + \frac{\partial^2 z^*}{\partial z^2}\frac{\partial T}{\partial z^*}$$

where

$$\frac{\partial z^*}{\partial x} = -\frac{H+\psi-z}{(H+\psi-\varphi)^2}\frac{\partial \varphi}{\partial x} - \frac{z-\varphi}{(H+\psi-\varphi)^2}\frac{\partial \psi}{\partial x}$$

$$\frac{\partial z^*}{\partial y} = -\frac{H+\psi-z}{(H+\psi-\varphi)^2}\frac{\partial \varphi}{\partial y} - \frac{z-\varphi}{(H+\psi-\varphi)^2}\frac{\partial \psi}{\partial y}$$

$$\frac{\partial z^*}{\partial z} = -\frac{1}{H+\psi-\varphi}$$

$$\frac{\partial^2 z^*}{\partial x^2} = -\frac{H+\psi-z}{(H+\psi-\varphi)^2}\frac{\partial^2 \varphi}{\partial x^2} - \frac{z-\varphi}{(H+\psi-\varphi)^2}\frac{\partial^2 \psi}{\partial x^2} + 2\frac{z-\varphi}{(H+\psi-\varphi)^3}\frac{\partial \psi}{\partial x}\left(\frac{\partial \psi}{\partial x} - \frac{\partial \varphi}{\partial x}\right)$$

$$+ 2\frac{H+\psi-z}{(H+\psi-\varphi)^3}\frac{\partial \varphi}{\partial x}\left(\frac{\partial \psi}{\partial x} - \frac{\partial \varphi}{\partial x}\right)$$

$$\frac{\partial^2 z^*}{\partial y^2} = -\frac{H+\psi-z}{(H+\psi-\varphi)^2}\frac{\partial^2 \varphi}{\partial y^2} - \frac{z-\varphi}{(H+\psi-\varphi)^2}\frac{\partial^2 \psi}{\partial y^2} + 2\frac{z-\varphi}{(H+\psi-\varphi)^3}\frac{\partial \psi}{\partial y}\left(\frac{\partial \psi}{\partial y} - \frac{\partial \varphi}{\partial y}\right)$$

$$+ 2\frac{H+\psi-z}{(H+\psi-\varphi)^3}\frac{\partial \varphi}{\partial y}\left(\frac{\partial \psi}{\partial y} - \frac{\partial \varphi}{\partial y}\right)$$

$$\frac{\partial^2 z^*}{\partial z^2} = 0$$

Similarly, the first- and second-order partial derivatives of variables $U(x, y, z^*, t)$, $V(x, y, z^*, t)$, and $W(x, y, z^*, t)$ are obtained. Thus, governing Eqs. (6.1) to (6.5) in the body-fitted coordinate system have the following forms:

Energy conservation equation:

$$\rho C_p\left(\frac{\partial T}{\partial t}+U\frac{\partial T}{\partial x}+V\frac{\partial T}{\partial y}+W_t\frac{\partial T}{\partial z*}\right)=\frac{\partial}{\partial x}\left(\lambda\frac{\partial T}{\partial x}\right)+\frac{\partial}{\partial y}\left(\lambda\frac{\partial T}{\partial y}\right)$$

$$+S\frac{\partial}{\partial z*}\left(\lambda\frac{\partial T}{\partial z*}\right)+\lambda C_t \qquad (6.20)$$

Momentum conservation equation:

$$\rho\left(\frac{\partial U}{\partial t}+U\frac{\partial U}{\partial x}+V\frac{\partial U}{\partial y}+W_1\frac{\partial U}{\partial z*}\right)=-\left(\frac{\partial P}{\partial x}+\frac{\partial P}{\partial z*}\frac{\partial z*}{\partial x}\right)$$

$$+\mu\left(\frac{\partial^2 U}{\partial x^2}+\frac{\partial^2 U}{\partial y^2}+S\frac{\partial^2 U}{\partial z*^2}\right)+C_u+F_x \qquad (6.21)$$

$$\rho\left(\frac{\partial V}{\partial t}+U\frac{\partial V}{\partial x}+V\frac{\partial V}{\partial y}+W_1\frac{\partial V}{\partial z*}\right)=-\left(\frac{\partial P}{\partial y}+\frac{\partial P}{\partial z*}\frac{\partial z*}{\partial y}\right)$$

$$+\mu\left(\frac{\partial^2 V}{\partial x^2}+\frac{\partial^2 V}{\partial y^2}+S\frac{\partial^2 V}{\partial z*^2}\right)+C_v+F_y \qquad (6.22)$$

$$\rho\left(\frac{\partial W}{\partial t}+U\frac{\partial W}{\partial x}+V\frac{\partial W}{\partial y}+W_1\frac{\partial W}{\partial z*}\right)=-\frac{\partial P}{\partial z*}\frac{\partial z*}{\partial x}$$

$$+\mu\left(\frac{\partial^2 W}{\partial x^2}+\frac{\partial^2 W}{\partial y^2}+S\frac{\partial^2 W}{\partial z*^2}\right)+C_w+F_z \qquad (6.23)$$

Continuity equation:

$$\frac{\partial U}{\partial x}+\frac{\partial V}{\partial y}+\frac{\partial W}{\partial z*}\frac{\partial z*}{\partial z}+C_m=0 \qquad (6.24)$$

Some terms in these governing equations are defined as follows:

$$W_t=U\frac{\partial z*}{\partial x}+V\frac{\partial z*}{\partial y}+W\frac{\partial z*}{\partial z}-\frac{\lambda}{\rho C_p}\left(\frac{\partial^2 z*}{\partial x^2}+\frac{\partial^2 z*}{\partial y^2}+\frac{\partial^2 z*}{\partial z^2}\right) \qquad (6.25)$$

$$W_1=U\frac{\partial z*}{\partial x}+V\frac{\partial z*}{\partial y}+W\frac{\partial z*}{\partial z}-\frac{\mu}{\rho}\left(\frac{\partial^2 z*}{\partial x^2}+\frac{\partial^2 z*}{\partial y^2}+\frac{\partial^2 z*}{\partial z^2}\right) \qquad (6.26)$$

$$S=\left(\frac{\partial z*}{\partial x}\right)^2+\left(\frac{\partial z*}{\partial y}\right)^2+\left(\frac{\partial z*}{\partial z}\right)^2 \qquad (6.27)$$

$$C_t = 2\left(\frac{\partial^2 T}{\partial z^* \partial x}\frac{\partial z^*}{\partial x} + \frac{\partial^2 T}{\partial z^* \partial y}\frac{\partial z^*}{\partial y}\right) \tag{6.28}$$

$$C_u = 2\mu\left(\frac{\partial^2 U}{\partial z^* \partial x}\frac{\partial z^*}{\partial x} + \frac{\partial^2 U}{\partial z^* \partial y}\frac{\partial z^*}{\partial y}\right) \tag{6.29}$$

$$C_v = 2\mu\left(\frac{\partial^2 V}{\partial z^* \partial x}\frac{\partial z^*}{\partial x} + \frac{\partial^2 V}{\partial z^* \partial y}\frac{\partial z^*}{\partial y}\right) \tag{6.30}$$

$$C_w = 2\mu\left(\frac{\partial^2 W}{\partial z^* \partial x}\frac{\partial z^*}{\partial x} + \frac{\partial^2 W}{\partial z^* \partial y}\frac{\partial z^*}{\partial y}\right) \tag{6.31}$$

$$C_m = \frac{\partial U}{\partial z^*}\frac{\partial z^*}{\partial x} + \frac{\partial V}{\partial z^*}\frac{\partial z^*}{\partial y} \tag{6.32}$$

Equations (6.20) to (6.24) are the conservation equations in the computing domain, and every term in these equations maintains its corresponding physical meaning in the Cartesian coordinate system. Although employment of the body-fitted coordinates can completely avoid the newly added boundaries resulting from the surface deformation, the governing equations in the body-fitted coordinate system have quite complex mixed partial derivative terms, which causes many difficulties in the discretization of the governing equations. Some special techniques are employed to overcome these difficulties.

6.2.4 Boundary Conditions in the Body-Fitted Coordinate System

6.2.4.1 Energy Boundary Conditions

As shown in Figure 6.1, the top surface of the baseplate is heated by a welding arc, and heat is conducted into the inner part of the baseplate at the same time; part of the heat is transferred into the ambient environment by convection, radiation, and evaporation. There is also radiation on other surfaces of the baseplate. But, compared to the energy boundary conditions in the Cartesian coordinate system, the energy boundary conditions in the body-fitted coordinate system are only altered on the top and bottom surface of the baseplate.

The welding arc deposits heat on the baseplate through a certain area. Generally, the Gaussian function is used to describe the heat flux distribution on the workpiece surface. On the top surface of the baseplate, take the arc centerline as the origin of the moving coordinate system and use r to refer to the distance away from the arc centerline. Thus, the heat flux distribution of the arc on the base metal can be written as

$$q_{arc} = \frac{\eta I U_a}{2\pi\sigma_q^2} \exp\left(-\frac{r^2}{2\sigma_q^2}\right), \quad r \le \sqrt{6}\sigma_q \qquad (6.33)$$

where $r = \sqrt{(x - v_0 t)^2 + y^2}$, η is the efficiency of the arc power, U_a is the arc voltage, I is the welding current, and σ_q is the distribution parameter of the arc heat flux.

In fact, if a Gaussian planar heat source is used, the predicted weld pool shape does not match well with the experimental one. For example, the predicted length of the weld pool is insufficient. So, an improved double-elliptical heat source mode is used here to describe the heat flux distribution, as shown in Figure 6.5.

Suppose the maximum value of heat density still locates at the arc centerline and consider the effect of welding speed on the heat flux distribution on the workpiece surface. Assume that heat flux distributions both ahead and behind the arc have an elliptical distribution and are symmetrical about the x-axis. Suppose 95% of heat energy locates in the two elliptical areas. After a series of derivations, the double-elliptical thermal flux distribution in a moving coordinate system can be expressed as

$$q_{arc}(x, y) = \frac{a_f}{a_f + a_r} \frac{6\eta I U_a}{\pi a_f b_h} \exp\left(-\frac{3x^2}{a_f^2}\right) \exp\left(-\frac{3y^2}{b_h^2}\right), \quad x \ge 0 \qquad (6.34)$$

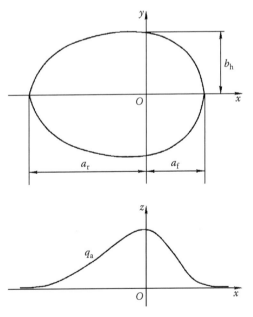

FIGURE 6.5
Schematic of the double-elliptical distribution.

$$q_{arc}(x,y) = \frac{a_r}{a_f + a_r} \frac{6\eta IU_a}{\pi a_r b_h} \exp\left(-\frac{3x^2}{a_r^2}\right) \exp\left(-\frac{3y^2}{b_h^2}\right), \quad x < 0 \tag{6.35}$$

where a_f, a_r, b_h are the heat flux distribution parameters that are related to the welding process parameters. Since the maximum heat density of the welding arc still deposits on the origin point, the relationship between the distribution parameters a_f, a_r, b_h in double-elliptical distribution and the parameter σ_q in Gaussian distribution can be obtained:

$$b_h(a_f + a_r) = 12\sigma_q^2 \tag{6.36}$$

In the stationary coordinate system shown in Figure 6.4, the double-elliptical distribution function is written as

$$q_{arc}(x,y) = \frac{a_f}{a_f + a_r} \frac{6\eta IU_a}{\pi a_f b_h} \exp\left[-\frac{3(x - v_0 t)^2}{a_f^2}\right] \exp\left(-\frac{3y^2}{b_h^2}\right), \quad (x - v_0 t) \geq 0 \tag{6.37}$$

$$q_{arc}(x,y) = \frac{a_r}{a_f + a_r} \frac{6\eta IU_a}{\pi a_r b_h} \exp\left[-\frac{3(x - v_0 t)^2}{a_r^2}\right] \exp\left(-\frac{3y^2}{b_h^2}\right), \quad (x - v_0 t) < 0 \tag{6.38}$$

The surface deformation of the weld pool causes the variation of the heat flux distribution on the top surface of the weld pool. But in TIG welding, there is comparatively little surface deformation, so it is acceptable to suppose that the energy boundary conditions do not change in the body-fitted coordinate system.

The heat losses on the workpiece surfaces, such as convection, radiation, and evaporation, can be expressed as

$$q_{conv} = \alpha_c(T - T_f) \tag{6.39}$$

$$q_{rad} = \sigma_s \varepsilon(T^4 - T_f^4) \tag{6.40}$$

$$q_{evap} = W_q L_q \tag{6.41}$$

where α_c is the convective heat transfer coefficient, σ_s is the Stefan-Boltzmann constant, ε is the radiation emissivity, W_q is the liquid-metal evaporation rate, T_f is the ambient temperature, and L_q is the latent heat of evaporation. For the stainless steel SS304 materials, an approximate equation was given for W_q in Eq. (6.41)[24,131]:

$$\log W_q = 2.52 + \left(6.121 - \frac{18836}{T}\right) - 0.5 \log T \tag{6.42}$$

So, the net heat transfer input at the top surface in a body-fitted coordinate system is as follows:

$$q = \lambda \frac{\partial T}{\partial z^*} \frac{\partial z^*}{\partial z} = q_{arc} - q_{conv} - q_{rad} - q_{evap} \tag{6.43}$$

At the symmetric surface, both sides have no net heat surplus:

$$\frac{\partial T}{\partial y} = 0 \tag{6.44}$$

At all other surfaces, there are only convection, radiation, and evaporation losses. Thus,

$$q = -q_{conv} - q_{rad} - q_{evap} \tag{6.45}$$

6.2.4.2 Momentum Boundary Conditions

When the weld pool is fully penetrated, the surface tension gradient (Marangoni force) will balance the viscous shear stress of fluid on both the top and the bottom surfaces of the weld pool. Because the body-fitted coordinate system is different from the Cartesian coordinate system only in the z-axis, in the computing domain just the momentum boundary conditions on the top and the bottom surfaces of the weld pool are changed. On the free surfaces of the weld pool, the momentum boundary conditions are

$$\mu \frac{\partial U}{\partial z^*} \frac{\partial z^*}{\partial z} = -\frac{\partial \gamma}{\partial T} \frac{\partial T}{\partial x}, \mu \frac{\partial V}{\partial z^*} \frac{\partial z^*}{\partial z} = -\frac{\partial \gamma}{\partial T} \frac{\partial T}{\partial y}, W = 0 \tag{6.46}$$

where γ is the surface tension, and $\partial \gamma / \partial T$ is the temperature coefficient of surface tension.

The magnitude of surface tension of liquid metal in the weld pool is a function of temperature. The surface-active elements in metals, such as sulfur and oxygen, also have an influence on the value of surface tension. Because both the temperature and the surface tension are not distributed uniformly on the weld pool surfaces, the value of $\partial \gamma / \partial T$ at each point is also different. This leads to the flow of liquid metal in the weld pool (i.e., Marangoni flow). Sahoo et al. gave the relationship between the surface tension and its temperature coefficient in the iron-sulfur binary alloy system[132]:

$$\gamma = \gamma^0 - A_c(T - T^0) - R_g T \Gamma_s \ln[1 + k_1 a_s \exp(-\Delta H^0 / R_g T)] \tag{6.47}$$

$$\frac{\partial \gamma}{\partial T} = -A_c - R_g \Gamma_s \ln(1 + K_{seg} a_s) - \frac{K_{seg} a_s \Delta H^0 \Gamma_s}{T(1 + K_{seg} a_s)} \tag{6.48}$$

where A_c is a constant, γ^0 is the surface tension when the temperature is T^0, R_g is the gas constant, Γ_s is the surface supersaturated parameter, a_s is the activity of sulfur in the alloy, ΔH^0 is the standard absorption heat, and k_1 is the segregation enthalpy. K_{seg} is the segregation balance parameter; it can be expressed as

$$K_{seg} = k_1 \exp(-\Delta H^0 / R_g T).$$

At the symmetric plane *xoz*, both sides do not have a net liquid surplus. So,

$$V = 0, \frac{\partial U}{\partial y} = 0, \frac{\partial V}{\partial y} = 0 \tag{6.49}$$

On the other boundaries of the weld pool, $U = 0$, $V = 0$, and $W = 0$.

The terms F_x, F_y, and F_z in momentum conservation equations (6.21) to (6.23) are body forces, which include electromagnetic forces and buoyancies caused by temperature variation[133]:

$$F_x = -\frac{\mu_m I^2}{4\pi^2\sigma_j^2 r}\exp\left(-\frac{r^2}{2\sigma_j^2}\right)\left[1-\exp\left(-\frac{r^2}{2\sigma_j^2}\right)\right]\left(1-\frac{z}{H}\right)^2 \frac{x}{r} \tag{6.50}$$

$$F_y = -\frac{\mu_m I^2}{4\pi^2\sigma_j^2 r}\exp\left(-\frac{r^2}{2\sigma_j^2}\right)\left[1-\exp\left(-\frac{r^2}{2\sigma_j^2}\right)\right]\left(1-\frac{z}{H}\right)^2 \frac{y}{r} \tag{6.51}$$

$$F_z = \frac{\mu_m I^2}{4\pi^2 H r^2}\left[1-\exp\left(-\frac{r^2}{2\sigma_j^2}\right)\right]\left(1-\frac{z}{H}\right)-\rho g\beta\Delta T \tag{6.52}$$

where H is the thickness of the workpiece, β is the thermal expansion coefficient, and $r = \sqrt{(x-v_0 t)^2 + y^2}$.

At the arc centerline as shown in Figure 6.4, since $x = v_0 t$ and $y = 0$, so $r = 0$. But Eqs. (6.50)–(6.52) are valid only if $r \neq 0$. Thus, take the limit to these equations at the point $(v_0 t, 0)$. The body force terms at $r = 0$ can be expressed as

$$F_x = -\frac{\mu_m I^2}{4\pi^2\sigma_j^2}\left(1-\frac{z}{H}\right)^2 \tag{6.53}$$

$$F_y = -\frac{\mu_m I^2}{4\pi^2\sigma_j^2}\left(1-\frac{z}{H}\right)^2 \tag{6.54}$$

$$F_z = \frac{\mu_m I^2}{4\pi^2 H}(1-\frac{z}{H})-\rho g\beta\Delta T \tag{6.55}$$

6.2.4.3 Initial Conditions

The initial conditions are as follows:

$$T = T_0, \quad U = V = W = 0, \quad t = 0 \tag{6.56}$$

6.3 Methods of Numerical Simulation

6.3.1 The Algorithm

This section discusses the algorithm used for numerical simulation. In this chapter, the finite difference method is utilized to analyze the 3-D transient problem. The calculation of the fluid flow field is the key to solving the governing equations. Here, a staggered grid system is used to achieve discretization of the governing equations. The temperature and pressure locate on the center of the main control volume, while the velocity components U, V, and W locate at the surfaces of the main control volume. So, there is a half grid step between the main control volume and the control volume of fluid velocity in the x-, y-, and z-directions, respectively.

In 1972, Patankar and Spalding proposed the SIMPLE (semi-implicit method for pressure-linked equations) algorithm.[134] Improved SIMPLER[135] and SIMPLEC[136] algorithms were developed in the 1980s. On the basis of the SIMPLE algorithm, Raithby et al.[137] provided the SIMPLEX algorithm, and Sheng[138] proposed the SIMPLET algorithms. All these algorithms are involved in improvement of some steps or some questions in the SIMPLE framework.

The convergence property of the SIMPLEC algorithm is much better than that of SIMPLE, even better than that of SIMPLER in some cases.[139] So, the SIMPLEC algorithm is used here. The basic steps are as follows:

1. Assume a velocity field, which is written as U^0, V^0, and W^0 in the x-, y-, and z-directions, respectively. Determine the coefficients and constant terms of the discretized momentum equations.

2. Assume a pressure field P^*.

3. Solve the momentum equations to produce estimated velocity fields U^*, V^*, and W^*.

4. Solve the pressure correction equation and get p'.

5. Update the velocity fields by substituting p' into the velocity correction equations.

6. Repeat these procedures using the updated velocity and pressure fields as the initial conditions of the next iteration step until a convergent solution has been obtained.

6.3.2 Establishment of Discretized Equations

The governing equations and their boundary conditions given in Section 6.2 have to be discretized through suitable difference schemes, and then they could be solved by the SIMPLEC algorithm. Generally, there are two considerations when difference schemes are selected. One is to ensure the physical meanings of the numerical solutions; another is to guarantee the accuracy of solutions under the condition of meeting the calculation economy requirement.

6.3.2.1 Grid System and Time Step

The first step of numerical simulation of flow fluid and heat transfer is to generate the grid system, which includes dividing the computing domain into control volumes or cells and discretizing the variables on the nodes. Actually, the final precision of calculated results and the efficiency of the calculating procedure are mainly determined by the divided grid system and the employed algorithm. The realization of a successful and effective numerical simulation just depends on a good match between grid generation and algorithm selection.

Figure 6.4 shows a schematic of the grid system in the body-fitted coordinate system. Because the workpieces used in TIG welding are usually thin, a uniform grid spacing scheme is adopted in the z-direction, and a nonuniform grid spacing scheme is used in the x- and y-directions. Finer and even grid spacing is used near the heat source, while coarser and gradually expanding grid spacing is used for the region far away from the heat source. In this way, the computation load will be decreased without decreasing computation precision. In fact, the coordinate transformation is only required in the z-direction, so that the finest grid spacing in the x-, y-, and z-directions is the same in the nondeformed and uniform grid region. In the deformed and uniform grid zone, the finest grid spacing in the x-, y-directions is unchanged, but this spacing in the z-direction varies with changing pool surface deformation. The spacing expanding parameters for nonuniform grids in both x- and y-directions are 1.2. The internal node method is employed to deal with the boundary conditions more easily, which means that the node lies in the center of the control volume and the grid lines form the interfaces of the control volume.

The selection of the time step is mostly limited to the stability of the difference scheme and the iterative method. Although an implicit difference scheme is applied to solve the algebraic equations (discretized equations), the alternation direction iterative (ADI) method is used to speed up the iteration convergence. During the calculation, the discretized equations of all nodes are first solved in the x-direction, then in the y-direction, and finally in the z-direction. But, it cannot ensure the implicit characteristic of the difference scheme for the previous two adjacent iterations. Therefore, the time step must satisfy the following rule[137]:

$$\frac{\lambda}{\rho C_p} \delta t \left(\frac{1}{\delta x^2} + \frac{1}{\delta y^2} + \frac{1}{\delta z^2} \right) \leq 1.5 \tag{6.57}$$

where δt is the time step; δx, δy, and δz are the grid spacing in the x-, y-, and z-directions, respectively. In this section, the time step is 0.001 s.

6.3.2.2 Discretization of Governing Equations

To eliminate the generation of an unreasonable pressure field, a 3-D staggered grid system is used to establish the discretization equations. The interfaces of control volumes consist of grid lines, and the node lies in the center of the control volume. As shown in Figure 6.6, the main node is denoted P, and its neighboring east, west, north, south, top, and bottom nodes are denoted E, W, N, S, B, and T, respectively. The corresponding interfaces are defined as e, w, n, s, b, and t. The temperature and pressure are arranged at the center of the control volume, while the velocity components are arranged on the interfaces. The positive directions of the coordinate axis are defined as the positive directions of velocity and pressure. The distance of two neighboring nodes or volume spacing along the x-, y-, and z-directions are defined as δx, δy, and δz^*, respectively. The items with 0 in the upper right corner represent the values of the last time step, and δt is denoted as the time step.

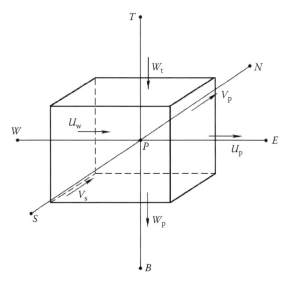

FIGURE 6.6
Schematic of 3-D staggered grid and main control volume.

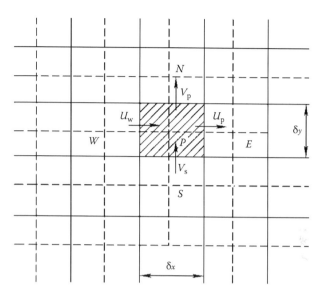

FIGURE 6.7
The control volume of T (*xoy* plane).

The 3-D energy conservation equation, Eq. (6.20), is discretized in the control volume as shown in Figure 6.7. After derivation, the discretization form in the uniform grid system is as follows:

$$f_P T_P = f_E T_E + f_W T_W + f_N T_N + f_S T_S + f_B T_B + f_T T_T + f_P^0 T_P^0 + f_1 \qquad (6.58)$$

where

$$f_P = f_E + f_W + f_N + f_S + f_B + f_T + f_P^0$$

$$f_E = D_e A(|P_{\Delta e}|) + \max(-F_e, 0)$$

$$f_W = D_w A(|P_{\Delta w}|) + \max(F_w, 0)$$

$$f_N = D_n A(|P_{\Delta n}|) + \max(-F_n, 0)$$

$$f_S = D_s A(|P_{\Delta s}|) + \max(F_s, 0)$$

$$f_B = D_b A(|P_{\Delta b}|) + \max(-F_b, 0)$$

$$f_T = D_t A(|P_{\Delta t}|) + \max(F_t, 0)$$

$$f_P^0 = \rho C_p \delta x \delta y \delta z^* / \delta t$$

$$f_1 = \lambda C_t \delta x \delta y \delta z^*$$

where the convection intensity F, the diffusion conductivity D, and the grid Peclet number P_Δ are written as

$$F_e = \rho C_p U_p \delta y \delta z^*, \ D_e = \lambda \delta y \delta z^* / \delta x, \ P_{\Delta e} = F_e / D_e$$

$$F_w = \rho C_p U_w \delta y \delta z^*, \ D_w = \lambda \delta y \delta z^* / \delta x, \ P_{\Delta w} = F_w / D_w$$

$$F_n = \rho C_p V_p \delta x \delta z^*, \ D_n = \lambda \delta x \delta z^* / \delta y, \ P_{\Delta n} = F_n / D_n$$

$$F_s = \rho C_p V_s \delta x \delta z^*, \ D_s = \lambda \delta x \delta z^* / \delta y, \ P_{\Delta s} = F_s / D_s$$

$$F_b = \rho C_p W_{1t} \delta x \delta y, \ D_b = \lambda S \delta x \delta y / \delta z^*, \ P_{\Delta b} = F_b / D_b$$

$$F_t = \rho C_p W_{1p} \delta x \delta y, \ D_t = \lambda S \delta x \delta y / \delta z^*, \ P_{\Delta t} = F_t / D_t$$

Here, the convection intensity F is the flow rate passing through interfaces, and the diffusion conductivity D is the reciprocal of diffusion resistance on interfaces. The grid Peclet number P_Δ is the ratio of convection intensity F to diffusion conductivity D. The term $A(|P_\Delta|)$ is defined as follows:

$$A(|P_\Delta|) = \max[0, (1.0 - 0.1|P_\Delta|)^5]$$

Figure 6.8 is the schematic of the control volume for variable U. After derivation, the 3-D momentum conservation equation (6.21) in the x-direction has a discretization form:

$$a_p U_p = a_e U_e + a_w U_w + a_n U_n + a_s U_s + a_b U_b + a_t U_t + a_p^0 U_p^0 + Add \qquad (6.59)$$

where

$$a_p = a_e + a_w + a_n + a_s + a_b + a_t + a_p^0$$

$$a_e = D_e A(|P_{\Delta e}|) + \max(-F_e, 0)$$

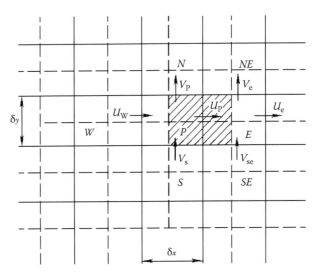

FIGURE 6.8
The control volume of U (*xoy* plane).

$$a_w = D_p A(|P_{\Delta p}|) + \max(F_p, 0)$$

$$a_n = D_{ne} A(|P_{\Delta ne}|) + \max(-F_{ne}, 0)$$

$$a_s = D_{se} A(|P_{\Delta se}|) + \max(F_{se}, 0)$$

$$a_b = D_{be} A(|P_{\Delta be}|) + \max(-F_{be}, 0)$$

$$a_t = D_{te} A(|P_{\Delta te}|) + \max(F_{te}, 0)$$

$$a_p^0 = \rho \delta x \delta y \delta z^* / \delta t$$

$$Add = (F_x + C_u)\delta x \delta y \delta z^* + (P_P - P_E)\delta x \delta z^* - z_x^* \delta x \delta y (P_B + P_{BE} - P_T - P_{TE}) / 4$$

where the convection intensity F, the diffusion conductivity D, and the grid Peclet number P_Δ are written as

$$F_e = \frac{1}{2}\rho(U_p + U_e)\delta y \delta z^*, \ D_e = \mu \delta y \delta z^* / \delta x, \ P_{\Delta e} = F_e / D_e$$

$$F_w = \frac{1}{2}\rho(U_w + U_p)\delta y \delta z^*, \ D_w = \mu \delta y \delta z^* / \delta x, \ P_{\Delta w} = F_w / D_w$$

$$F_{ne} = \frac{1}{2}\rho(V_p + V_e)\delta x \delta z^*, \ D_{ne} = \mu \delta x \delta z^* / \delta y, \ P_{\Delta ne} = F_{ne} / D_{ne}$$

$$F_{se} = \frac{1}{2}\rho(V_s + V_{se})\delta x \delta z^*, \ D_{se} = \mu \delta x \delta z^* / \delta y, \ P_{\Delta se} = F_{se} / D_{se}$$

$$F_{be} = \frac{1}{2}\rho(W_{1p} + W_{1e})\delta x \delta y, \ D_{be} = \mu S \delta x \delta y / \delta z^*, \ P_{\Delta be} = F_{be} / D_{be}$$

$$F_{te} = \frac{1}{2}\rho(W_{1t} + W_{1te})\delta x \delta y, \ D_{te} = \mu S \delta x \delta y / \delta z^*, \ P_{\Delta te} = F_{te} / D_{te}$$

After derivation, the 3-D momentum conservation equation (6.22) in the y-direction has a discretization form in the uniform grid system as shown in Figure 6.9 as follows:

$$b_p V_p = b_e V_e + b_w V_w + b_n V_n + b_s V_s + b_b V_b + b_t V_t + b_p^0 V_p^0 + Bdd \qquad (6.60)$$

where

$$b_p = b_e + b_w + b_n + b_s + b_b + b_t + b_p^0$$

$$b_e = D_{en} A(|P_{\Delta en}|) + \max(-F_{en}, 0)$$

$$b_w = D_{wn} A(|P_{\Delta wn}|) + \max(F_{wn}, 0)$$

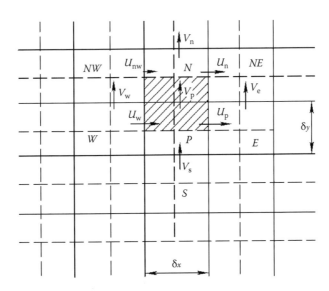

FIGURE 6.9
The control volume of V (xoy plane).

$$b_n = D_n A(|P_{\Delta n}|) + \max(-F_n, 0)$$

$$b_s = D_p A(|P_{\Delta p}|) + \max(F_p, 0)$$

$$b_b = D_{bn} A(|P_{\Delta bn}|) + \max(-F_{bn}, 0)$$

$$b_t = D_{tn} A(|P_{\Delta tn}|) + \max(F_{tn}, 0)$$

$$b_P^0 = \rho C_p \delta x \delta y \delta z^* / \delta t$$

$$Bdd = (F_y + C_v)\delta x \delta y \delta z^* + (P_P - P_N)\delta x \delta z^* - z_y^* \delta x \delta y(P_B + P_{NB} - P_{TN} - P_T)/4$$

where the convection intensity F, the diffusion conductivity D, and the grid
Peclet number P_Δ are written as

$$F_{en} = \frac{1}{2}\rho(U_p + U_n)\delta y \delta z^*, \; D_{en} = \mu \delta y \delta z^* / \delta x, \; P_{\Delta en} = F_{en} / D_{en}$$

$$F_{wn} = \frac{1}{2}\rho(U_{nw} + U_w)\delta y \delta z^*, \; D_{wn} = \mu \delta y \delta z^* / \delta x, \; P_{\Delta wn} = F_{wn} / D_{wn}$$

$$F_n = \frac{1}{2}\rho(V_p + V_n)\delta x \delta z^*, \; D_n = \mu \delta x \delta z^* / \delta y, \; P_{\Delta n} = F_n / D_n$$

$$F_p = \frac{1}{2}\rho(V_s + V_p)\delta x \delta z^*, \; D_p = \mu \delta x \delta z^* / \delta y, \; P_{\Delta p} = F_p / D_p$$

$$F_{bn} = \frac{1}{2}\rho(W_{1p} + W_{1n})\delta x \delta y, \; D_{bn} = \mu S \delta x \delta y / \delta z^*, \; P_{\Delta bn} = F_{bn} / D_{bn}$$

$$F_{tn} = \frac{1}{2}\rho(W_{1t} + W_{1tn})\delta x \delta y, \; D_{tn} = \mu S \delta x \delta y / \delta z^*, \; P_{\Delta tn} = F_{tn} / D_{tn}$$

For variable W, the 3-D momentum conservation equation (6.23) in the z-direction has a discretization form in the uniform grid system as shown in Figure 6.10 as follows:

$$c_p W_p = c_e W_e + c_w W_w + c_n W_n + c_s W_s + c_b W_b + c_t W_t + c_p^0 W_p^0 + Cdd \tag{6.61}$$

where

$$c_p = c_e + c_w + c_n + c_s + c_b + c_t + c_p^0$$

$$c_e = D_{eb} A(|P_{\Delta eb}|) + \max(-F_{eb}, 0)$$

$$c_w = D_{wb} A(|P_{\Delta wb}|) + \max(F_{wb}, 0)$$

$$c_n = D_{nb} A(|P_{\Delta nb}|) + \max(-F_{nb}, 0)$$

$$c_s = D_{bs} A(|P_{\Delta bs}|) + \max(F_{bs}, 0)$$

$$c_b = D_b A(|P_{\Delta b}|) + \max(-F_b, 0)$$

$$c_t = D_p A(|P_{\Delta p}|) + \max(F_p, 0)$$

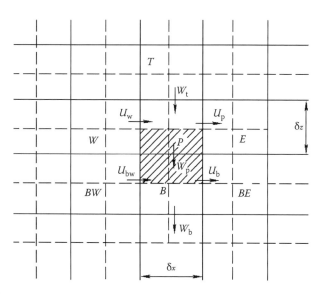

FIGURE 6.10
The control volume of W (*xoz* plane).

$$c_P^0 = \rho \delta x \delta y \delta z^* / \delta t$$

$$Cdd = (F_z + C_w)\delta x \delta y \delta z^* + (P_P - P_B)\delta x \delta y\, z_z^*$$

where the convection intensity F, the diffusion conductivity D, and the grid Peclet number P_Δ are written as

$$F_{eb} = \frac{1}{2}\rho(U_p + U_b)\delta y \delta z^*,\ D_{eb} = \mu\delta y\delta z^* / \delta x,\ P_{\Delta eb} = F_{eb} / D_{eb}$$

$$F_{wb} = \frac{1}{2}\rho(U_{bw} + U_w)\delta y \delta z^*,\ D_{wb} = \mu\delta y\delta z^* / \delta x,\ P_{\Delta wb} = F_{wb} / D_{wb}$$

$$F_{nb} = \frac{1}{2}\rho(V_p + V_b)\delta x \delta z^*,\ D_{nb} = \mu\delta x\delta z^* / \delta y,\ P_{\Delta nb} = F_{nb} / D_{nb}$$

$$F_{sb} = \frac{1}{2}\rho(V_{sb} + V_s)\delta x \delta z^*,\ D_{sb} = \mu\delta x\delta z^* / \delta y,\ P_{\Delta sb} = F_{sb} / D_{sb}$$

$$F_b = \frac{1}{2}\rho(W_{1p} + W_{1b})\delta x \delta y,\ D_b = \mu S\delta x\delta y / \delta z^*,\ P_{\Delta b} = F_b / D_b$$

$$F_p = \frac{1}{2}\rho(W_{1p} + W_{1t})\delta x \delta y,\ D_p = \mu S\delta x\delta y / \delta z^*,\ P_{\Delta p} = F_p / D_p$$

The solution of the pressure field is a key to the method of separation variables. Because the pressure field is implied in the continuity equation, when a correct pressure field is substituted into the momentum equations, the obtained velocity fields should satisfy the continuity equation. By using the backward difference scheme, the continuity equation (6.24) is discretized in the main control volume as shown in Figure 6.7:

$$(U_p - U_w)\delta y\delta z^* + (V_p - V_s)\delta x\delta z^* + (W_p - W_t)\delta x\delta y = -C_m\delta x\delta y\delta z^* \quad (6.62)$$

Meanwhile, some terms are defined based on the discretized momentum equations:

$$U_p = \left[\sum a_{np}U_{np} + a_{1np} + (P_P - P_E)\delta y\delta z^*\right] / a_{pp}$$

$$U_w = \left[\sum a_{nw}U_{nw} + a_{1nw} + (P_W - P_P)\delta y\delta z^*\right] / a_{wp}$$

$$V_p = \left[\sum b_{np}V_{np} + b_{1np} + (P_P - P_N)\delta x\delta z^*\right] / b_{pp}$$

$$V_s = \left[\sum b_{ns} V_{ns} + b_{1ns} + (P_S - P_P)\delta x \delta z^* \right] / b_{sp}$$

$$W_p = \left[\sum c_{np} W_{np} + c_{1np} + z_x^*(P_P - P_B)\delta x \delta y \right] / c_{pp}$$

$$W_t = \left[\sum c_{nt} W_{nt} + c_{1nt} + z_x^*(P_T - P_P)\delta x \delta y \right] / c_{tp}$$

The term a_{pp} is the a_p in calculation of U_p in P volume. Likewise, a_{wp} is the a_p in calculation of U_w in W volume, and so on for the other terms. By substituting these six equations mentioned into Eq. (6.62), we get the pressure equation:

$$h_P P_P = h_E P_E + h_W P_W + h_N P_N + h_S P_S + h_B P_B + h_T P_T + h_0 \tag{6.63}$$

where

$h_E = (\delta y \delta z^*)^2 / a_{pp}$, $h_W = (\delta y \delta z^*)^2 / a_{wp}$,

$h_N = (\delta x \delta z^*)^2 / b_{pp}$, $h_S = (\delta x \delta z^*)^2 / b_{sp}$,

$h_B = (z_x^* \delta x \delta y)^2 / c_{pp}$, $h_T = (z_x^* \delta x \delta y)^2 / c_{tp}$,

$h_P = h_E + h_W + h_N + h_S + h_T + h_B$,

$h_0 = (U_w^0 - U_p^0)\delta y \delta z^* + (V_s^0 - V_p^0)\delta x \delta z^* + z_x^*(W_t^0 - W_p^0)\delta x \delta y - C_m \delta x \delta y \delta z^*$.

Based on the SIMPLEC algorithm, a pressure correction equation can also be obtained:

$$h_{1P} P_{1P} = h_{1E} P_{1E} + h_{1W} P_{1W} + h_{1N} P_{1N} + h_{1S} P_{1S} + h_{1B} P_{1B} + h_{1T} P_{1T} + h_{10} \tag{6.64}$$

where

$h_{1E} = (\delta y \delta z^*)^2 / a_{pp}$, $h_{1W} = (\delta y \delta z^*)^2 / a_{wp}$,

$h_{1N} = (\delta x \delta z^*)^2 / b_{pp}$, $h_{1S} = (\delta x \delta z^*)^2 / b_{sp}$,

$h_{1B} = (z_x^* \delta x \delta y)^2 / c_{pp}$, $h_{1T} = (z_x^* \delta x \delta y)^2 / c_{tp}$,

$h_{1P} = h_{1E} + h_{1W} + h_{1N} + h_{1S} + h_{1T} + h_{1B}$,

$h_{10} = (U_W^* - U_P^*)\delta y \delta z^* + (V_S^* - V_P^*)\delta x \delta z^* + (W_T^* - W_P^*)\delta x \delta y - C_m \delta x \delta y \delta z^*$.

The velocity and pressure correction equations are as follows:

$$U_p = U_p^* + (P_{1P} - P_{1E})\delta y \delta z^* / a_{PP} \tag{6.65}$$

$$V_p = V_p^* + (P_{1P} - P_{1N})\delta x \delta z^* / b_{pp} \tag{6.66}$$

$$W_p = W_p^* + (P_{1P} - P_{1T})z_x^* \delta x \delta y / c_{pp} \tag{6.67}$$

$$P_p = P_p^* + P_{1P} \tag{6.68}$$

where U^*, V^*, W^*, and P^* are the velocity components and pressure before corrected, and P_1 is the value of pressure correction.

There is no mass exchange on the symmetrical plane, solid-liquid interface, and top and bottom surfaces of the weld pool, so that the flow velocity in the normal direction of all these surfaces is zero. Taking the control volume P on the top surface of the weld pool as an example, it is impossible to obtain the coefficient C_t corresponding to W_t, so let the relevant efficient h_t and h_{1t} in the pressure equation and pressure correction equation be zero, and the velocity does not need to be corrected.

6.3.2.3 Discretization of the Surface Deformation Equation

When the workpiece is fully penetrated, the surface deformation equations (6.11) and (6.12) on the top and bottom surfaces of the weld pool and their constraint condition (6.13) are discretized.

By using the central difference scheme in the main control volume, the first- and second-order partial differential derivatives of the top surface deformation equation can be written as follows:

$$\frac{\partial \varphi}{\partial x} = \frac{1}{2}(\varphi_E - \varphi_W) / \delta x, \quad \frac{\partial \varphi}{\partial y} = \frac{1}{2}(\varphi_N - \varphi_S) / \delta y$$

$$\frac{\partial^2 \varphi}{\partial x^2} = (\varphi_E - 2\varphi_P + \varphi_W) / (\delta x)^2, \quad \frac{\partial^2 \varphi}{\partial y^2} = (\varphi_N - 2\varphi_P + \varphi_S) / (\delta y)^2$$

$$\frac{\partial^2 \varphi}{\partial x \partial y} = \frac{1}{4}(\varphi_{NE} - \varphi_{SE} - \varphi_{NW} + \varphi_{SW}) / (\delta x \delta y)$$

Let

$$a_1 = (1 + \varphi_x^2 + \varphi_y^2)^{\frac{3}{2}} / \gamma, \, a_2 = 2\varphi_x \varphi_y \varphi_{xy}, \, a_3 = 1 + \varphi_y^2, \, a_4 = 1 + \varphi_x^2$$

Then, substituting these algebraic expressions into Eq. (6.11), we can obtain

$$e\varphi_P = e_1 \varphi_E + e_2 \varphi_W + e_3 \varphi_N + e_4 \varphi_S + e_0 \tag{6.69}$$

where
$e = e_1 + e_2 + e_3 + e_4 + \rho g a_1$
$e_1 = a_3 / (\delta x)^2$
$e_2 = a_3 / (\delta x)^2$

$e_3 = a_4/(\delta y)^2$

$e_4 = a_4/(\delta y)^2$

$e_0 = a_1(P_{arc} + C_2) - a_2$

Similarly, the discretized forms of partial differential derivatives of the bottom surface deformation equation are as follows:

$$\frac{\partial \psi}{\partial x} = \frac{1}{2}(\psi_E - \psi_W)/\delta x, \quad \frac{\partial \psi}{\partial y} = \frac{1}{2}(\psi_N - \psi_S)/\delta y$$

$$\frac{\partial^2 \psi}{\partial x^2} = (\psi_E - 2\psi_P + \psi_W)/(\delta x)^2, \quad \frac{\partial^2 \psi}{\partial y^2} = (\psi_N - 2\psi_P + \psi_S)/(\delta y)^2$$

$$\frac{\partial^2 \psi}{\partial x \partial y} = \frac{1}{4}(\psi_{NE} - \psi_{SE} - \psi_{NW} + \psi_{SW})/(\delta x \delta y)$$

Let

$$b_1 = (1 + \psi_x^2 + \psi_y^2)^{\frac{3}{2}}/\gamma, \quad b_2 = 2\psi_x \psi_y \psi_{xy}, \quad b_3 = 1 + \psi_y^2, \quad b_4 = 1 + \psi_x^2$$

Substitute these algebraic expressions into Eq. (6.12), we can get:

$$e\psi_P = e_1\psi_E + e_2\psi_W + e_3\psi_N + e_4\psi_S + e_0 \tag{6.70}$$

where

$e = e_1 + e_2 + e_3 + e_4 - \rho g b_1$

$e_1 = b_3/(\delta x)^2$

$e_2 = b_3/(\delta x)^2$

$e_3 = b_4/(\delta y)^2$

$e_4 = b_4/(\delta y)^2$

$e_0 = b_1[\rho g(H-\varphi) + C_2] - b_2$

The surplus pressure C_2 is an important parameter that is given by Eq. (6.13). Integrate Eqs. (6.11) and (6.12) in surface deformation areas Ω_1 and Ω_2, respectively, and set

$$AP = \iint_{\Omega1} P_{arc}dxdy, \quad AN = \iint_{\Omega1} dxdy,$$

$$AX = \iint_{\Omega1} -\gamma \frac{(1+\varphi_y^2)\varphi_{xx} - 2\varphi_x\varphi_y\varphi_{xy} + (1+\varphi_x^2)\varphi_{yy}}{(1+\varphi_x^2+\varphi_y^2)^{3/2}} dxdy$$

$$BP = \rho g \iint_{\Omega2} (H-\varphi)dxdy, \quad BN = \iint_{\Omega2} dxdy$$

$$BX = \iint_{\Omega 2} -\gamma \frac{(1+\psi_y^2)\psi_{xx} - 2\psi_x\psi_y\psi_{xy} + (1+\psi_x^2)\psi_{yy}}{(1+\psi_x^2+\psi_y^2)^{3/2}} dxdy$$

So, we can obtain

$$\rho g \iint_{\Omega 1} \varphi dxdy = AP + C_2 AN - AX$$

$$\rho g \iint_{\Omega 2} \psi dxdy = -BP - C_2 BN + BX$$

Substituting them into Eq. (6.11), we have

$$C_2 = (AX + BX - AP - BP)/(AN + BN) \qquad (6.71)$$

6.3.3 Discretization of Boundary Conditions

6.3.3.1 Energy Boundary Conditions

It can be seen from the previous section that the energy boundary conditions are the second or the third category boundary conditions. The additional source term method (ASTM) is an effective method for dealing with such boundary conditions in the finite difference method. The basic idea of ASTM[140,141] is to convert the interface heat flux caused by the second or the third category boundary conditions into an inner heat source in the control volume for an inner node adjacent to the boundary and at the same time to take this interface as the adiabatic boundary. In this way, the conservation characteristic of the equation is kept, and the unknown temperature at the boundary node is successfully eliminated from the governing equations of the inner node.

For a generalized source term S_{ad} in control volume P, it can be expressed as

$$S_{ad} = S_{c,ad} + S_{p,ad} T_p \qquad (6.72)$$

The specific forms of the additional source term are defined as[31]
For the second category boundary condition,

$$S_{c,ad} = q_B / \delta x, \, S_{p,ad} = 0 \qquad (6.73)$$

For the third category boundary condition,

$$S_{c,ad} = \frac{1}{\delta x} \times \frac{T_f}{1/\alpha_c + \delta x/\lambda}, \, S_{p,ad} = \frac{1}{\delta x} \times \frac{1}{1/\alpha_c + \delta x/\lambda} \qquad (6.74)$$

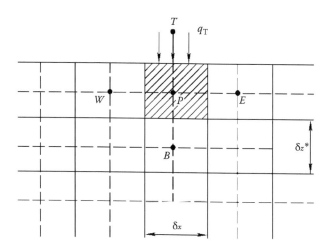

FIGURE 6.11
The ASTM on the top energy boundary condition.

Taking the top surface of the weld pool as the example, the energy boundary condition is the second category boundary condition, as shown in Figure 6.11. To exclude the unknown boundary temperature T_T from the algebraic expression, subtract $f_T T_p$ from both sides of Eq. (6.58). So, we get

$$(f_P - f_T)T_P = f_E T_E + f_W T_W + f_N T_N + f_S T_S + f_B T_B + f_T(T_T - T_P) + f_P^0 T_P^0 + f_1 \quad (6.75)$$

Because the convection flux on free surfaces is zero, from the derivation process of Eq. (6.58) we can know that

$$F_t = 0, P_{\Delta t} = 0, f_T = D_t = \lambda S \delta x \delta y / \delta z^*.$$

So

$$f_T(T_T - T_P) = \frac{\lambda S \delta x \delta y}{\delta z^*}(T_T - T_P) = S \delta x \delta y \frac{\lambda(T_T - T_P)}{\delta z^*}.$$

According to Fourier's law, on the top surface of the weld pool, there is

$$q = \lambda \frac{\partial T}{\partial z^*}\frac{\partial z^*}{\partial z} = \frac{\partial z^*}{\partial z}\frac{\lambda(T_T - T_P)}{\delta z^*} = Z_z^* \frac{\lambda(T_T - T_P)}{\delta z^*}, \quad (6.76)$$

Therefore,

$$f_T(T_T - T_P) = Sq \delta x \delta y / Z_z^*$$

Thus, Eq. (6.75) can be expressed as

$$(f_P - f_T)T_P = f_E T_E + f_W T_W + f_N T_N + f_S T_S + f_B T_B + Sq \delta x \delta y / Z_z^* + f_P^0 T_P^0 + f_1 \quad (6.77)$$

6.3.3.2 Momentum Boundary Conditions

The velocity of liquid metal in the weld pool is zero on the liquid-solid interfaces, which belongs to the first category boundary conditions. But, the momentum boundary conditions on free surfaces are very special, and it is hard to deal with them. Based on the idea used in the energy boundary condition, the ASTM can also be applied in the momentum boundary conditions.

Taking velocity U as an example, on the top surface of the weld pool its momentum boundary condition is

$$\mu \frac{\partial U}{\partial z^*} \frac{\partial z^*}{\partial z} = -\frac{\partial \gamma}{\partial T} \frac{\partial T}{\partial x}.$$

To exclude the unknown velocity U_t, subtract $a_t U_p$ from both sides of Eq. (6.59). We can obtain

$$(a_p - a_t)U_p = a_e U_e + a_w U_w + a_n U_n + a_s U_s + a_b U_b + a_t(U_t - U_p) + a_p^0 U_p^0 + Add \quad (6.78)$$

On the top surface, $W = 0$, therefore

$$F_{te} = 0, \ P_{\Delta te} = 0, \ a_t = D_{te} = \mu S \delta x \delta y / \delta z^*.$$

So,

$$a_t(U_t - U_p) = \mu S \delta x \delta y \frac{U_t - U_p}{\delta z^*} \quad (6.79)$$

And since

$$\mu \frac{\partial U}{\partial z^*} \frac{\partial z^*}{\partial z} = \mu \frac{U_t - U_p}{\delta z^*} Z_z^* = -\frac{\partial \gamma}{\partial T} \frac{\partial T}{\partial x},$$

then

$$\mu \frac{U_t - U_p}{\delta z^*} = -\frac{\partial \gamma}{\partial T} \frac{\partial T}{\partial x} \frac{1}{Z_z^*} = -\frac{1}{Z_z^*} \frac{\partial \gamma}{\partial T} \frac{T_{TE} - T_E}{\delta x},$$

Substituting it into Eq. (6.79), we obtain

$$a_t(U_t - U_p) = \frac{\partial \gamma}{\partial T} \frac{S}{Z_x^*} \delta y(T_{TE} - T_E) \quad (6.80)$$

Thus, the discretized equation of control volume P on the top surface of the weld pool can be expressed as

$$(a_p - a_t)U_p = a_eU_e + a_wU_w + a_nU_n + a_sU_s + a_bU_b$$

$$+ \frac{\partial\gamma}{\partial T}\frac{S}{Z_x^*}\delta y(T_{TE} - T_E) + a_p^0U_p^0 + Add \qquad (6.81)$$

Similarly, other discretized equations on different surfaces can be derived.

6.4 Calculation Procedure and Programming

6.4.1 The Main Program of the 3-D Transient Numerical Analysis

The surface deformation of the weld pool makes the calculations of fluid flow and heat transfer in the transient state more complex than those in steady or quasi-steady cases. First, the weld pool shape and size must be obtained from the calculations of the temperature field. Then, the surface deformation can be determined within the weld pool. The temperature and fluid flow fields are coupled with the surface deformation through the body-fitted coordinates and the thermophysical properties of the workpiece material. Figure 6.12 demonstrates the coupling relationship among the surface deformation, fluid flow field, and temperature profiles. Thus, it is a nonlinear and strong-coupling problem of numerical solution.

Based on the SIMPLE algorithm, the surface deformation, fluid flow field, and temperature field are solved in turn using the method of variable separation. At each time step, the value of every variable is improved once until

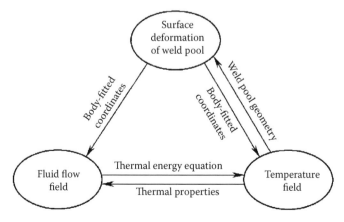

FIGURE 6.12
Coupling relationship among surface deformation, heat, and fluid flow fields.

all satisfy the convergence criteria. Thus, for every time step, the calculation process can be divided into the following steps:

1. Calculate the temperature profiles on the workpiece.
2. Determine the 3-D shape and size of the weld pool based on the results calculations of the temperature field and then calculate the surface deformation of the weld pool.
3. Conduct the coordinate system transformation. Calculate the fluid flow field in the body-fitted coordinate system and obtain convergent results.
4. Calculate the temperature field in the body-fitted coordinate system and obtain convergent results.
5. Repeat steps 2 to 4, improve the results of surface deformation, fluid flow field, and temperature field calculation continuously until they satisfy their convergent criteria.

Figure 6.13 shows the flowchart of the main program. In the calculating process, the grid system is first generated. Although the grid system is fixed in the workpiece, the grid system in every time step is quite different because the surface deformation of the weld pool changes with the moving arc. At each time step, calculation procedures 1 to 5 above are carried out, and the convergent criteria are as follows:

$$|T_{N+1} - T_N| / T_N \leq 0.005 \qquad (6.82)$$

$$|U_{N+1} - U_N| / U_N \leq 0.005 \qquad (6.83)$$

$$|V_{N+1} - V_N| / V_N \leq 0.005 \qquad (6.84)$$

$$|W_{N+1} - W_N| / W_N \leq 0.005 \qquad (6.85)$$

The contradiction between calculating speed and calculating accuracy is taken into account for selecting convergent conditions. On the premise of meeting the requirement of analysis accuracy, the cost of calculation is fully considered.

Some thermophysical properties, such as the specific heat, thermal conductivity, and dynamic viscosity, are temperature dependent. Their relations with temperature are expressed in subroutines, which are called by the main program.

6.4.2 Subroutine of Pool Surface Deformation

As shown in Figure 6.14, the deformation of the top and bottom surfaces of the weld pool are calculated by the Gauss-Seidel iteration method. First, the

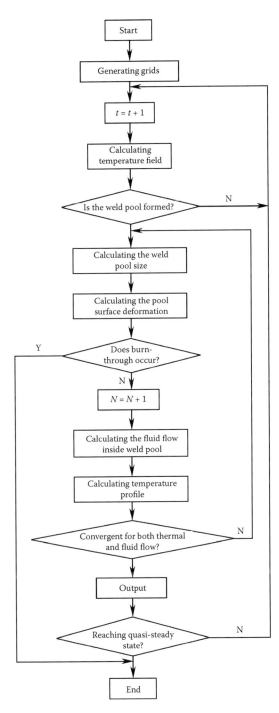

FIGURE 6.13
Flowchart of the main program.

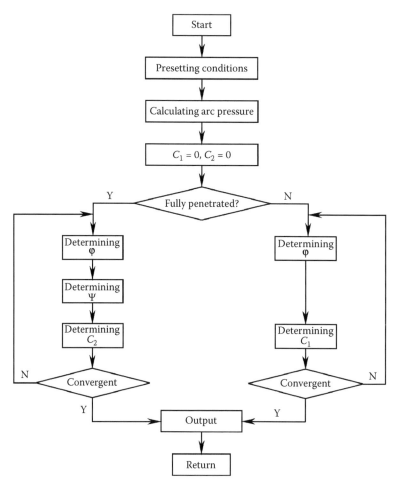

FIGURE 6.14
Flowchart of the weld pool surface deformation subroutine.

boundaries of the weld pool are determined on the basis of the calculated temperature field. Then, the arc pressure is calculated, and whether the workpiece is penetrated is judged. According to the penetrating condition, appropriate surface deformation equations are employed. When the pool surface deformation is obtained, convergence is judged. The calculation is repeated until it satisfies the convergent condition.

6.4.3 Calculating Heat and Fluid Flow in the Weld Pool

As mentioned, the SIMPLEC algorithm is used to calculate the fluid flow field. The calculation of the temperature field is only involved in solving the energy conservation equation and its boundary condition. At each time step, a stricter convergent criterion is used in the heat and fluid flow subroutines:

$$|T_{N+1} - T_N| \leq 1\,(\text{K}) \tag{6.86}$$

$$|U_{N+1} - U_N| / U_N \leq 0.001 \tag{6.87}$$

$$|V_{N+1} - V_N| / V_N \leq 0.001 \tag{6.88}$$

$$|W_{N+1} - W_N| / W_N \leq 0.001 \tag{6.89}$$

6.5 Material Properties and Workpiece Dimension

Numerical simulations are performed for bead-on-plate welding. The workpiece materials are SS304 stainless steel plate 3-mm thick and low-carbon steel Q235 plate 2-mm thick. A half-workpiece with a domain 200 × 50 × 3 mm for SS304 or 200 × 50 × 2 mm for Q235 is divided into a mesh of 352 × 60 × 10 or 384 × 64 × 10 grid nodes, respectively. Finer grid spacing is utilized in the molten region. Some physical parameters are given in Table 6.1. The

TABLE 6.1

Other Thermophysical Properties and Parameters Used in the Calculation

Symbol	Property or Parameter	Units	Value for SS304	Value for Q235
a_s	Activity of sulfur	wt%	0.22	0.22
A_v	Evaporation constant	—	2.52	2.52
β	Thermal expansion coefficient	K^{-1}	10^{-4}	10^{-4}
γ	Surface tension	N m^{-1}	1.0	1.0
σ_s	Stefan-Boltzmann constant	W m^{-2} k^{-4}	5.67×10^{-8}	5.67×10^{-8}
ε	Surface radiation emissivity	—	0.4	0.4
g	Gravitational acceleration	m s^{-2}	9.8	9.8
α_c	Convective heat transfer coefficient	W m^{-2}K^{-1}	80	80
ΔH^0	Standard absorption heat	J g^{-1}mol^{-1}	1.463×10^3	1.463×10^3
k_1	Segregation enthalpy		-1.38×10^{-2}	-1.38×10^{-2}
L_q	Latent heat of vaporization	J kg^{-1}	73.43×10^5	73.43×10^5
ρ	Density	Kg m^{-3}	7,200	6,900
T_m	Melting point	K	1,723	1,789
$T\infty$	Ambient temperature	K	293	293
η	Arc power efficiency	—	0.65	0.65

specific heat C_p, dynamic viscosity μ, and thermal conductivity λ are temperature dependent and can be expressed as follows:

For SS 304 stainless steel plates,

$$\lambda = \begin{cases} 10.717 + 0.014955T & T \leq 780K \\ 12.076 + 0.013213T & 780K \leq T \leq 1672K \\ 217.12 - 0.1094T & 1672K \leq T \leq 1727K \\ 8.278 + 0.0115T & 1727K \leq T \end{cases} \quad (\text{W m}^{-1} \text{ K}^{-1}) \quad (6.90)$$

$$\mu = \begin{cases} 37.203 - 0.0176T & 1713K \leq T \leq 1743K \\ 20.354 - 0.008T & 1743K \leq T \leq 1763K \\ 34.849 - 0.0162T & 1763K \leq T \leq 1853K \\ 13.129 - 0.0045T & 1853K \leq T \leq 1873K \end{cases} \quad (10^{-3} \text{ kg m}^{-1} \text{ s}^{-1}) \quad (6.91)$$

$$C_p = \begin{cases} 438.95 + 0.198T & T \leq 773K \\ 137.93 + 0.59T & 773K \leq T \leq 873K \\ 871.25 - 0.25T & 873K \leq T \leq 973K \\ 555.2 + 0.0775T & 973K \leq T \end{cases} \quad (\text{J kg}^{-1}) \quad (6.92)$$

For Q235 steel plates,

$$\lambda = \begin{cases} 60.719 - 0.027857T & T \leq 851K \\ 78.542 - 0.0488T & 851K \leq T \leq 1082K \\ 15.192 + 0.0097T & 1082K \leq T \leq 1768K \\ 349.99 - 0.1797T & 1768K \leq T \leq 1798K \end{cases} \quad (\text{W m}^{-1} \text{ K}^{-1}) \quad (6.93)$$

$$\mu = \begin{cases} 119.00 - 0.061T & 1823K \leq T \leq 1853K \\ 10.603 - 0.025T & 1853K \leq T \leq 1873K \\ 36.263 - 0.0162T & 1873K \leq T \leq 1973K \end{cases} \quad (10^{-3} \text{ kg m}^{-1} \text{ s}^{-1}) \quad (6.94)$$

$$C_p = \begin{cases} 513.76 - 0.335T + 6.89 \times 10^{-4}T^2 & T \le 973K \\ -10539 + 11.7T & 973K \le T \le 1023K \\ 11873 - 10.2T & (J \, kg^{-1}) \quad 1023K \le T \le 1100K \quad (6.95) \\ 644 & 1100K \le T \le 1379K \\ 354.34 + 0.21T & 1379K \le T \end{cases}$$

6.6 The Transient Development of the Weld Pool Shape and Fluid Flow Field

A complete TIG weld process consists of a few stages: arc ignition, weld pool formation and expansion, quasi-steady state, and contraction and diminution of the weld pool after arc extinguishment. By employing the developed model and algorithm for TIG welding, the dynamic behaviors of the weld pool are analyzed numerically from the partially penetrated to the fully penetrated conditions. The dynamic behaviors of the weld pool include the variations of the pool shape and size and pool surface deformation with time, as well as the evolution of the fluid dynamics inside the weld pool. The development of the weld pool is divided into several stages, such as formation of the weld pool, partial penetration, full penetration, and the quasi-steady state. This section focuses on the study of the transient development of the weld pool shape and size, pool surface deformation, and fluid flow field from partial penetration to full penetration.

For the case of process condition A (TIG welding, Q235 steel plate of 2-mm thick, 16-V arc voltage, 110-A welding current, 160-mm/min welding speed), the calculated results show that the weld pool appears at 1.62 s after arc ignition. Then, the size of the weld pool expands gradually. The workpiece gets penetrated at 2.88 s, and the quasi-steady state of the weld pool and the temperature profile are achieved at 4.20 s. Although a fixed coordinate is used in numerical simulation, a moving coordinate is employed to compare the 3-D shape of the weld pool at different times. Both Figure 6.15 and Figure 6.16 demonstrate the calculated results under process condition A.

Figure 6.15 shows the transient development of the weld pool geometry and fluid flow field inside the pool. Figures 6.15(a), 6.15(b), and 6.15(c) are the top surface, longitudinal cross section, and transverse cross section, respectively of the weld pool. The width and length of the weld pool at the top surface increase quickly after arc ignition, as shown in Figure 6.15(a). When the weld pool reaches the quasi-steady state, the pool geometry is kept

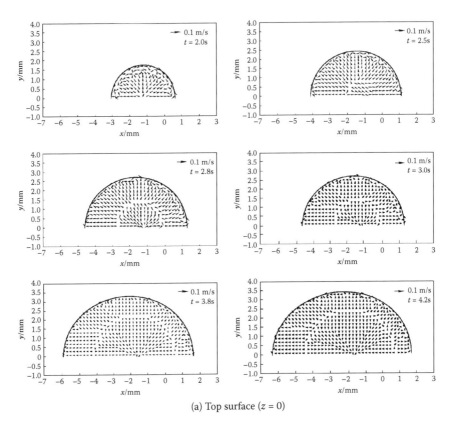

(a) Top surface ($z = 0$)

FIGURE 6.15
Transient variation of weld pool shape and fluid flow field (TIG welding with condition A) (from Zhao P C, Wu C S, and Zhang Y M, Proc. Instn. Mech. Eng., Part B: *J. Eng. Manuf.*, 219: 99–110, 2005): (a) top surface ($z = 0$).

nearly constant. When the workpiece is partially penetrated, the penetration depth increases with time. After the workpiece is fully penetrated, the lower part of the weld pool expands quickly, and the expanding rate of both pool width and pool length at the top surface slows, as shown in Figure 6.15(b). Figure 6.15(c) is the transverse cross section of the weld pool behind the arc centerline ($x = -12$ mm).

The fluid flow state does not change much before and after the workpiece is fully penetrated. The maximum fluid velocity appears at the region near the electrode centerline ($x = 0$), and the value is 0.007 m/s at $t = 2.0$ s and 0.04 m/s at $t = 4.12$ s.

Figure 6.16 shows the variations of the pool width and length at both top and bottom surfaces as well as the weld penetration versus time. For the welding conditions used (process condition A), the weld pool emerges at $t = 1.62$ s, the workpiece is penetrated at $t = 2.88$ s, and the quasi-steady state is reached at $t = 4.20$ s.

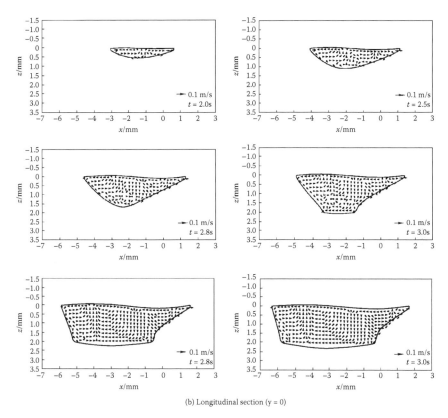

(b) Longitudinal section (y = 0)

FIGURE 6.15 (Continued)
Transient variation of weld pool shape and fluid flow field (TIG welding with condition A) (From Zhao P C, Wu C S, and Zhang Y M, Proc. Instn. Mech. Eng., Part B: *J. Eng. Manuf.*, 219: 99–110, 2005): (b) longitudinal section ($y = 0$).

6.7 Transient Variation of Weld Pool Surface Deformation

For another case of process condition B (TIG welding, SS304 steel plate 3-mm thick, 14-V arc voltage, 100-A welding current, 125-mm/min welding speed), the computed results show that the weld pool emerges at $t = 0.82$ s, then expands continuously, is fully penetrated at $t = 3.54$ s, and reaches the quasi-steady state at $t = 4.24$ s.

Figure 6.17 shows the transient development of the pool surface deformation, that is, the maximum values of the depression at both sides and the hump at the top side versus time. As shown in Figure 6.17(a), after the weld pool is formed at $t = 0.82$ s, the pool surface deformation is produced. As the pool volume expands with increasing time, the extent of the pool surface deformation gets bigger, and both maximum depression and hump at the top side increase with time. The test plate is completely penetrated

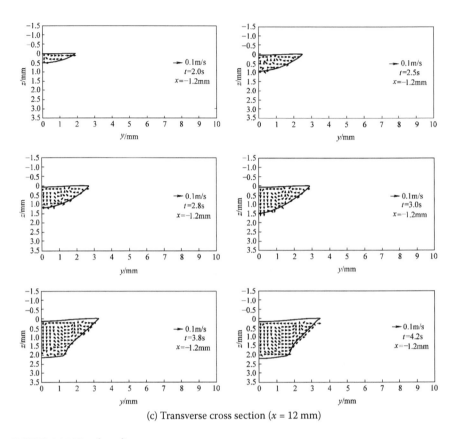

(c) Transverse cross section (x = 12 mm)

FIGURE 6.15 (Continued)
Transient variation of weld pool shape and fluid flow field (TIG welding with condition A) (From Zhao P C, Wu C S, and Zhang Y M, Proc. Instn. Mech. Eng., Part B: *J. Eng. Manuf.*, 219: 99–110, 2005): (c) transverse cross section (x = –12 mm).

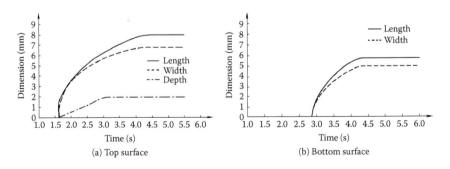

FIGURE 6.16
The weld pool dimensions versus time (TIG welding with condition A) (From Zhao P C, Wu C S, and Zhang Y M, Proc. Instn. Mech. Eng., Part B: *J. Eng. Manuf.*, 219: 99–110, 2005): (a) top surface; (b) bottom surface.

at t = 3.54 s. In the meantime, the bottom surface of the weld pool starts to deform, so the whole weld pool is depressed. Then, the hump at the top side decreases, while the depressions at both sides rise with a higher rate. When the thermal process reaches the quasi-steady state at t = 4.24 s, the weld pool geometry keeps constant, the hump at the top side becomes zero, and the depressions of the weld pool at both sides attain their maximum and do not vary anymore with time. It can be seen that the increasing rate of the pool surface depressions is quite different before and after the pool is completely penetrated.

Figures 6.17(b), 6.17(c), 6.17(d), and 6.17(e) illustrate the transient development of weld pool surface deformation at both the top and the bottom surfaces of the weld pool. In this figure, (b) and (c) are the longitudinal cross sections (side view), while (d) and (e) are the transverse cross sections (front view). Compared to the top surface of the weld pool, the bottom surface gets depressed more seriously and quickly. The maximum depression at the bottom surface increases from 0 mm at t = 3.54 s (the moment when the pool is just completely penetrated) to 0.26 mm at t = 4.24 s (the instant when the quasi-steady state is reached). The increasing rate is 0.371 mm/s. As shown in Figure 6.17(e), there is a minor oscillation of the pool surface deformation at the bottom side after the weld pool geometry reaches quasi-steady state. But, the amplitude of such oscillation is so low that the bottom surface contours at t = 4.2 s and t = 4.4 s are nearly identical to each other. For the top surface depression, the increasing rates of maximum depression are 0.031 mm/s before complete penetration (from 0 mm at t = 0.82 s to 0.098 mm at t = 4.0 s) and 0.117 mm/s after complete penetration (from 0.098 mm at t = 4.0 s to 0.126 mm at t = 4.24 s), respectively.

Since the variation rate of the top surface depression of the weld pool has a marked increase after the pool is completely penetrated, it can be taken as an indicator to judge whether the plate is penetrated. On the other hand, the pool length and width at the top side are also changed after complete penetration is achieved. To quantitatively describe the correlation of the top-side surface depression with the extent of penetration, two characteristic variables are used to reflect the variation of the whole weld pool geometry, that is, the ratio of the maximum depression Dd_{max} to the pool width W (Dd_{max}/W), and the ratio of Dd_{max} to the pool length L (Dd_{max}/L).

Figure 6.18 shows the ratios of Dd_{max}/W and Dd_{max}/L versus time. The three-segment curves of such ratios reflect information on the penetration. During expanding of the nonpenetrated weld pool, the values of Dd_{max}/W and Dd_{max}/L rise slowly with time. At the moment the weld pool is fully penetrated (t = 3.54 s), the rising rates of Dd_{max}/W and Dd_{max}/L are suddenly increased, that is, the slopes of two curves increase in a marked way. The first kink point on the curves corresponds to the moment when the weld pool is fully penetrated. When the quasi-steady state is obtained at t = 4.24 s, the weld pool geometry is in a relatively stable condition, and Dd_{max}/W and Dd_{max}/L are nearly constant, so the curves are just straight lines after 4.24 s. The second kink point on

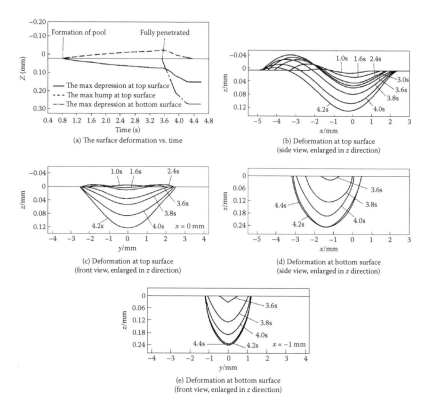

FIGURE 6.17

The transient development of the weld pool surface deformation (From Wu C S, Zhao P C, and Zhang Y M, *Weld. J.*, 83(12): 330s–335s, 2004) (TIG welding with condition B): (a) surface deformation versus time; (b) deformation at the top surface (side view, enlarged in the z-direction); (c) deformation at the top surface (front view, enlarged in the z-direction); (d) deformation at the bottom surface (side view, enlarged in the z-direction); (e) deformation at the bottom surface (front view, enlarged in the z-direction).

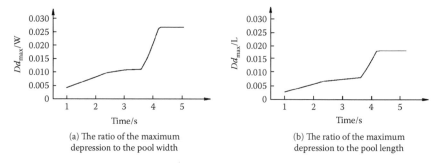

FIGURE 6.18

The ratios of Dd_{max}/W and Dd_{max}/L versus time (From Wu C S, Zhao P C, and Zhang Y M, *Weld. J.*, 83(12): 330s–335s, 2004) (TIG welding with condition B): (a) ratio of the maximum depression to the pool width; (b) ratio of the maximum depression to the pool length.

the curves corresponds to the moment when the weld pool reaches the quasi-steady state. Because the depression of the weld pool surface at the top side has the characteristics mentioned, it can be employed as an indicator of the extent of weld penetration. In practice, the top-side sensor can be developed to measure the weld pool surface depression for weld penetration control.

6.8 Transient Variation of Weld Pool Shape and Size after Arc Extinguishment

The transient behaviors of the weld pool when the arc is extinguished are also calculated. This process is actually the cooling and solidifying of the weld pool. For the process condition C (workpiece 3-mm thick SS304 stainless steel plate, 14-V arc voltage; 100-A welding current, 120-mm/s welding speed) used, the quasi-steady state is achieved at $t = 4.0$ s; then, the arc is intentionally extinguished, and both the welding current and the speed become zero. Because there is no heat input, the weld pool decreases quickly.

The calculation results show that the weld pool disappears totally at $t = 4.6$ s. Figure 6.19 illustrates the transient contraction behavior of the weld pool geometry at the top surface, bottom surface, longitudinal cross section, and transverse cross section. During the cooling process, the weld pool moves backward relative to the electrode centerline ($x = 0$). The penetration depth changes slowly at first, and then it varies more rapidly than the pool width and length.

Figure 6.20 illustrates the transient variation of pool depth as well as pool length and width at both top and bottom surfaces of the plate with time, which is a detailed expression of the variation of pool length, width, and depth with time in Figure 6.19.

Figure 6.21 shows the transient variation of the fluid flow field after the arc has been extinguished. Because of the disappearance of the arc pressure and electromagnetic forces, the fluid flow is driven only by the surface tension gradient and buoyancy; therefore, it lasts a short time, and the amplitude of flow velocity is much lower than that in the quasi-steady state.

6.9 Experimental Verifications

Generally, experimental measurements must be performed to validate the numerical model. A transient TIG welding process is analyzed in this chapter. The 3-D shape and size, pool surface deformation, temperature profile, and fluid flow field are all time dependent. If the transient information on weld

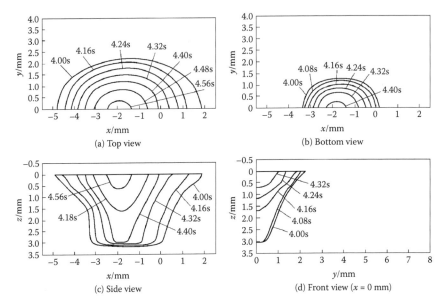

FIGURE 6.19
Transient variation in the three-dimensional shape of a weld pool after the arc has been extinguished (From Zhao P C, Wu C S, and Zhang Y M, Proc. Instn. Mech. Eng., Part B: *J. Eng. Manuf.*, 219: 99–110, 2005) (TIG welding with condition C): (a) top view; (b) bottom view; (c) side view; (d) front view ($x = 0$ mm).

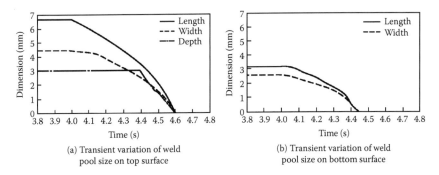

FIGURE 6.20
Transient variation of weld pool size after the arc has been extinguished at $t = 4.0$: (a) transient variation of weld pool size on the top surface; (b) transient variation of weld pool size on the bottom surface.

pool behaviors can be obtained in real time during the TIG welding process, a complete validation of the simulation results can be offered. But, limited to the measuring and experimental conditions, only the geometry of the weld pool on the top surface can be obtained in the current experiments. Moreover, because the final weld shape and size are consequent to fluid flow and heat transfer inside the weld pool, measurement of weld dimensions at the transverse cross

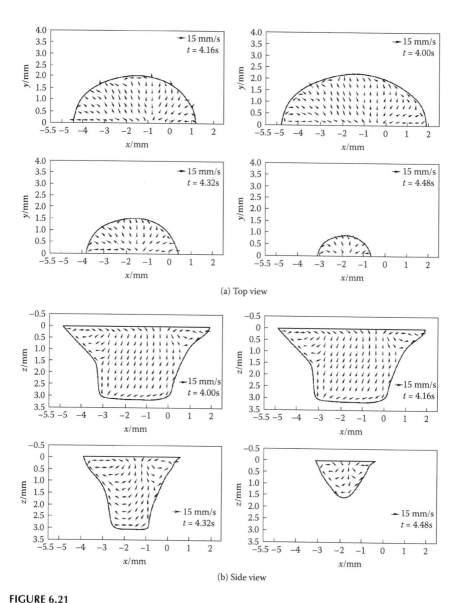

(a) Top view

(b) Side view

FIGURE 6.21

Transient variation in the fluid flow field after the arc has been extinguished (From Zhao P C, Wu C S, and Zhang Y M, Proc. Instn. Mech. Eng., Part B: *J. Eng. Manuf.*, 219: 99–110, 2005) (TIG welding with condition C): (a) top view; (b) side view.

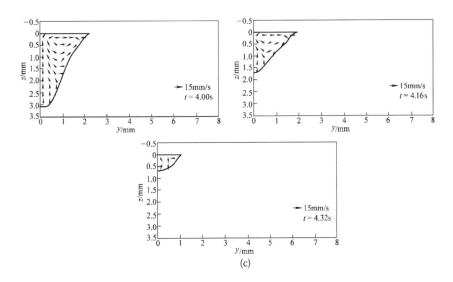

FIGURE 6.21 (Continued)
Transient variation in the fluid flow field after the arc has been extinguished (From Zhao P C, Wu C S, and Zhang Y M, Proc. Instn. Mech. Eng., Part B: *J. Eng. Manuf.*, 219: 99–110, 2005) (TIG welding with condition C): (c) front view ($x = 0$ mm).

section of the weld bead can reflect comprehensive performance of the weld pool behaviors. Therefore, the experiments are conducted to detect the shape and size of the weld pool at the top surface, the cross-sectional profile of the weld, and the pool surface deformation on both the top and the bottom sides.

The common commercial charge-coupled device (CCD) camera combined with a special narrowband filter is used to capture the images of the weld pool during the TIG welding process. A detailed description of this system can be found in Chapter 10.

During the TIG welding operation, once the image of the weld pool captured by the camera [Figure 6.22(a)] is digitized through a frame grabber, it is stored in a computer as a matrix in which one element represents a dot of the image. A special image-processing algorithm has been developed to extract the weld pool edges so that the weld pool geometry of the top side is determined [Figure 6.22(b)]. After welding, the weld bead is sectioned, and metallographic samples of the weld beam on the cross section are made. Then, macrophotographs of welds contain information on the weld shape and size and depression of the weld pool.

6.9.1 Comparison of Weld Pool Shape on the Top Surface

Figure 6.23 and Table 6.2 show a comparison of the predicted and measured weld pool shape and size on the top surface. It can be seen that the predicted weld pool surface geometry generally agrees with the measured geometry except for the trailing part. Because the latent heat is not considered in the model, the calculated weld pool trail is not elongated sufficiently.

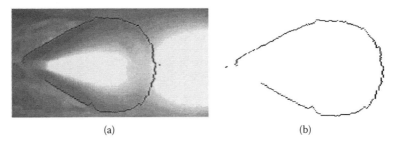

(a)   (b)

FIGURE 6.22
The image of the weld pool and its geometry: (a) the image of the weld pool captured by the camera; (b) geometry of weld pool extracted.

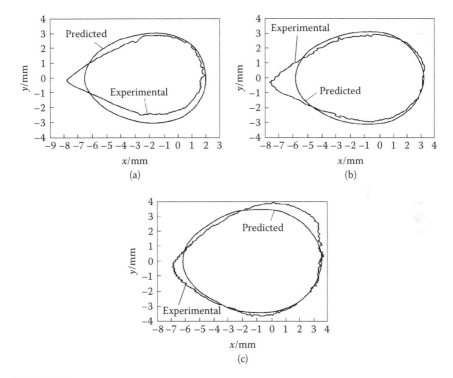

FIGURE 6.23
Comparison between the predicted and experimental surface geometry of weld pools: (a) Q235, 2-mm thickness, 110 A, 16 V, 160 mm/min; (b) SS304, 3-mm thickness, 100 A, 16 V, 125 mm/min; (c) SS304, 3-mm thickness, 100 A, 16 V, 113 mm/min.

6.9.2 Comparison of Surface Depression

When TIG welding reaches the quasi-steady state, the shape of the weld pool keeps relatively stable. At this time, solid metal in front of the weld pool melts continuously, while molten metal in the tail of the weld pool solidifies. So, the maximum width of the weld pool corresponds to the weld width, and

TABLE 6.2

Comparison of the Predicted and Experimental Weld Pool Surface Geometry

Figure Number	Welding Parameters			Experimental Results			Predicted Results		
	Current (A)	Voltage (V)	Speed (mm/min)	Length (mm)	Width (mm)	Area (mm²)	Length (mm)	Width (mm)	Area (mm²)
Figure 6.23(a)	110	16	160	9.55	5.33	34.84	8.56	6.02	40.36
Figure 6.23(b)	100	16	125	10.91	5.89	43.24	9.06	6.16	43.63
Figure 6.23(c)	100	16	113	10.43	7.39	55.27	9.75	6.74	51.11

the pool surface deformation corresponds to the surface depression of the weld seam. Therefore, the macrophotographs of welds contain information on the pool width and surface depression. Tables 6.3 and 6.4 provide comparisons of the predicted and experimental results.

6.10 Dynamic Characteristics of TIG Weld Pool Geometry with Step Changes of Welding Parameters

Welding process parameters have a critical influence on the shape and size of the weld pool. For development of an intelligent control system in TIG welding, it is necessary to understand the response of weld pool behaviors to the step changes of welding process parameters. Based on the numerical model established, the dynamic behaviors of a fully penetrated weld pool are simulated numerically after a sudden step change of process parameters. The main idea focuses on the study of the dynamic response of 3-D geometry and surface depression of the weld pool to a step change in welding current or welding speed, which could provide basic data for the establishment of an intelligent control model.

6.10.1 Adjustment of the Main Program

The dynamic response of the weld pool to the step change of welding current or welding speed is essentially a transient transformation of the weld pool from an original quasi-steady state to a new one. The step change of welding current has great influence on many subroutines in the main program, such as temperature field, surface deformation, and flow field. For example, the subroutines for the heat input and the arc pressure on the top surface of the weld pool and the electromagnetic force in the fluid flow field need to be changed. The step change of the welding speed mainly affects the heat input and the position of the projection point of the tungsten electrode. Consequently, some subroutines, such as the fluid flow field, temperature field, and data output, need to be adjusted.

6.10.2 Dynamic Response of TIG Weld Pool Geometry to a Step Change of the Welding Current

For TIG welding on a Q235 plate 2-mm thick with a 110-A welding current, 16-V arc voltage, and 160-mm/min welding speed, the weld pool and the temperature field achieve the quasi-steady state at $t = 4.2$ s. To analyze the dynamic variation of the weld pool, the welding current is suddenly changed from 110 A to 90 A at $t = 5$ s. Because of this step change of the welding

TABLE 6.3

Comparison of the Maximum Depression of the Weld Pool on the Top Surface

Metals	Thickness (mm)	Welding Parameters			Experimental Results		Predicted Results	
		Current (A)	Voltage (V)	Speed (mm/min)	Width (mm)	Depression (mm)	Width (mm)	Depression (mm)
SS304	3	110	12	125	6.5	0.12	5.4	0.14
Q235	2	100	16	160	4.7	0.16	5.3	0.14
Q235	2	110	16	160	6.2	0.24	7.4	0.28

TABLE 6.4

Comparison of the Maximum Depression of the Weld Pool on the Bottom Surface

Metals	Thickness (mm)	Welding Parameters			Experimental Results		Predicted Results	
		Current (A)	Voltage (V)	Speed (mm/min)	Width (mm)	Depression (mm)	Width (mm)	Depression (mm)
SS304	3	110	12	125	1.7	0.3	1.9	0.27
Q235	2	100	16	160	3.0	0.3	3.2	0.23
Q235	2	110	16	160	4.8	0.56	5.2	0.49

current, there are dynamic variations of the weld pool geometry, temperature, and fluid flow fields, so a transient transformation process starts. When this transformation process ends, a new quasi-steady state is reached while the welding process is at the new condition. In this case, the transformation process starts at about $t = 5$ s and ends at about $t = 9$ s. The waveform of the welding current is shown in Figure 6.24.

6.10.2.1 Dynamic Variation of the Weld Pool Shape

The dynamic responses of the 3-D shape of the weld pool after a 20-A step drop of welding current are shown in Figure 6.25, where (a), (b), (c), and (d) are the weld pool geometry at the top surface ($z = 0$), at the bottom surface ($z = H$), on the longitudinal cross section ($y = 0$), and on the transverse cross section ($x = -1.2$ mm), respectively. During the welding process, the arc travels at the welding speed and so does the weld pool. To compare the weld pool geometries at different instants, the weld pool geometry in the moving coordinate system with the origin located at the intersection between the electrode centerline and the top surface of the workpiece is shown in Figure 6.25. As shown in Figures 6.25(a) and 6.25(b), after the sudden change of welding current from 110 to 90 A, the pool length at the top and bottom surfaces shortens immediately at the front of the weld pool, while it extends slightly at the rear of the weld pool. The reason is that the temperature gradient is much steeper at the front of the weld pool. The electrical parameters (welding current) can be decreased suddenly, but the temperature field changes with a stagnation because of the time delay resulting from thermal diffusion. This results in the quick contraction at the front of the weld pool and a little prolongation at the rear of the weld pool.

The dynamic responses of the length and width of the weld pool after a 20-A step drop of welding current are shown in Figure 6.26. As time goes on, due to the decrease in welding current, the whole length of the weld pool is gradually shortened as the rear edge of the weld pool moves forward but the

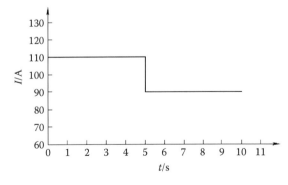

FIGURE 6.24
The step change of the welding current.

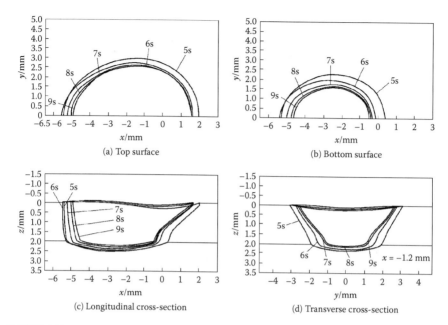

FIGURE 6.25

The dynamic response of the three-dimensional weld pool geometry to the sudden decrease of welding current from 110 to 90 A (from Zhao P C, Wu C S, and Zhang Y M, *Model. Simul. Mater. Sci. Eng.*, 12: 765–780, 2004) (workpiece: Q235, 2-mm thickness, 16 V, 160 mm/min; the welding current is changed from 110 to 90 A at $t = 5$ s): (a) top surface; (b) bottom surface; (c) longitudinal cross section; (d) transverse cross section.

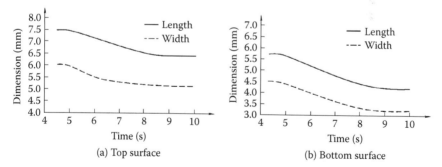

FIGURE 6.26

The dynamic response of length and width on the top and bottom surfaces of the weld pool to the sudden decrease of welding current from 110 to 90 A (workpiece: Q235, 2-mm thickness, 16 V, 160 mm/min; the welding current is changed from 110 to 90 A at $t = 5$ s): (a) top surface; (b) bottom surface.

front edge hardly moves relative to the electrode centerline ($x = 0$). The rate of change of the weld pool width is different at different moments. At the initial stage of the transformation period, the pool width decreases quickly. Then, its rate of change slows, until the transformation period is finished. There is a larger final decrease of weld pool width at the bottom surface than that at the top surface. When the new quasi-steady state is reached, the whole weld pool contracts due to the decreasing welding current. After the transformation period (from 4 to 9 s) ends, the pool width and length at the top surface change from 6.02 to 5.16 mm and from 7.5 to 6.4 mm, respectively, while those at the bottom surface change from 4.49 to 3.25 mm and from 5.74 to 4.25 mm, respectively.

6.10.2.2 Dynamic Variations of Pool Surface Deformation

The dynamic variations of pool surface deformation on the top and bottom surfaces at different times are shown in Figure 6.27. To depict more clearly the dynamic responses of pool surface deformation to a step change of welding current, a finer scale is used in the z-direction, which is equal to an enlargement of surface deformation in the z-direction.

Figure 6.27(a) represents that, as welding current suddenly drops, both the arc pressure and the weld pool volume decrease, which leads to a rapid drop in the maximum depression on the top surface. Figure 6.27(b) indicates that there is a relatively even change of the maximum depression on the bottom surface. Figures 6.27(c) and 6.27(d) show the variation of pool surface

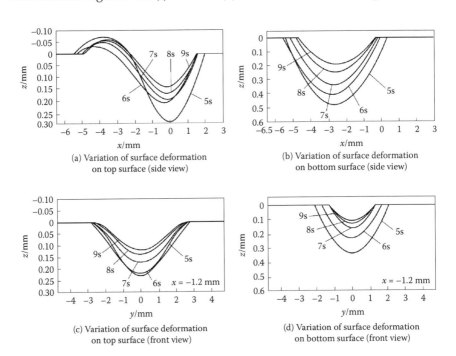

(a) Variation of surface deformation on top surface (side view)

(b) Variation of surface deformation on bottom surface (side view)

(c) Variation of surface deformation on top surface (front view)

(d) Variation of surface deformation on bottom surface (front view)

FIGURE 6.27
Responses of surface deformation to a step change in welding current (enlarged in the z-direction): (a) variation of surface deformation on the top surface (side view); (b) variation of surface deformation on the bottom surface (side view); (c) variation of surface deformation on the top surface (front view); (d) variation of surface deformation on the bottom surface (front view).

deformation on the top and bottom surfaces at $x = -1.2$ mm. It can be seen that there is a larger variation of surface depression on the bottom surface than that on the top surface, and the responses of surface deformation on both the top and bottom surfaces to step change of the welding current are quite obvious.

The maximum deformations on both the top and bottom surfaces, such as the maximum depression on the top surface, the maximum hump on the top surface, and the maximum depression on the bottom surface, are shown in Figure 6.28. It can be seen that the maximum depression on the bottom surface is almost linearly decreased with the time, while the maximum depression on the top surface is irregular and decreases rapidly in the first second after the step change and slows in the subsequent 3 s. Figure 6.29 indicates the ratios of Dd_{max}/L and Dd_{max}/W on the top surface versus time. It can be seen that, in the first 4 s after the step change in welding current (from the fifth second to the ninth second), the curve is composed of two sections of lines. The line at 5 to 6 s is quite steep with a large slope, while the line at 6 to 9 s is relatively straight with a small slope. This means a rapid change of

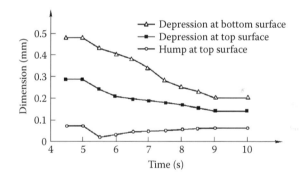

FIGURE 6.28
The maximum deformation on both the top and the bottom surface versus time.

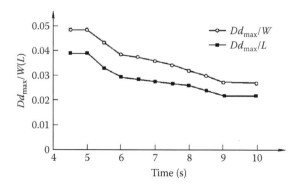

FIGURE 6.29
Ratios Dd_{max}/L and Dd_{max}/W on the top surface versus time.

weld pool profile and surface depression on the top surface at the initial time after the step change and a slow change at the subsequent time.

6.10.3 Dynamic Responses of TIG Weld Pool Geometry with Step Changes of Welding Speed

By using the numerical model, the dynamic behavior of the weld pool is simulated when the welding speed undergoes a step change. Because the same welding conditions as mentioned are used, the weld pool reaches quasi-steady state at $t = 4.2$ s. Then, the welding speed is suddenly changed from 160 to 180 mm/min at $t = 4.5$ s. The weld pool attains its new quasi-steady state at $t = 6.5$ s. The step change in welding speed is shown in Figure 6.30.

6.10.3.1 Dynamic Variation of Weld Pool Shape

The response of the 3-D shape of the weld pool is shown in Figure 6.31, where 6.31(a), 6.31(b), 6.31(c), and 6.31(d) are the transient weld pool geometries at the top surface ($z = 0$), at the bottom surface ($z = H$), on the longitudinal cross section ($y = 0$), and on the transverse cross section ($x = -1$ mm), respectively. It is clear that the weld pool geometry varies quickly with the step increase in welding speed. During the initial 0.5 s after the change, the weld pool moves 1 mm backward with respect to the electrode centerline. Then, further movement backward is slowed. The increase in welding speed causes a decrease in heat input. Thus, the whole weld pool contracts. The variation in the weld pool width is larger than that in the weld pool length. The variation trends are almost the same for both the top and the bottom surfaces of the weld pool, but the top surface changes more quickly than the bottom surface because of thermal diffusion resulting in a time delay before the bottom surface reacts.

The dynamic responses of the length and width of the weld pool after a 20-mm/min step jump of welding speed are shown in Figure 6.32. Obviously,

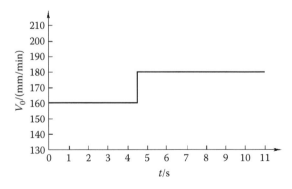

FIGURE 6.30
The step change of the welding speed.

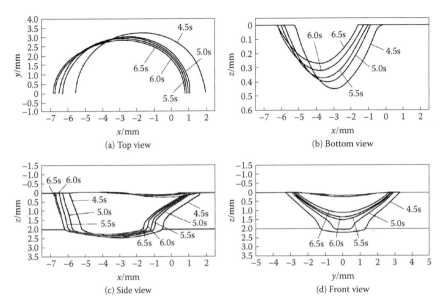

FIGURE 6.31
The dynamic response of three-dimensional weld pool geometry to the sudden increase of welding speed from 160 to 180 mm/min (From Zhao P C, Wu C S, and Zhang Y M, *Model. Simul. Mater. Sci. Eng.*, 12: 765–780, 2004): (a) top view; (b) bottom view; (c) side view; (d) front view.

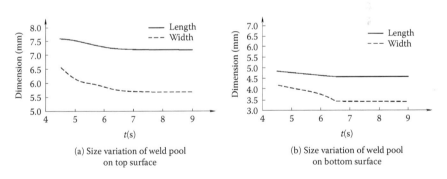

FIGURE 6.32
The transient variation of length and width on both the top and the bottom surfaces: (a) size variation of the weld pool on the top surface; (b) size variation of the weld pool on the bottom surface.

variation in width is relatively larger than that in length of the weld pool after the step change in welding speed. On the top surface of the weld pool, the width of the weld pool decreases rapidly, while the length decreases slowly. On the bottom surface, the variation speed of the width is smaller than that on the top surface because of the delay caused by thermal diffusion.

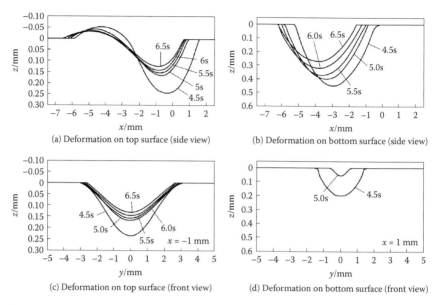

(a) Deformation on top surface (side view)

(b) Deformation on bottom surface (side view)

(c) Deformation on top surface (front view)

(d) Deformation on bottom surface (front view)

FIGURE 6.33
Responses of surface deformation to a step change in welding speed (enlarged in the z-direction): (a) deformation on the top surface (side view); (b) deformation on the bottom surface (side view); (c) deformation on the top surface (front view); (d) deformation on the bottom surface (front view).

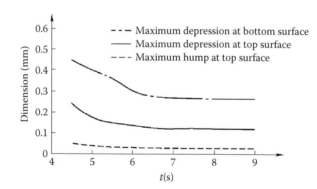

FIGURE 6.34
The maximum deformation on both the top and the bottom surface versus time.

6.10.3.2 Dynamic Variation of Pool Surface Deformation

The dynamic variations of surface deformation on the top and bottom surfaces at different times are shown in Figure 6.33. On the top surface, as shown in Figures 6.33(a) and 6.33(c), the weld pool moves backward and shrinks its surface area because of the increase of welding speed, which results in a decrease of heat input. During the first 0.5 s after the step change in welding

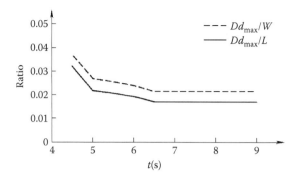

FIGURE 6.35
Ratios Dd_{max}/L and Dd_{max}/W on the top surface versus time.

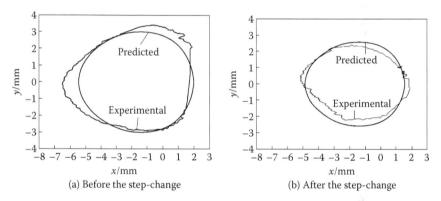

(a) Before the step-change

(b) After the step-change

FIGURE 6.36
Comparison of the calculated and experimental geometries of the weld pool on the top surface before and after a step change in the welding current.

speed, the depression on the top surface diminishes immediately, but its variation trend becomes slower during the next 2 s. The hump on the top surface reduces gradually, but it does not decrease much. As the weld pool volume shrinks, the depression on the bottom surface as seen in Figures 6.33(b) and 6.33(d) decreases continually and fades away in the end. On the transverse cross section where $x = -1$ mm, the loss of deformation is because of the backward moving of weld pool rather than shrinkage of the weld pool. Consequently, the responses of surface deformation to a step change in the welding speed are also in evidence.

The transient variation of the maximum deformation (the maximum depression and hump on the top surface, the maximum depression on the bottom surface) is shown in Figure 6.34. After the step change in welding speed, variations of the maximum depression on both the top and bottom surfaces are comparatively bigger. At the beginning after the step change, all maximum deformations decrease rapidly and trend to be stable when

FIGURE 6.37
Macrograph of the weld in cross section after a step change in the welding current.

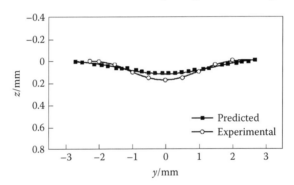

FIGURE 6.38
Comparison of the calculated and experimental weld depression on the top surface.

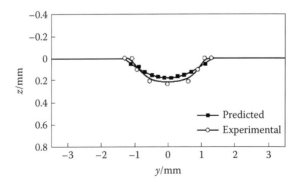

FIGURE 6.39
Comparison of the calculated and experimental weld depression on the bottom surface.

the weld pool reaches the second quasi-steady state. Ratios of Dd_{max}/L and Dd_{max}/W on the top surface versus time are indicated in Figure 6.35. The three-segment curves of such ratios reflect the information on the step change in the welding speed. At the moment the step change in the welding speed is given ($t = 4.2$s), the slopes of Dd_{max}/L and Dd_{max}/W are suddenly changed, that is, the slopes of two curves increase in a marked way. When the quasi-steady state is obtained at $t = 6.5$ s, the weld pool geometry is in a relatively stable condition, and Dd_{max}/W and Dd_{max}/L are nearly constant. Thus, they can clearly represent the step change in the welding speed.

6.10.4 Experimental Verification

The same experimental method is performed to verify the model. Figure 6.36 shows a comparison of the calculated and experimentally observed geometries of the weld pool at the top surface in the quasi-steady state before and after the step change in the welding current. The experimental and the predicted pool surface configurations are consistent with each other.

The macrograph of the weld at the transverse cross section after the step change in the welding current is shown in Figure 6.37. Figure 6.38 and Figure 6.39 show a comparison of the calculated and experimentally observed weld surface depression on the top and bottom surfaces, respectively. It can be seen that the predicted maximum deformations on the top and bottom surfaces are a little smaller than the experimental ones, and the calculated width on the top surface is also smaller than the experimental result.

7

Analysis of Dynamic Process of Metal Transfer in GMAW

7.1 Introduction

Gas metal arc welding (GMAW) has become the most popular type of arc welding processes because of its variety of advantages, such as high efficiency, good weld quality, low production cost, and ease of automation. This process uses consumable wire as the electrode. During the welding process, the wire melts to form molten drops at the end of the wire; these drops are detached and transferred into the weld pool to provide filler metal and heat. Metal transfer is one of the important factors affecting the stability of the welding process and weld quality. In addition, stable metal transfer is the basis for automatic welding and intelligent control welding. The welding process parameters affect the metal transfer mode and features, which in turn have great influence on the stability of the welding process parameters. Therefore, it is of great significance to establish the quantitative relation between metal transfer mode and welding parameters. So far, such quantitative relation is mainly established using experimental methods. This results in costs in terms of labor and materials, and the application range of the results is limited to the experimental conditions. New experiments are required when the base metal and wire vary. By modeling and simulation, the correlation of metal transfer mode and welding parameters can be established for any process conditions. It can achieve rational accuracy with a limited amount of experimental verifications. Numerical analysis of metal transfer not only can provide a database for generating welding procedure specifications but also can provide powerful tools to study some complicated physical phenomena (e.g., the evolution of drop size and profile, temperature field and fluid flow inside a drop, etc.) that are useful for modeling and simulation of weld pool behaviors. Therefore, it is of both theoretical and practical significance to conduct numerical simulation of the metal transfer in GMAW.

7.2 Mathematical Model for Metal Transfer in GMAW

Metal transfer is a process in which events in a series occur continuously, such as the tip of the filler wire melts; a droplet forms, grows, detaches; and then the droplet is transferred into the molten pool. It is similar to water dripping from a faucet. So, we can assume that the molten metal flows from a dummy pipe at a certain speed and gradually forms a globular droplet because of the combined action of gravity, surface tension, electromagnetic force, and plasma drag. The droplet geometry and the fluid flow field inside it change with time, resulting in a pinching neck. Finally, the droplet detaches from the tip of the filler wire. This is a system involving free boundaries. Some boundary profiles of the object to be solved change with time. This kind of boundary needs to be traced.

7.2.1 Governing Equations

For simplification, we make the following reasonable assumptions:

1. The droplet remains axisymmetric during its formation, evolution, detachment, and movement from the electrode tip to the molten pool, so the metal transfer system is reduced to a two-dimensional problem.
2. The molten metal is incompressible.
3. The physical properties of the molten metal are constants.
4. The temperature field in the droplet is uniform.
5. The interface between the molten droplet and the electrode is a plane perpendicular to the axis of electrode.

Based on these assumptions, the velocity field in the droplet can be described by Navier-Stokes equations:

$$\frac{\partial u}{\partial t}+u\frac{\partial u}{\partial r}+v\frac{\partial u}{\partial z}=-\frac{1}{\rho}\frac{\partial p}{\partial r}+\mu\left(\frac{\partial^2 u}{\partial r^2}+\frac{\partial^2 u}{\partial z^2}+\frac{1}{r}\frac{\partial u}{\partial r}-\frac{u}{r^2}\right)+\frac{F_v}{\rho} \tag{7.1}$$

$$\frac{\partial v}{\partial t}+u\frac{\partial v}{\partial r}+v\frac{\partial v}{\partial z}=-\frac{1}{\rho}\frac{\partial p}{\partial y}+\mu\left(\frac{\partial^2 v}{\partial r^2}+\frac{\partial^2 v}{\partial z^2}+\frac{1}{r}\frac{\partial v}{\partial r}\right)+\frac{F_v}{\rho} \tag{7.2}$$

where u and v are fluid velocity components in the radial (r) and axial (z) directions, respectively; t is time; μ is the coefficient of dynamic viscosity; ρ is the molten metal density; p is the pressure; and F_v is the body force that acts on the molten metal.

As an incompressible fluid, molten metal satisfies the continuity equation

$$\frac{\partial u}{\partial r} + \frac{\partial v}{\partial z} + \frac{u}{r} = 0 \tag{7.3}$$

7.2.2 Tracing Free Boundary: Volume of Fluid Method

A variety of methods has been developed for tracing free boundaries. Of all the methods, the volume of fluid method (VOF) is the most popular. For implementing VOF, a function $F(i, j, t)$ is introduced. A cell with $F(i, j, t) = 1$ is full of fluid, whereas a zero value of $F(i, j, t)$ corresponds to a cell that contains no fluid. A cell with an $F(i, j, t)$ value between 0 and 1 contains a free surface. By calculating $F(i, j, t)$, we can determine the free surface cell and the normal direction to the boundary. The normal direction to the boundary is the direction in which the value of $F(i, j, t)$ changes most rapidly.

The time dependence of $F(i, j, t)$ is governed by the equation

$$\frac{\partial F}{\partial t} + u\frac{\partial F}{\partial r} + v\frac{\partial F}{\partial z} = 0 \tag{7.4}$$

where u and v are the fluid velocity components in the radial (r) and axial (z) directions, respectively. As Eq. (7.4) implies, F moves with the fluid. With standard finite difference approximation, it is difficult to reach convergence, so special approximation should be adopted.

7.2.3 Forces Acting on a Molten Drop

During growth and detachment of a molten droplet, the following forces act on it: gravity, electromagnetic force, surface tension, plasma drag, and recoil pressure due to metal evaporation. Because there is a lack of data for describing recoil pressure, the body forces in Eqs. (7.1) and (7.2) only include gravity, electromagnetic force, surface tension, and plasma drag.

7.2.3.1 Electromagnetic Force

The electromagnetic force results from the interaction of the welding current with its own magnetic field. It can be expressed as

$$F_m = \vec{J} \times \vec{B} \tag{7.5}$$

where $\vec{J}$ is the current density, and $\vec{B}$ is the magnetic field vector.

The magnetic field is calculated as follows:

$$B_\theta = \frac{\mu_m}{r} \int_0^r J_z r dr \tag{7.6}$$

The current density is calculated from Ohm's law:

$$J_r = -\sigma_e \frac{\partial \phi}{\partial r} \tag{7.7}$$

$$J_z = -\sigma_e \frac{\partial \phi}{\partial z} \tag{7.8}$$

where ϕ is the electric potential, σ_e is the electric conductivity, μ_m is the magnetic permeability, and J_r and J_z are the components of current density in the radial (r) and axial (z) directions, respectively. The electric potential is governed by the current continuity equation:

$$\frac{1}{r} \frac{\partial}{\partial r} \left(r \frac{\partial \phi}{\partial r} \right) + \frac{\partial^2 \phi}{\partial z^2} = 0 \tag{7.9}$$

Because of the difficulty in performing measurement, the experimental data concerning current density distribution in the molten drop are not yet available, so a rational assumption is made. The current density on the interface between the molten drop and the electrode is assumed to remain in the same distribution in the electrode (i.e., uniform distribution), while the current density on the surface of the drop is assumed by taking a Gaussian distribution:

$$J_z = \frac{k_d I z}{2\pi H_D} \exp(-k_d r^2) + \frac{I}{2\pi R_w^2} \left(1 - \frac{z}{H_D} \right) \tag{7.10}$$

where k_d is a coefficient indicating the current concentration in the molten drop, I is the welding current, H_D is the length of the molten drop, z is the distance between the surface cell and the solid-liquid interface of the electrode, and R_w is the diameter of the electrode.

From Eqs. (7.10) and (7.9), we get the current density in the radial direction:

$$J_r = \frac{I}{2\pi r H_D} \left[1 - \exp(-k_d r^2) - \frac{k_d r}{2\pi H_D} \right] \tag{7.11}$$

The magnetic field is calculated as follows:

$$B_\theta = \frac{\mu_m I z}{2\pi r H_D}[1 - \exp(-k_d r^2)] + \frac{\mu_m k_d r I}{4\pi}\left(1 - \frac{z}{H_D}\right) \tag{7.12}$$

From Eqs. (7.10) and (7.12), we get the electromagnetic force component in the radial direction:

$$F_m^r = -B_\theta J_z \tag{7.13}$$

Based on Eqs. (7.11) and (7.12), we get the electromagnetic force component in the axial direction:

$$F_m^z = -B_\theta J_r \tag{7.14}$$

7.2.3.2 Surface Tension

The surface tension acting on the drop surface is described by the surface tension coefficient and the curvature radius of the surface:

$$P_\sigma = \gamma(\frac{1}{R_1} + \frac{1}{R_2}) \tag{7.15}$$

where γ is the surface tension coefficient, and R_1 and R_2 are the principal radii of the curvature of the free surface.

The pressure in the free surface cell must be interpolated with the pressure on the free surface and the neighbor cell (this cell is called the interpolation cell). In general, a free surface cell has more than one neighbor cell; the interpolation cell must be selected based on the direction of the free surface.

7.2.3.3 Plasma Drag

The plasma drag acting on a molten drop is similar to the force acting on a spherical particle immersed in a liquid stream. The drag acting on a spherical particle immersed in a liquid moving at constant speed is expressed as[66]

$$F_p = C_D A_p\left(\frac{\rho_f v_f^2}{2}\right) \tag{7.16}$$

where C_D is the dimensionless drag coefficient, A_p is the projected area of the droplet on the plane perpendicular to the direction of fluid flow, ρ_f is the density of plasma, and v_f is the velocity of plasma. For the plasma drag on a molten drop, C_D depends on the Renault number of shielding gas. The Renault number can be determined using the velocity of plasma. Since experimental data concerning the velocity of plasma in GMAW are not available, it is assumed to be 100 m/s, which is the same as velocity of plasma in gas tungsten arc welding (GTAW) (i.e., the corresponding Renault number is 0.44).[89] ρ_f takes the density of argon (6×10^{-6} kg/m³).[81]

7.2.4 Boundary Conditions

The boundary and initial conditions for solving these mentioned equations are shown in Figure 7.1.

1. The molten metal is assumed to flow from a dummy pipe at a certain speed. The velocity is assumed to be the same as the wire feed speed S_m, that is, $u(i, 1) = 0$, $v(i, 1) = S_m$.
2. The drop is assumed to be axisymmetric. So, only the right side of the centerline is taken as the calculation domain.
3. There is no slip at the solid boundary, and the other boundaries are all slip boundaries.
4. For simplicity and minimization of computing time, the initial shape of the molten drop is assumed to be a semisphere.

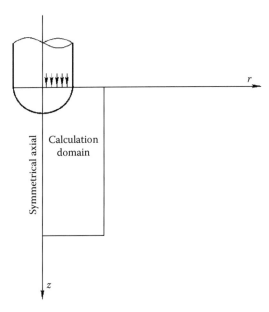

FIGURE 7.1
Boundary and initial conditions.

7.3 Algorithm and Program

For this problem, it is convenient to adopt rectangle mesh in the Eulerian coordinate. The arrangement of variables is shown in Figure 7.2. The volume of fluid function F is used to distinguish different mesh cells. There are three kinds of mesh cells: (1) a free surface cell, which is defined as a cell that has a nonzero value of F and has at least one neighboring cell $(i + 1, j)$, $(i - 1, j)$, $(i, j + 1)$, or $(i, j - 1)$ containing a zero value of F; (2) an empty cell, which has a zero value of F; (3) a cell full of fluid, which has a unit value of $F = 1$ and no empty neighboring cells.

For the cell shown in Figure 7.2, convert Eqs. (7.1)–(7.3) into a finite difference approximation. The basic procedure for advancing a solution through one time step δt is as follows:

1. Using the initial conditions or previous time-level values of velocity and pressure fields, explicit approximations of Eqs. (7.1) and (7.2) are used to compute the new time-level values of velocity field.

2. To satisfy the continuity equation, Eq. (7.3), pressure in every cell is iteratively adjusted, and the velocity variations caused by pressure change are added to the velocities computed in step 1. An iteration is required because the change in pressure needed in one cell to satisfy the continuity equation will upset the balance in the four neighboring cells.

3. Finally, update the value of F in each cell to determine the new surface configuration of the droplet.

Repeat these steps until the desired time is reached. At each step, proper boundary conditions must be imposed at all elements and free surfaces.

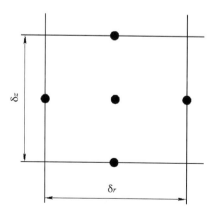

FIGURE 7.2
Arrangement of variables in a typical cell.

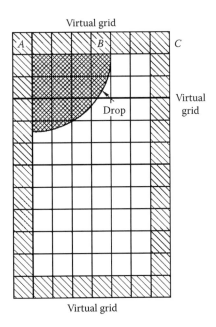

FIGURE 7.3
Boundary conditions and dummy cells.

Besides free surface boundary conditions, the computation domain boundary condition must be determined. For simplification, a layer of virtual cells is added to the side of the real computation domain, as seen in Figure 7.3.

To improve accuracy, the values of δr and δz shall be chosen small enough, but the smaller are δr and δz, the longer is the computing time, and the more computer memory is required. Therefore, moderate δr and δz are often chosen. After the mesh sizes are chosen, a suitable time increment δt shall be chosen to guarantee stability.

First, the displacement of the fluid in a cell (i, j) within one time step should not exceed the size of the same cell. Thus, δt must satisfy the inequality

$$\delta t < \min\left(\frac{\delta r_i}{ABS(u_{i,j})}, \frac{\delta z_i}{ABS(v_{i,j})}\right) \tag{7.17}$$

where ABS means the absolute value. δt must be computed for every cell. Generally, δt is chosen equal to one-fourth to one-third of the minimum cell transit time.

In the case of a nonzero value of dynamic viscosity, momentum must not diffuse more than one cell in one time step. Therefore, δt must satisfy the inequality

$$\delta t < \frac{1}{2\mu} \frac{\delta r_i^2 \delta z_j^2}{\delta r_i^2 + \delta z_j^2} \qquad (7.18)$$

where μ is the dynamic viscosity of the fluid.

When surface tension is included, the capillary waves must not travel more than one cell in one time step δt. So, δt must satisfy the inequality

$$\delta t < \sqrt{\frac{\rho \delta r_{min}^3}{8\gamma}} \qquad (7.19)$$

where ρ is the density of fluid, and δr_{min} is the minimum size in mesh (δr_i or δz_i).

The computation program contains 13 subroutines. The flowchart is shown in Figure 7.4. This program provides capabilities for a wide range of applications. For a specific problem, only a few subroutines need to be adjusted based on particular conditions. The subroutines and brief descriptions are as follows:

1. Main program
   a. Input and output data.
   b. Call other subroutines and control the iteration.
   c. $t = t + \delta t°$.
   d. CYCLE = CYCLE + 1°.
   e. Stop the program and provide a shutdown procedure.
2. Subroutine BC (boundary condition)
   a. Set the variable values at the boundaries of the computation domain.
   b. Set and adjust the parametric values of cells near the free surface.
3. Subroutine TSA (time step adjustment)
   a. Compute maximum allowable δt using Eqs. (7.17)–(7.19).
   b. Adjust δt according to the fluid velocity variations.
   c. Adjust the relaxation factors for pressure.
4. Subroutine SOUTW (set output window)
   a. Set the features of the output window.
   b. Convert the physical coordinate to the window coordinate.

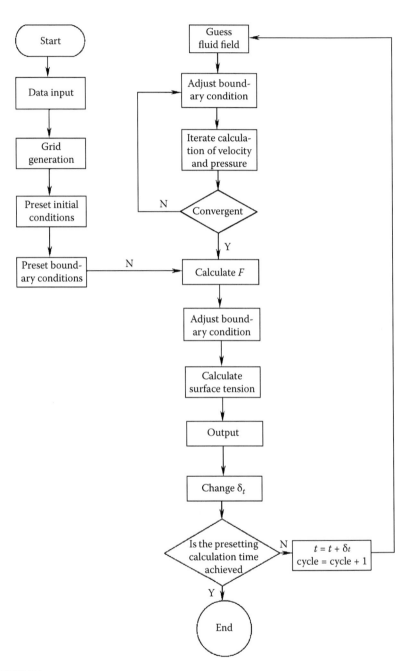

FIGURE 7.4
Flowchart of the program.

5. Subroutine OUTMESH: Output mesh in the window.
6. Subroutine OUTVEF: Output velocity field.
7. Subroutine OUTDROP: Output the profile of the calculated molten drop.
8. Subroutine CBFORCE: Calculate the body forces.
9. Subroutine MESHSET (mesh setup)
   a. Generate the computing mesh.
   b. Predict the necessary geometric variables.
10. Subroutine INTFC (interpolation factor calculation)
   a. Determine the slope of the free surface.
   b. Determine the cell flag NF for each surface cell.

   NF = 1, the neighbor to the left is the interpolation cell.

   NF = 2, the neighbor to the right is the interpolation cell.

   NF = 3, the neighbor to the bottom is the interpolation cell.

   NF = 4, the neighbor to the top is the interpolation cell.

   Determine the orientation of the surface using this information.
   c. Compute the pressure in every surface cell.
11. Subroutine PRESSIT (pressure iteration)
   a. Iterate the velocity and pressure fields until the continuity equation is satisfied.
   b. Adjust the pressure in every surface cell based on the applied surface pressure.
12. Subroutine SETUP
   a. Initialize all the constants needed in computation.
   b. Set and compute the initial $F(i, j)$.
   c. Set and compute the initial $P(i, j)$.
   d. Set and compute the initial $u(i, j)$ and $v(i, j)$.
   e. Compute the relaxation factors.
13. Subroutine TVEC (temporary velocity calculation): Compute new time-level velocity field using an explicit approximation of the momentum equation.
14. Subroutine VFCONV (volume fraction convection)
   a. Compute new time level $F(i, j)$ using Eq. (7.4).
b. Compute and store $F(i, j)$.

7.4 Numerical Simulation Results

The model and program were used to conduct numerical analysis of the metal transfer in GMAW of low-carbon steel plates. The diameter of electrode wire was 1.6 mm. The wire feed speed was 70 mm/s to 125 mm/s when the welding current varied from 150 to 320 A.[142] The physical properties of base metal are listed in Table 7.1.

7.4.1 Evolution of Molten Drop Profile

Figure 7.5 shows the molten drop profile evolution at a current level of 150 A. The equivalent diameter of the detached molten drop is 4.2 mm. The frequency of droplet transfer is 6 drops a second. At the beginning, the drop at the tip of the electrodes takes a semispherical shape and gradually changes to a pear shape. A pinching neck forms near the solid-liquid interface under

TABLE 7.1

Physical Properties of Wire Metal
Used in Calculation

Density ρ	7860 kg/m³
Viscosity μ	7.76 kg/(m s)
Surface tension γ	1.2 N/m

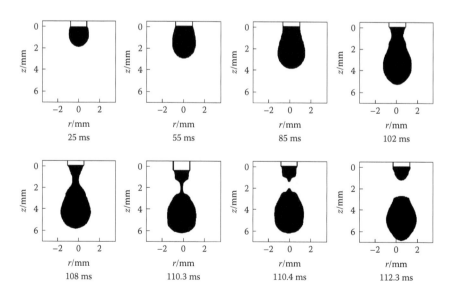

FIGURE 7.5
The formation, growth, and detachment of a drop (150-A welding current). (From Chen M A and Wu C S, *Acta Metall. Sinica*, 40(11): 1227–1232, 2004.)

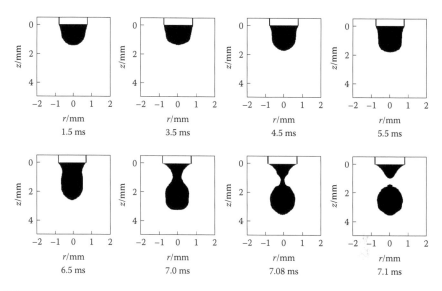

FIGURE 7.6
The formation, growth, and detachment of a drop (320-A welding current 320 A). (From Chen M A and Wu C S, *Acta Metall. Sinica*, 40(11): 1227–1232, 2004.)

the action of electromagnetic force and plasma drag. As time passes, the electromagnetic force at the pinching neck increases rapidly, and the neck pinching becomes more and more severe; at last, the drop detaches from the pinching neck.

Figure 7.6 shows the drop profile evolution at a current level of 320 A. The equivalent diameter of the detached molten drop is 1.4 mm. The frequency of droplet transfer is 140 drops per second. The electromagnetic force increases with increasing welding current. The drop moves toward the weld pool at a higher speed. The detached drop size becomes smaller, and detachment frequency is much greater.

7.4.2 Velocity Field in a Molten Drop

Figure 7.7 shows the velocity field inside a molten drop with a welding current of 150 A. A vortex is observed near the surface of the molten drop during the period of drop growth. The vortex is caused by the surface tension force. In the area near the centerline, the gravity and electromagnetic force dominate the fluid flow, while the surface tension has little effect. So, the liquid metal in this area moves along the axial direction. In the area near the surface, the surface tension has great effect on the fluid velocity. The fluid at the side surface is driven by surface tension with an upward component, causing the vortex.

With growth of the molten drop, the center of the vortex moves downward, and the vortex itself becomes weak. When the vortex moves out of the area

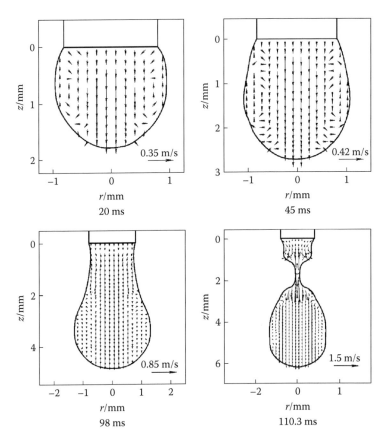

FIGURE 7.7
Velocity field in molten drop (150-A welding current). (From Chen M A and Wu C S, *Acta Metall. Sinica*, 40(11): 1227–1232, 2004.)

near the solid–liquid interface, a pinching neck forms under the action of electromagnetic force. The fluid gradually turns to flow axially, while the vortex zone decreases. When the critical necking is reached, the vortex zone disappears. The molten metal at the middle of the neck moves toward the centerline of the drop; the molten metal at the upper part of the neck moves upward, and that at the bottom part of the neck moves downward. Thus, the neck breaks quickly, and the droplet is detached rapidly. The maximum velocity appears at the location of the drop center. In a detached droplet, the fluid direction is along the wire axis.

With larger welding current, a vortex is also observed at the initial stage of drop formation, as shown in Figure 7.8. Due to the higher level of welding current, the electromagnetic force is much greater than the surface tension, and the vortex is weaker than that in the situation with lower current.

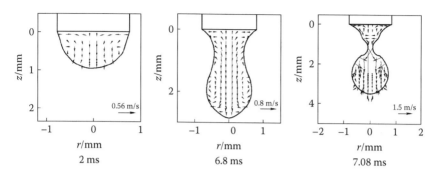

FIGURE 7.8

Velocity field in molten drop (320-A welding current). (From Chen M A and Wu C S, *Acta Metall. Sinica*, 40(11): 1227–1232, 2004.)

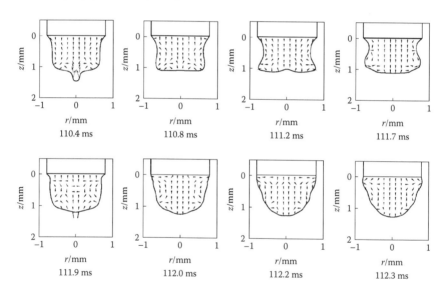

FIGURE 7.9

Evolution of retained drop profile at 150 A. (From Chen M A and Wu C S, *Acta Metall. Sinica*, 40(11): 1227–1232, 2004.)

7.4.3 Evolutions of the Retained Drop Profile

When a drop is detached from the wire, some molten metal remains at the wire end and is referred to as the retained drop. Both the retained drop and the detached drop change their profiles with time. The retained drop tends to retract to the solid-liquid interface and changes its shape to a semi-spherical one gradually, as shown in Figure 7.9. On detachment, the fluid at the bottom of the retained drop moves upward quickly. The upward speed decreases rapidly to zero after a short time. Then, the fluid at the bottom

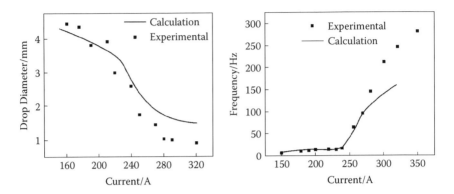

FIGURE 7.10
Comparison between the calculated and experimental drop diameter and frequency. (From Chen M A and Wu C S, *Acta Metall. Sinica*, 40(11): 1227–1232, 2004.)

of the retained drop starts to move downward and accelerates. After the downward speed reaches a certain value, it decreases gradually to zero.

7.4.4 Effect of Welding Current on Detached Drop Size

Figure 7.10 shows the relationship between the calculated drop size and welding current. This figure also gives the comparison between the measured and calculated drop sizes. The measured data were taken from the literature.[89] It can be seen that the drop size decreases and the frequency of droplet transfer increases with increasing welding current. The drop sizes vary sharply in the current range of 230–260 A, indicating that the transitional current lies in this range. In the lower current range, the calculated drop sizes are in good agreement with experimental ones, while in the higher-current range, the calculated drop size is much greater than the experimental one at a particular current level. The possible reasons are as follows: (1) In the computing model, the surface tension coefficient was assumed to be a constant in spite of the current level. In fact, it decreases with increasing current. (2) The solid-liquid interface is a cone when the welding current is beyond the transitional value.[81] The assumed initial condition of a planar solid-liquid interface is far from a practical situation.

7.5 Dynamic Model for Metal Transfer Based on the "Mass-Spring" Theory

A variety of different mathematical models has been established to investigate metal transfer in GMAW. In the classical theories, that is, pinch instability theory (PIT) and static force balance theory (SFBT), only the static balance or wavelength instability at the onset of detachment is considered, while the effect of drop evolution and melting rate is not included, not to say the oscillation speed of the droplet during its growth and detachment. Thus, the predicted values are far from the experimental ones.[79–83] Great progress has been made in the simulation of metal transfer phenomena, and the predicted results tend to approach the experimental ones. Currently, efforts in this field have been focusing on the "mass-spring" theory and fluid dynamics theory. Sections 7.2 to 7.3 present the practical application of fluid dynamics theory in analysis of metal transfer. Based on this theory, a two-dimensional dynamic model is developed for metal transfer in GMAW. The computation is time consuming because a group of differential equations has to be solved. In addition, this theory cannot describe the oscillation of the droplet during its evolution. Therefore, fluid dynamics theory is only fit for simulating metal transfer at a low welding current but not for the simulation of combined transfer and spray transfer at a high current level. So far, studies of simulation of metal transfer based on mass-spring theory are scarce, and no work on simulation of metal transfer in pulsed GMAW has been done to realize the ideal metal transfer mode of one drop for one pulse.

In this section, based on the feature of metal transfer, a dynamic model for metal transfer is developed using the mass-spring theory, which lays a basis for simulation of metal transfer in GMAW and one-drop for one-pulse metal transfer in active controlled pulsed GMAW.

7.5.1 Development of the Mathematical Model

Based on the mass-spring theory,[82] the droplet at the tip of the electrode is regarded as a mass-spring system. The dummy spring is assumed to be attached to the electrode at one side and attached to the droplet at the other side (see Figure 7.11). For simplification, the following assumptions are made:

1. The physical properties of the molten metal are constant.
2. The melting rate of the electrode is constant.
3. The mass-spring system is axisymmetric.
4. The oscillation velocity in the direction perpendicular to the electrode axis is negligible.

Based on these assumptions, the growth and detachment of the droplet can be described with the following differential equations:

$$m_D \frac{d^2x}{dt^2} + kx + b\frac{dx}{dt} = F_T \tag{7.20}$$

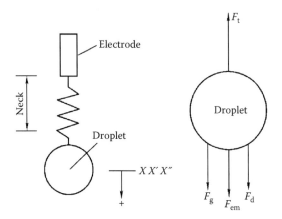

FIGURE 7.11
Forces acting on the droplet. (From Watkins A D, Smartt H B, and Johnson J A. A dynamic model of droplet growth and detachment in GMAW. *Proceedings of the 4th International Conference, Gatlinburg Tennessee, USA, 5–8 June 1995.)*

$$\frac{dm_D}{dt} = V_M \qquad (7.21)$$

where x is the droplet displacement, m_D is the time-varying mass of the droplet, t is the time, b is the damping coefficient, k is the spring constant, V_M is the increasing rate of droplet weight, and F_T is the sum of the external force acting on the droplet. The spring force is $f_k = kx$. The damping force acting on the droplet is $f_b = b(dx/dt)$.

Figure 7.11 shows the forces acting on the droplet during welding. The droplet mass can be approximately assumed as varying with time linearly, and the droplet displacement varies during the droplet oscillation. The spring force represents the surface tension force and varies with the time step. When the displacement reaches a critical value x_c, the droplet mass is reduced suddenly and is detached from the electrode. The retained droplet oscillation still follows Eq. (7.20), and the droplet mass follows Eq. (7.21).

In Eq. (7.20), F_T can be expressed as

$$F_T = F_g + F_{em} + F_d$$

where F_g is the gravitational force, F_{em} is the electromagnetic force, and F_d is the plasma drag force. Thus, the dynamic force equilibrium in Eq. (7.20) can be expressed as

$$F_T - f_i - f_k - f_b = 0 \qquad (7.22)$$

where f_i is the inertia force.

The droplet growth and detachment occur continuously. Assume the critical droplet mass at critical displacement is M_c, and the mass of the detached droplet is ΔM. ΔM has the forms of

$$\Delta M = \ell_1 M_c \tag{7.23}$$

$$\Delta M = \ell_2 V_c \tag{7.24}$$

$$\Delta M = \ell_3 M_c V_c \tag{7.25}$$

where l_1, l_2, and l_3 are coefficients to be determined; V_c is the oscillation velocity of the droplet tip. The droplet detachment is the result of combined action of individual forces and inertia momentum. Equations (7.23) and (7.24) do not include all the factors, while Eq. (7.25) includes the effect of critical droplet mass and oscillation velocity. Here, Eq. (7.25) is adopted.

Figure 7.12 is a schematic of droplet detachment. The bigger sphere is the detached droplet, and the smaller one is the retained droplet. In the model developed, the critical displacement of the mass center of the system is x_c, and the initial displacement of the retained droplet can be expressed as

$$x_0 = x_c - R_D \frac{\Delta M}{M_c} \tag{7.26}$$

where $R_D = \left(3\frac{\Delta M}{4\pi\rho} \right)^{1/3}$, ρ is the droplet density, and R_D is the droplet radius.

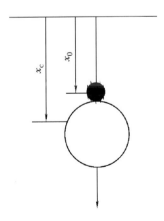

FIGURE 7.12
Schematic of droplet detachment.

7.5.2 Modeling of Forces Acting on Droplet

The forces acting on the droplet during its growth and detachment include gravity, electromagnetic force, plasma drag, surface tension force, and recoil pressure due to metal evaporation. The recoil pressure due to evaporation can be neglected. The surface tension force is represented by the spring force and can be used as a criterion for droplet detachment. The forces need to be determined for the dynamic model of metal transfer.

7.5.2.1 Electromagnetic Force F_{em}

The electromagnetic force results from the interaction of the welding current with its own magnetic field. It can be expressed as

$$F_{em} = \vec{J} \times \vec{B} \tag{7.27}$$

where $\vec{J}$ is the current density, and $\vec{B}$ is the magnetic field. When the welding current flows into a droplet, its conduction area varies. Since the current path either converges or diverges, the electromagnetic force has two components, that is, axial and radial. When the current diverges in the droplet, the electromagnetic force acts as a detaching force:

$$F_{em} = \int_V (J \times B) \sin \varphi \, dV \tag{7.28}$$

where φ is the angle between the force vector and the axial direction. During the growth process of the droplet, the radial component of the electromagnetic force can be neglected as the detaching force. Thus, the axial component of electromagnetic force can be described as[80]

$$F_{em} = \frac{\mu_m I^2}{4\pi} \left[\ln \frac{r_D \sin \theta_d}{r_w} - \frac{1}{4} - \frac{1}{1 - \cos \theta_d} + \frac{2}{(1 - \cos \theta_d)^2} \ln \frac{2}{1 + \cos \theta_d} \right] \tag{7.29}$$

where I is the welding current, μ_m is the permeability of free space, r_D is the droplet radius, r_w is the wire radius, and θ_d is the angle shown in Figure 7.13. When θ_d is greater than $60°$, its effect can be neglected.[78] Thus, for simplification, the θ_d is assumed to be $150°$.

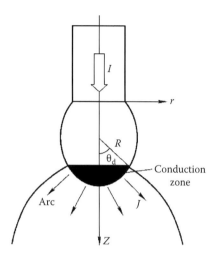

FIGURE 7.13
Arc-covered ellipsoidal drop to calculate the electromagnetic force.

7.5.2.2 *Plasma Drag Force F_d*

The aerodynamic drag force on the droplet is similar to the force acting on a spherical particle immersed in a fluid stream. The drag force may be calculated by[79]

$$F_d = C_d A_p \left(\frac{\rho_f v_f^2}{2} \right) \tag{7.30}$$

where C_d is the dimensionless drag coefficient, A_p is the projected area of plasma on the plane perpendicular to the direction of fluid flow, ρ_f is the density of the plasma, and v_f is the relative fluid flow to particle velocity in the plasma. The area A_p can be calculated by

$$A_p = \pi(r_D^2 - r_w^2) \tag{7.31}$$

7.5.2.3 *Surface Tension Force and Gravity*

The surface tension force F_σ, which acts to retain the droplet on the electrode, is given by

$$F_\sigma = 2\pi r_w \gamma \tag{7.32}$$

where γ is the surface tension for the liquid metal, and r_w is the electrode radius.

The gravity of the droplet is

$$F_g = \frac{4}{3}\pi r_D^3 \rho g \qquad (7.33)$$

where ρ is the density of the droplet, g is the gravitational acceleration, and r_D is the droplet radius.

7.5.3 Solution Techniques

7.5.3.1 Calculation Method

A four-order Runge-Kutta procedure[143] is used to solve the differential equation. This method features fourth-order accuracy.

Equations (7.20) and (7.21) are second-order differential equations. They can be first converted to first-order differential equations. Then, it is easy to obtain the solution based on the initial conditions. The derivation is as follows:

Based on

$$m_D \frac{d^2x}{dt^2} + kx + b\frac{dx}{dt} = F_T$$

we set

$$y_1(t) = x(t) \qquad (7.34)$$

$$\frac{dy_1}{dt} = y_2 \qquad (7.35)$$

Then,

$$\frac{dy_2}{dt} = \frac{d^2y_1}{dt^2} = \frac{d^2x}{dt^2} \qquad (7.36)$$

we can get

$$\begin{cases} \dfrac{dy_1}{dt} = y_2 \\[2mm] \dfrac{dy_2}{dt} = \dfrac{F_T - b\dfrac{dy_1}{dt} - ky_1(t)}{m_D(t)} \end{cases} \qquad (7.37)$$

The solution based on the fourth-order Runge-Kutta procedure is as follows :

$$\begin{cases} y_{1k+1} = y_{1k} + \dfrac{\Delta t}{6}\left[K_{11} + 2K_{12} + 2K_{13} + K_{14}\right] \\[2mm] y_{2k+1} = y_{2k} + \dfrac{\Delta t}{6}\left[K_{21} + 2K_{22} + 2K_{23} + K_{24}\right] \end{cases} \tag{7.38}$$

$$\begin{cases} K_{11} = y_{2k} \\[1mm] K_{21} = f(t_k, y_{1k}, y_{2k}) \\[1mm] K_{12} = y_{2k} + \dfrac{\Delta t}{2}K_{21} \\[1mm] K_{22} = f\left(t_k + \dfrac{\Delta t}{2}, y_{1k} + \dfrac{\Delta t}{2}K_{11}, y_{2k} + \dfrac{\Delta t}{2}K_{21}\right) \\[1mm] K_{13} = y_{2k} + \dfrac{\Delta t}{2}K_{22} \\[1mm] K_{23} = f\left(t_k + \dfrac{\Delta t}{2}, y_{1k} + \dfrac{\Delta t}{2}K_{12}, y_{2k} + \dfrac{\Delta t}{2}K_{22}\right) \\[1mm] K_{14} = y_{2k} + \Delta t K_{23} \\[1mm] K_{24} = f(t_k + \Delta t, y_{1k} + \Delta t K_{13}, y_{2k} + \Delta t K_{23}) \end{cases} \tag{7.39}$$

where Δt is the time step.

Through selecting the appropriate time step, time step number, and initial conditions, this method can be used to simulate the dynamic growth and detachment process of metal transfer. The physical property data should be adjusted slightly until convergence is reached. During the calculation, the mass of the droplet increases linearly with time.

7.5.3.2 Calculation of Droplet Mass

According to Eq. (7.21), the droplet mass increases with time and is proportional to the melting rate of electrode S_m, which is a function of welding current I and the wire extension L_e. The droplet mass can be expressed as

$$\frac{dm}{dt} = \rho A_w S_m \tag{7.40}$$

where ρ is the density of the electrode, A_w is the cross-sectional area of the electrode, and S_m is the melting rate of the electrode. The melting rate of the electrode can be expressed as[144]

$$S_m = \tau_1 I + \tau_2 L_e I^2 \tag{7.41}$$

TABLE 7.2

Melting Constants[80,145]

Wire Extension L_e (mm)	τ_1 (mm·A^{-1}·s^{-1})	τ_2 (A^{-2}·s^{-1})
26	0.1347	1.413×10^{-5}
36	0.1342	1.4266×10^{-5}

Source: From Kim Y S and Eagar T W, *Weld. J.*, 72(6): 269s–277s, 1993, and Choi S, Kim Y S, and Yoo C D, *J. Phys. D: Appl. Phys.*, 32: 326–334, 1999.

where τ_1, τ_2 are the melting constants corresponding to the arc and Joule heat, respectively; I is the welding current. The melting constants are listed in Table 7.2.

In the case of a constant welding current and wire extension, the melting rate of the electrode is constant; then,

$$\frac{dm_D}{dt} = \rho A_w S_m = V_M \tag{7.42}$$

7.5.3.3 Determination of the Spring Constant

The spring constant k is the most important parameter for analysis of metal transfer. Unfortunately, there is neither detailed description nor specific data of the spring constant k for application in welding. The droplet oscillation results in an oblate or prolate droplet shape.[92,94] According to the surface tension acting on the liquid metal droplet and its surface potential energy, the spring constant k may be described as follows[92]:

$$dU_d = \gamma dS = F_s dz \tag{7.43}$$

$$k = dF_s / dz \tag{7.44}$$

where U_d is the surface potential energy, γ is the surface tension coefficient, S is the droplet surface area, F_s is the spring force, and z is the mass center displacement of the droplet. Figure 7.14 shows the geometry of the droplet.

According to thermodynamic theory, the surface potential of a liquid boundary can be described by integrating within the whole surface. The surface potential energy on a infinitesimal area dS is

$$dU_d = \gamma \cdot dS \tag{7.45}$$

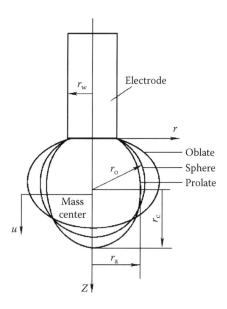

FIGURE 7.14
Geometry of the droplet. (From Jae H C, Jihye L, and Choong D Y, *J. Phys. D: Appl. Phys.*, 34: 2658–2664, 2001.)

The total surface potential energy on the droplet is

$$U_d = \oint_S \gamma \cdot dS \tag{7.46}$$

During oscillation, the shape of the droplet at the tip of the electrode is oblate. The total surface potential energy on the droplet is calculated as[92]

$$S = \pi r_a \left[r_a + \alpha_d \arcsin\left(\frac{r_c}{\alpha_d}\right) + \frac{\beta_d}{\alpha_d}\sqrt{\alpha_d^2 - \beta_d^2} + \alpha_d \arcsin\left(\frac{\beta_d}{\alpha_d}\right) \right] \tag{7.47}$$

$$\alpha_d = r_c^2 / (r_c^2 - r_a^2)^{\frac{1}{2}}, \beta_d = \left(\frac{r_c}{r_a}\right)(r_a^2 - r_w^2)^{\frac{1}{2}} \tag{7.48}$$

where r_a and r_c are principal radii of the droplet (see Figure 7.14), r is the radius of the electrode, and $r_w = r_e$.

When the droplet shape changes from sphere to prolated spheroid, its volume remains unchanged. Neglecting the variation of the mass center displacement of the droplet, the spherical droplet radius can be derived in the series by the Taylor expansion with elimination of the high-order terms:

$$r_o^3 \cong r_c r_a^2 \tag{7.49}$$

where r_o is the spherical droplet radius at the equilibrium state. The mass center position of a spheroid droplet can be express as[92]

$$Z_G = (V_U Z_U + V_L Z_L)/(V_U + V_L) \tag{7.50}$$

where Z_G is the mass center position of a spheroid droplet; V_U and Z_U are the volume and mass center position of the upper half of the spheroid droplet, respectively; and V_L and Z_L are the volume and mass center position of the lower half of the spheroid droplet, respectively.

The principal radius r_c is interpolated using the polynomial as[92]

$$r_c = 1.355 Z_G - 1.957 (r_e^2 / r_0^3) Z_G^2 + 5.19 (r_e^4 / r_o^6) Z_G^3 \tag{7.51}$$

Substituting Eqs. (7.49) and (7.51) into Eq. (7.47), we get

$$F_s = \gamma \frac{dS}{dz} = \pi \gamma p_1 z + \pi \gamma r_o p_1 \left(p_2 - \frac{z_o}{r_o} \right) \tag{7.52}$$

where

$$p_1 = 219.5 - 1242.2(r_e / r_0) + 2345.9(r_e / r_0)^2 - 1443.5(r_e / r_0)^3 \tag{7.53}$$

$$p_2 = 1.06 - 0.26(r_e / r_0) \tag{7.54}$$

where z_0 is the mass center position of the spherical droplet.

The spring force can be approximately obtained using Eq. (7.52). The second term in Eq. (7.52) is negligible compared to the first term. Then, the spring constant of the spherical droplet k can be expressed approximately as

$$k = \pi \gamma p_1 \tag{7.55}$$

Experimental results and theoretical prediction show that the spring constant decreases linearly with increasing droplet mass during the growth and detachment. The spring constant can be calibrated using experimental data. It can be considered as a function of droplet mass:

$$k = C_1 - C_2 m(t) \tag{7.56}$$

where C_1 and C_2 are the undetermined constants that depend on the initial conditions and critical displacement. The droplet mass m varies with time.

In the case of constant current, the spring constant varies slightly with time. It can be considered a constant provided that relevant parameters are optimized. Of course, through adjusting the spring constant k during calculation, the calculation results will be closer to the experimental ones.

7.5.3.4 Determination of Damping Coefficient

It has been found that the damping coefficient b has a minor effect on the metal transfer compared with other factors, such as surface tension and gravity.[92] The damping coefficient b can be expressed as

$$b = 3\mu \frac{V_d}{x^2} \qquad (7.57)$$

where μ is the dynamic viscosity, V_d is the droplet volume, and x is the droplet displacement.

Because the damping coefficient has a very low order of magnitude, it can take an appropriate assumed value based on the properties of the molten droplet during practical calculation. Here, referring to the literature,[92,94] take a constant value for the damping coefficient.

7.5.3.5 Determining Critical Displacement

In the model developed, the critical displacement and detaching size of the droplet are calculated separately, and the time dependence of the critical displacement is also analyzed. The critical displacement is calculated using SFBT. When the critical displacement is reached, the droplet comes into a detachment stage. The droplet radius at this moment is the critical radius of the droplet. When the droplet mass is suddenly reduced, the droplet is detached to transfer into the weld pool. In SFBT, the critical radius is considered as the detaching radius. In fact, when the detaching force balances the retaining force, the droplet just comes into a detachment stage; a pinching neck forms, but it is not really detached. In SFBT, effects of both the oscillation velocity and the pinching neck on the metal transfer are not considered.

7.5.4 Calculation Method and Selected Physical Properties

For simulating the dynamic process of metal transfer, the forces acting on the droplet must be calculated first, and the initial condition (including initial displacement and velocity) and critical instability condition need to be determined. During calculation, the droplet mass increases linearly with time, while the spring constant is dependent on the detaching mass of the droplet. The droplet will eventually reach a dynamic balance under the action of various forces and oscillation velocity. The droplet displacement increases with increasing droplet mass. When the droplet reaches the critical size, the forces on the droplet reach a dynamic balance, and the droplet mass is suddenly reduced, representing a detaching droplet.

The spring constant is the most important parameter for calculating the droplet displacement versus time relation and detaching droplet size using a fourth-order Runge-Kutta procedure. When the model is calibrated, the spring constant is adjusted and optimized by comparing the calculated results with the experimental ones. Figure 7.15 shows the flowchart of the computation program developed.

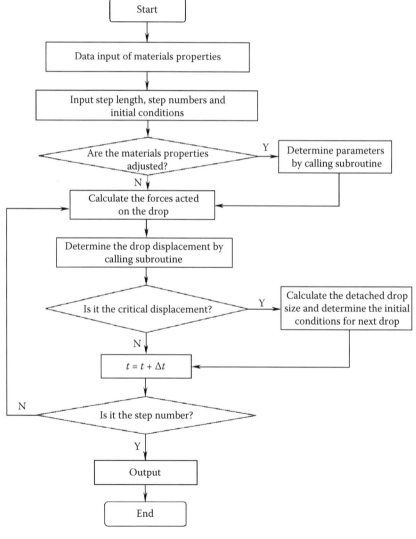

FIGURE 7.15
Flowchart of the developed program.

TABLE 7.3

Physical Properties Used in Calculation

Symbol	Value	Properties
μ_m	1.256×10^{-6} (H/m)	Magnetic permeability
C_d	0.44	Plasma drag coefficient
ρ_f	0.06 (kg/m³)	Density of the plasma
ρ	7,800 (kg/m³)	Density of electrode
g	9.807 (m/s²)	Gravity acceleration
v_f	100 (m/s)	Velocity of the plasma
θ_d	150(°)	Half-angle projection
γ	1.22 (N/m)	Surface tension coefficient
b	0.0028 (N·s/m)	Damping coefficient

The values of physical properties directly affect the accuracy of the calculation results. Table 7.3 gives the physical properties used in calculation of droplet size and displacement.

Using the model developed, numerical simulations were carried out for a mild steel electrode with a diameter of 1.6 mm. The shielding gas was 95% argon plus 5% CO_2. The wire extension was 26 mm. The welding current range was from 180 to 320 A.

7.5.5 Droplet Oscillation and Detachment at Different Current Levels

Using the mass-spring model developed, the oscillation and detachment were numerically simulated. When the dummy spring reaches the critical length, that is, the displacement of the droplet reaches the critical value, the droplet mass is reduced suddenly, and the droplet is detached and transferred to the weld pool at a certain speed. The retained droplet oscillates at a certain speed at the end of the electrode, and the dummy spring extends gradually until another droplet is detached.

Figure 7.16 shows the predicted droplet displacement as a function of time under different levels of welding current. The vertical line segment in the curves represents the droplet detachment. At a current level of 200 A, the critical displacement is large, and the transfer period is long [see Figure 7.16(a)]. This means that globular transfer occurs at this level of current (200 A). It is also noticed that the transfer period of each droplet is almost the same. After the detachment of the first droplet, the initial oscillating amplitude of the retained droplet is larger because the detachment results in a sudden change of the forces acting on the retained droplet, and the retaining force is much higher than the detaching forces. Figure 7.16(b) shows the droplet oscillation at a current level of 260 A. As compared with that in Figure 7.16(a), the oscillation amplitude and the critical displacement are much lower. The time interval of the transfer varies randomly. Following a globular transfer, one or more little droplets are detached as project transfer. The transfer is

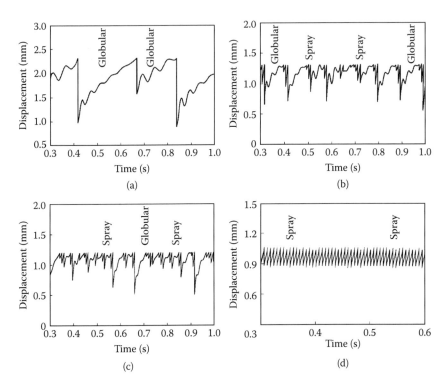

FIGURE 7.16
Droplet oscillation and detachment: (a) $I = 200$ A; (b) $I = 260$ A; (c) $I = 270$ A; (d) $I = 320$ A.

not regular, indicating a mixed transfer of globules and spray. Figure 7.16(c) shows the droplet oscillation at a current level of 270 A. It can be seen that the spray transfer predominates, with globular transfer occurring occasionally at this current level. As compared with Figures 7.16(a) and 7.16(b), the frequency of droplet transfer is remarkably higher. Figure 7.16(d) shows the droplet oscillation at a current value of 320 A. The growth and detachment of droplets are very stable. The transfer period is much shorter. The oscillation amplitude is very low, indicating the spray transfer mode at such higher current.

It can be seen from Figure 7.16 that under the condition of constant welding current, the elastic displacement of the droplet increases with time until it reaches the critical displacement. The detachment of a droplet causes the oscillation of the retained droplet at the tip of the electrode and affects the size and transfer period of the subsequent detached droplet. During the growth phase, the oscillation amplitude decreases with time due to the damping force. For controlling metal transfer, suitable use of momentum resulting from the downward oscillation speed can reduce the critical current for spray transfer. The initial displacement and velocity of the retained droplet is dependent on the oscillating velocity and size of the detached droplet. It is noticed that the initial displacements of individual droplets are different. This indicates that the influence of oscillation on metal transfer is important. It can also be seen that the droplet transfer frequency increases and the critical displacement decreases with increasing welding current. The calculations revealed that there exists a mixed-mode transfer between globular transfer and spray transfer. When welding current value is close to the transition current, the transfer period becomes irregular, globular and spray transfer occur alternately, and one globule transfer is accompanied with one or a few projection transfers. With the welding current increasing in the transition zone, the amount of globular detachment decreases, and the amount of spray detachment increases gradually until complete spray transfer appears.

It can be concluded that there are several factors affecting the metal transfer mode, of which the welding current is the most important one. With the other conditions constant, the transfer mode varies with the welding current. At low welding current, the droplet is detached in a globular mode. At higher current beyond the transition current, the droplet is detached in the spray mode. In the transition zone, the spray and globular transfer coexist.

7.5.6 Dynamic Process of Metal Transfer

Most previous models of metal transfer do not consider the oscillation velocity of the droplet in predicting the droplet size. But, the droplet detachment is due not only to the action of various forces but also to the action of oscillation velocity. In this section, the effect of oscillation velocity on the detaching size of the droplet is discussed.

Figure 7.17 shows the variations of the droplet mass center displacement, oscillation velocity, and relevant electromagnetic force with time at a current level of 220 A. The spring constant takes the value of $k = 2.2$ N/m through parameter optimization. It can be seen that the time interval of droplet growth and detachment is relatively long, which indicates that it is globular transfer. During the globular transfer process, the oscillation velocity decreases gradually due to the damping force. At the onset of detachment, the oscillation velocity is low; thus, the detachment momentum is also low. At a current level of 220 A, the electromagnetic force is up to 3.06×10^{-3} N when the displacement reaches the critical value [see Figure 7.17(c)]. Therefore, the droplet detachment is mainly ascribed to the gravity and electromagnetic forces, and the oscillation velocity plays little role in it. In this situation, SFBT is applicable.

According to Eq. (7.29), the electromagnetic force is a function of droplet radius and welding current if other conditions are kept constant. For a certain welding current, this force is dependent on the droplet radius. It can be seen from Figure 7.17(c) that the electromagnetic force decreases sharply after it reaches its peak value. The reason is that the initial size of the retained droplet is small after droplet detachment, so that a smaller droplet

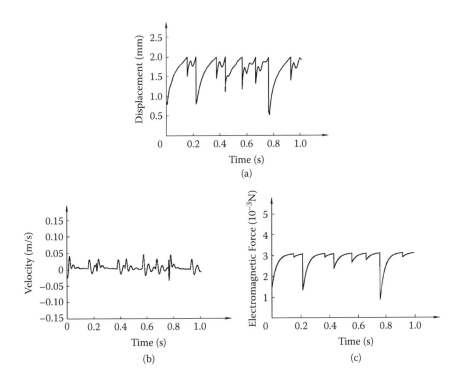

FIGURE 7.17
Displacement, oscillation velocity, and electromagnetic force versus time (220-A current). (From Wu C S, Chen M A, and Li S K, *Chinese J. Mech. Eng.*, 42(2): 76–81, 2006.)

radius corresponds to lower electromagnetic force. The electromagnetic force exerted on each droplet at the initial stage is not regular.

Figure 7.18 shows the droplet displacement, velocity, and electromagnetic force at a current level of 265 A. The spring constant increases with decreasing detached droplet mass. In this case, it takes a value of 4.2 N/m. It can be seen from Figure 7.18 that the time interval of droplet detachment is irregular. Some droplets have a longer detachment time, while others have a shorter detachment time. Figure 7.18(b) shows that when a droplet oscillates for a longer time, it has a lower detaching velocity, and globular transfer occurs easily. If a droplet oscillates for a shorter time, it has a higher detaching velocity, and spray transfer takes place. It can be seen from Figure 7.18(c) that the electromagnetic force is also irregular during droplet growth and detachment due to the irregularity of the retained droplet.

Figure 7.19 shows the droplet displacement, velocity, and electromagnetic force with a current value of 280 A. The spring constant takes the value of 4.6 N/m. Figure 7.19(a) demonstrates that the frequency of droplet transfer is much higher than that in Figure 7.18(a), and the droplet is detached and transferred in a very short time. The droplet oscillation is relatively stable, and the oscillation velocity changes little during droplet growth [see Figure 7.19(b)].

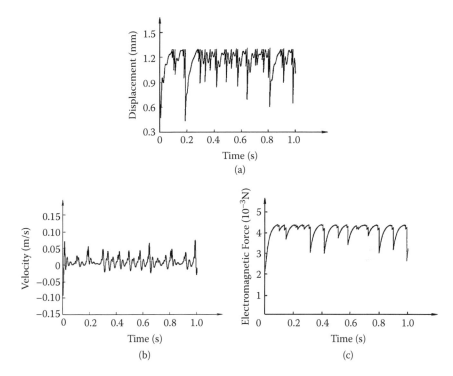

FIGURE 7.18
Displacement, oscillation velocity, and electromagnetic force versus time (265-A current). (From Wu C S, Chen M A, and Li S K, *Chinese J. Mech. Eng.*, 42(2): 76–81, 2006.)

The electromagnetic force is 4.56×10^{-3} N at the instant when the critical displacement is reached. It is the predominant detaching force, resulting in spray transfer.

Figure 7.20 shows the electromagnetic force versus the welding current at the detachment instant. As welding current increases, the electromagnetic force exerted on the droplet increases. When the welding current is up to 280 A, the electromagnetic force no longer changes with the current, which indicates stable spray transfer.

Figure 7.21 shows the relation between the welding current and the droplet gravity at the detachment instant. As the welding current increases, the gravity force acting on the droplet decreases due to the decreased size of the droplet. When the welding current increases to 280 A, the gravity force does not change with the current any more, which indicates the mode of spray transfer.

Figure 7.22 shows the plasma drag versus the welding current at the detachment instant. The plasma velocity is assumed to be 100 m/s.[78] The plasma drag force rises with increasing welding current. It can be seen that the plasma drag contributes little to the metal transfer.

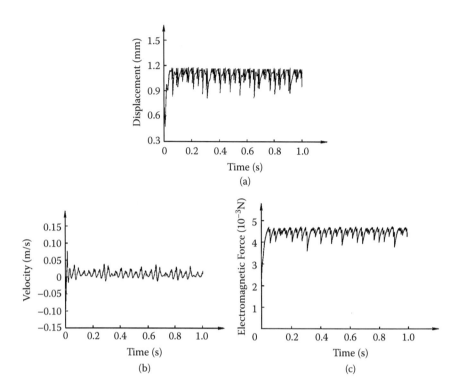

FIGURE 7.19
Displacement, oscillation velocity, and electromagnetic force versus time (280-A current). (From Wu C S, Chen M A, and Li S K, *Chinese J. Mech. Eng.*, 42(2): 76–81, 2006.)

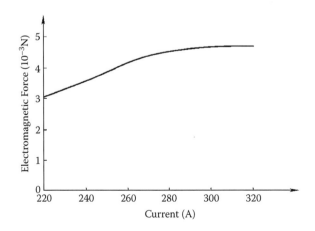

FIGURE 7.20
Electromagnetic force as a function of welding current.

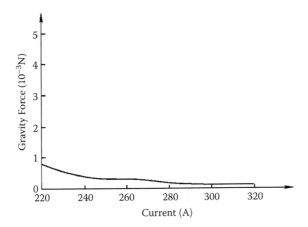

FIGURE 7.21
Gravity force as a function of welding current.

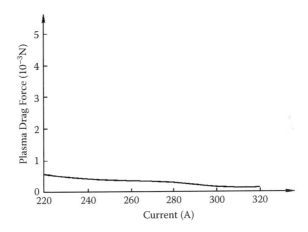

FIGURE 7.22
Plasma drag as a function of welding current.

It can be concluded that the electromagnetic force varies continuously during the growth and detachment period of the droplet even if the welding current is constant. As the welding current increases, the electromagnetic force on a droplet at the detaching instant increases gradually. The electromagnetic force is the key factor determining the transfer mode. It is the variation of electromagnetic force that determines the relationship between the welding current and the transfer mode. For cases of low welding current, the gravity force cannot be neglected. For high welding current conditions, the droplet is very small; thus, the electromagnetic force is the predominant detaching force.

7.5.7 Prediction of Droplet Size

Reaching critical displacement does not mean that the droplet has been detached. During calculation, the critical displacement is determined using SFBT. Only when the droplet mass is reduced suddenly can the droplet be considered detached. Figure 7.23 shows that the critical displacement decreases with increasing welding current. When the welding current is above 280 A, the critical displacement is nearly constant; when lower, the droplet is detached mainly by electromagnetic force, and it is the spray transfer mode. When the welding current is below 240 A, the droplet is detached in a globular mode, the droplet is prolonged by the electromagnetic and gravity forces, and the critical displacement is relatively high. With the welding current ranging from 240 to 280 A, the droplet is detached in a mixed mode.

Using the developed model that includes the oscillation velocity, the detaching condition was analyzed. The droplet size and frequency of droplet transfer were calculated.

Figure 7.24 shows the predicted droplet size and transfer frequency. With increasing welding current, the droplet size decreases, and the frequency of droplet transfer increases. With a current level of 200 or 220 A, the droplet size is large, but it is within a narrow range. For a case of 200 A, the average droplet radius is 1.52 mm, much greater than the electrode radius of 0.8 mm, so it is globular transfer. At a current value of 240 A, the average droplet radius is 1.1 mm, still higher than that of the electrode. The droplet radius falls in a wide range. At a current level of 260 A, the average droplet radius is 0.9 mm, close to that of the electrode. Most of the droplets have radius close to that of the electrode, but there are also some big droplets. This means that the transfer at this current is a mixed one, with spray transfer as the predominant mode. When the current is up to 280 A, the droplet radius is 0.75 mm, slightly less than that of the electrode, so it is spray transfer.

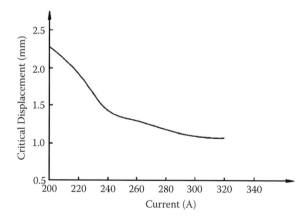

FIGURE 7.23
Critical displacement as a function of welding current.

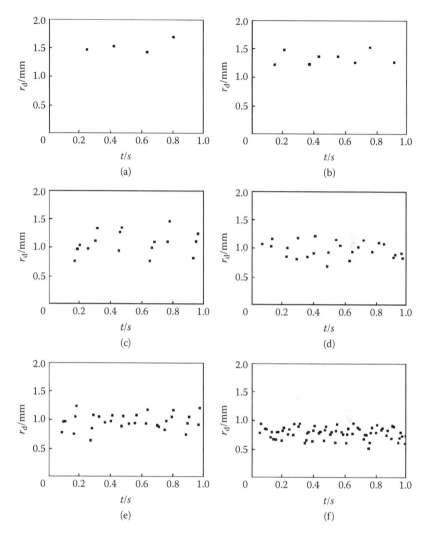

FIGURE 7.24

Calculated droplet radius and transfer frequency (from Wu C S, Chen M A, and Li S K, *Chinese J. Mech. Eng.*, 42(2): 76–81, 2006): (a) *I* = 200 A; (b) *I* = 220 A; (c) *I* = 240 A; (d) *I* = 250 A; (e) *I* = 260 A; (f) *I* = 280 A.

It can be concluded that the mass-spring model developed here can be used to predict not only the averaged droplet radius but also the detached droplet radius and transfer frequency at different instants.

The critical radius and detached radius were calculated using the model (see Figure 7.25). It can be seen that the critical radius of the droplet is greater than the detached radius of a droplet at a certain current level. This is because the oscillation velocity and momentum are considered in computing

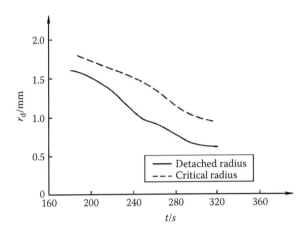

FIGURE 7.25
Comparison between critical radius and detached radius.

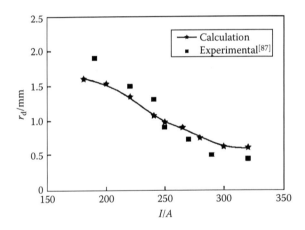

FIGURE 7.26
Comparison between the predicted and measured droplet radii. (From Wu C S, Chen M A, and Li S K, *Chinese J. Mech. Eng.*, 42(2): 76–81, 2006.)

detached droplet size. When the droplet reaches the critical size, a pinching neck forms gradually. There is a decreasing process of droplet mass during the extreme short time from droplet necking to its complete detachment.

7.5.8 Comparison between the Calculated and Experimental Results

Figure 7.26 compares the calculated detached droplet radii and the experimental ones. The experimental data are from the literature.[89] It can be seen that the droplet radius decreases with increasing welding current. The predicted values are in good agreement with the experimental ones, with the

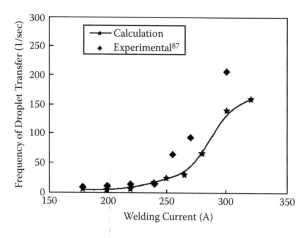

FIGURE 7.27
Comparison between the calculated and measured frequency of droplet transfer. (From Wu C S, Chen M A, and Li S K, *Chinese J. Mech. Eng.*, 42(2): 76–81, 2006.)

highest calculation accuracy achieved at about 250 A. When the welding current was greater than 300 A, the droplet was detached in spray mode, and the droplet size changed little with the welding current. However, the predicted size in this current range was greater than the experimental ones. The possible reason is that the tip of the electrode takes a shape like a pencil tip when the welding current is greater than 300 A, and finer droplets are detached in spray mode.

Figure 7.27 shows the comparison between the predicted frequency of droplet transfer and measured ones. The measured data are cited from the literature.[87] It can be seen that the predicted values are in good agreement with the experimental ones. But, the predicted transfer frequency was lower than the experimental one when the welding current was above the transition current.

8

Numerical Simulation of Weld Pool Behaviors in MIG/MAG Welding

8.1 Introduction

Metal inert gas (MIG) welding and metal active gas (MAG) welding are widely used in automatic and robotic welding of various kinds of metallic materials.[146] Determining the relationships between the welding process parameters and the weld quality not only is a prerequisite for realizing automatic production and intelligent control but also is a significant basis for generating welding process procedures in MIG/MAG welding. To this end, it is essential to conduct numerical simulation of fluid flow and heat transfer in MIG/MAG weld pools.

For modeling the MIG/MAG welding process, an appropriate and realistic description of the heat input deposited on the baseplate must be made first. The heat input in MIG/MAG welding consists of two parts: one is the arc heat flux acting on the workpiece surface; another is the enthalpy delivered into the weld pool by transferred droplets.[147] Usually, the Gaussian distribution mode was employed to depict the heat flux distribution on the weld pool surface in almost all previous studies of MIG/MAG weld pool behaviors. Because there is a larger surface deformation of the MIG/MAG weld pool, the arc behaviors will be affected by the deformed weld pool surface.[22] Thus, the distribution of arc heat flux on the severely deformed weld pool surface is obviously altered, and the Gaussian distribution mode is no longer suitable. The distribution of droplet enthalpy inside the weld pool is another important problem, which is usually treated with some unreasonable assumptions,[59,60,62,63] In this chapter, the heat input distribution mode in MIG/MAG welding is established based on the basic principles of arc physics; this is then applied to the numerical simulation of weld pool behaviors.

8.2 Model of MIG/MAG Weld Pool Behaviors

Figure 8.1 is a schematic of the MIG/MAG welding process. The small-diameter electrode wire is fed continuously into the arc from a coil. When the arc moves along the x-direction at a speed of v_0, the filler wire is melted at a velocity S_m (wire feed rate or wire melting rate) and transferred into the weld pool as droplets. The o-xyz coordinate system, which has an origin at the intersection point of the electrode centerline and the bottom surface of the workpiece, moves with the same speed of arc. As droplets are continuously transferred from the wire tip to the weld pool, they also bring energy, momentum, and mass into the weld pool. This extra heat and the arc heat constitute the total heat input deposited on the workpiece and determine the temperature distribution in the weld pool and its neighboring heat-affected zone. On the other hand, transferring droplets exert an impact (momentum) on the pool surface. Thus, severe surface deformation of the weld pool occurs under the combined effects of both arc pressure and droplet impingement in MIG/MAG welding.

MIG/MAG welding has the following features: (1) Wire melting and metal transfer cause the increasing molten metal volume in the weld pool, and weld reinforcement forms after solidification; (2) droplets deliver extra heat content into the weld pool; (3) droplets exert an impact on the weld pool surface; and (4) a severe surface deformation of the weld pool exists. The numerical analysis model of weld pool behaviors in MIG/MAG welding must account for these features.

8.2.1 Governing Equations in Cartesian Coordinate System

Suppose the liquid metal in the weld pool is incompressible fluid with a laminar flow mode. In the moving coordinate system o-xyz, the energy conservation

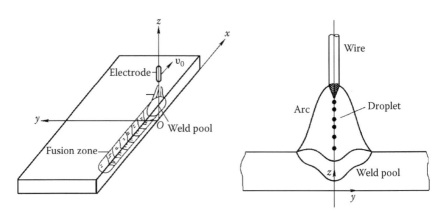

FIGURE 8.1
Schematic of MIG/MAG welding process.

equation, momentum conservation equation, and continuity equation that describe weld pool behaviors take the following unified forms:

$$AE_x + AF_y + AG_z = AS \tag{8.1}$$

where

$$AE = \begin{Bmatrix} \rho u \\ \rho(u-v_0)u - \mu u_x \\ \rho(u-v_0)v - \mu v_x \\ \rho(u-v_0)w - \mu w_x \\ \rho C_p(u-v_0)T - \lambda T_x \end{Bmatrix}, AF = \begin{Bmatrix} \rho v \\ \rho vu - \mu u_y \\ \rho vv - \mu v_y \\ \rho vw - \mu w_y \\ \rho C_p vT - \lambda T_y \end{Bmatrix} \tag{8.2}$$

$$AG = \begin{Bmatrix} \rho w \\ \rho wu - \mu u_z \\ \rho wv - \mu v_z \\ \rho ww - \mu w_z \\ \rho C_p wT - \lambda T_z \end{Bmatrix}, AS = \begin{Bmatrix} 0 \\ -P_x + F_x \\ -P_y + F_y \\ -P_z + F_z \\ Q_{vd} \end{Bmatrix} \tag{8.3}$$

where T is temperature; u, v, and w are the velocity component of fluid flow in the x-, y-, and z-direction, respectively; P is pressure; F_x, F_y, and F_z are body force (see Section 8.4); ρ is the density of metal; μ is the dynamic viscosity coefficient; C_p is the specific heat capacity; λ is thermal conductivity; and Q_{vd} is the internal heat source term in the energy equation (which reflects the droplet enthalpy transferred into the weld pool in welding, with its magnitude and distribution defined in Section 8.6). The symbols with subscripts x, y, and z denote the first-order partial derivatives of the relevant variables with respect to x, y, and z, respectively.

During the welding process, phase transition, including melting and solidification, occurs in a temperature range, which means two phases coexist. Here, an effective heat capacity method[54,148,149] is used in the treatment of the two-phase region. The latent heat is treated as a very large heat capacity in the temperature range of phase transformation. Thus, the governing equation and its boundary conditions in different regions can be converted into a single nonlinear heat conduction equation in the whole region. When the temperature field is calculated, the location of the two-phase interface can be determined. The effective heat capacity method uses temperature as the undetermined function rather than the enthalpy, so the calculation speed can be improved greatly. However, the distribution of specific heat capacity and thermal conductivity must be constructed separately in the temperature range of the phase transformation.

Let T_L and T_s denote the temperature of the liquidus and solidus of the base metal, respectively, and assume that the thermal conductivity and liquid fraction change linearly with temperature in the two-phase region when $T_s \leq T \leq T_L$. Let

$$T_m = (T_s + T_L) / 2 \qquad (8.4)$$

$$\Delta T = (T_L - T_s) / 2 \qquad (8.5)$$

Then, the specific heat capacity C_p and thermal conductivity λ in the whole domain, including liquid, mushy (liquid-solid), and solid regions, take the forms:

$$C_p(T) = \begin{cases} c_s & T < T_s \\ \dfrac{Q_m}{T_L - T_s} + \dfrac{c_s + c_L}{2} & T_s \leq T \leq T_L \\ c_L & T > T_L \end{cases} \qquad (8.6)$$

$$\lambda(T) = \begin{cases} \lambda_s & T < T_s \\ \lambda_s + \dfrac{\lambda_L - \lambda_s}{T_L - T_s}(T - T_s) & T_s \leq T \leq T_L \\ \lambda_L & T > T_L \end{cases} \qquad (8.7)$$

where Q_m is the latent heat of fusion (kJ/kg); c_s and c_L are specific heat capacity of the solid phase and liquid phase, respectively; λ_s and λ_L are the thermal conductivity of the solid phase and liquid phase, respectively.

8.2.2 Surface Deformation of MIG/MAG Weld Pool and Weld Reinforcement

The droplet impact on the weld pool surface in MIG/MAG welding produces a concave weld pool surface. When the welding current is lower, metal transfer is in the globular transfer mode; the cavity due to the prior droplet impact may disappear because of the backflow of liquid metal before the impact of the next droplet. When the welding current increases and spay transfer occurs, several hundred droplet impacts per second continuously act on the weld pool surface, so there is not enough time for backflowing of liquid metal. The weld pool surface is depressed severely. Therefore, an approximately stable surface deformation occurs in the weld pool, as shown in Figure 8.2.

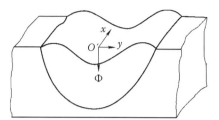

FIGURE 8.2
Schematic of surface deformation and the Φ-*xyz* coordinate system.

The surface deformation of the MIG/MAG weld pool is related to various forces, such as arc pressure, droplet impact, gravity and surface tension of liquid metal, and fluid flow in the weld pool. Studies indicated that the influence of fluid flow on weld pool shape is rather small compared with that of arc pressure, so it could be neglected.[23] As shown in Figure 8.2, the shape function φ(*x*, *y*) of the weld pool surface in the *o-xyz* coordinate system satisfies the dynamic balance condition in the gravity field under the interaction of both droplet impact and arc pressure. The system energy in a functional form takes the following form:

$$E_t = \iint_\Omega \left(\gamma\sqrt{1+\Phi_x^2+\Phi_y^2} + \frac{1}{2}\rho g\Phi^2 - P_a\Phi - P_d\Phi \right) dxdy = \iint_s Fdxdy \quad (8.8)$$

where Ω is the enclosure region of the weld pool boundary at the top surface, P_a is the arc pressure, P_d is droplet impact, and ρ and γ are the density and surface tension of liquid metal, respectively. Subscripts x and y denote the first-order partial derivatives of variable Φ with respect to x and y, respectively. The first term in brackets denotes the corresponding energy caused by surface change of the weld pool, the second term is the potential energy of liquid metal in the gravity field, and the third and fourth terms are the work that arc pressure and droplet impact do when they act on the weld pool surface, respectively.

Equation (8.8) must satisfy this constraint condition that the volume addition of the weld pool is equal to the volume of filler metal ΔV. Thus,

$$\iint_s \Phi dxdy = \iint_s Gdxdy = \Delta V \quad (8.9)$$

Using the Euler-Lagrangian criterion[150] in the calculus of variations, we get

$$\frac{\partial}{\partial \Phi}(F+C_3G) - \frac{\partial}{\partial x}\left[\frac{\partial}{\partial \Phi_x}(F+C_3G)\right] - \frac{\partial}{\partial y}\left[\frac{\partial}{\partial \Phi_y}(F+C_3G)\right] = 0 \quad (8.10)$$

After a series of derivations, it takes the following form:

$$P_a - \rho g \Phi + P_d + C_3 = -\gamma \frac{\left(1+\Phi_y^2\right)\Phi_{xx} - 2\Phi_x\Phi_y\Phi_{xy} + \left(1+\Phi_x^2\right)\Phi_{yy}}{\left(1+\Phi_x^2+\Phi_y^2\right)^{3/2}} \qquad (8.11)$$

where C_3 is a constant that can be calculated by the constraint condition Eq. (8.9). The droplet impact on the weld pool surface P_d is expressed as follows[63]:

$$P_d = \frac{4 v m_D v_D}{\pi d_D^2} \qquad (8.12)$$

where droplet parameters such as transfer frequency v, droplet mass m_D, velocity v_D, and diameter d_D are obtained from experiments.[147,151]

Because of the filler metal added into the weld pool, weld reinforcement forms after solidification of the weld pool. The volume of filler metal is equal to the product of the cross-sectional area of the weld reinforcement and length of the weld bead. The cross-sectional area of reinforcement is affected by the wire feeding speed S_m, wire diameter d_w, and welding speed v_0. Assuming the metal mass conservation, the cross-sectional area of reinforcement may be written as follows:

$$A_R = \frac{S_m \pi d_w^2}{4 v_0} \qquad (8.13)$$

The geometry shape of the weld reinforcement is described by the shape function $\Phi(x, y)$ determined in the surface deformation equation, Eq. (8.11).

8.2.3 Governing Equations in a Body-Fitted Coordinate System

Under the interaction of arc pressure, droplet impact, gravity of the liquid metal, and surface tension, severe surface deformation is generated in the MIG/MAG weld pool. On the other hand, the weld reinforcement is formed behind the weld pool. For such curved surfaces, the body-fitted coordinate system is quite applicable[152]; this system is classified as an orthogonal system and a nonorthogonal system. In the orthogonal body-fitted coordinate, no mixed partial derivatives arise, but the coordinate system requires redefinition after every iteration step. This means that variables on all grid nodes need to be reset before every iteration step, which will slow the speed of convergence, lengthen the calculating time, and require large computer memory. So, the nonorthogonal body-fitted coordinate system is employed here, and the coordinate transformation takes the following forms:

$$x = x, \, y = y, \, z^* = \frac{z}{H - \Phi(x,y)} \tag{8.14}$$

where H is the thickness of the workpiece. Obviously, on the top surface of the workpiece, $z = H - \Phi(x, y)$, so $z^* = 1$; on the bottom surface, $z = 0$, so $z^* = 0$. In this way, the coordinate transformation from Cartesian coordinates x-y-z to the body-fitted coordinates x-y-z^* is realized.

If this body-fitted coordinate system is employed in the whole workpiece domain, a staggered grid in the SIMPLER algorithm may make the discretization of governing equations extremely complex, with more than 100 items of mixed partial derivatives, which slows the solution speed. Therefore, the nonorthogonal body-fitted coordinate system is only applied on the workpiece surface to deal with the curved surface of both the weld pool and the weld reinforcement, while the Cartesian coordinate system is still used inside the workpiece to improve calculation speed and simplify the problem. At the juncture of two coordinate systems, coordinate transformation is carried out to convert the values in one coordinate system into those in another.[44] Solution practice has proved that this is an economic, simple, high-speed, practical method.

In the nonorthogonal body-fitted coordinate system defined by Eq. (8.14), the energy conservation equation, momentum conservation equation, and continuity equation that describe weld pool behaviors can be written as follows:

$$AR_x + AB_y + AQ_{z^*} = AD \tag{8.15}$$

where,

$$AR = \begin{cases} \rho u \\ \rho(u - v_0)u - \mu u_x \\ \rho(u - v_0)v - \mu v_x \\ \rho(u - v_0)w - \mu w_x \\ \rho C_p(u - v_0)T - \lambda T_x \end{cases}, \, AB = \begin{cases} \rho v \\ \rho v u - \mu u_y \\ \rho v v - \mu v_y \\ \rho v w - \mu w_y \\ \rho C_p v T - \lambda T_y \end{cases} \tag{8.16}$$

$$AQ = \left\{ \begin{array}{l} \rho w \dfrac{\partial z^*}{\partial z} \\[2mm] \rho w_1 u - \mu S u_{z^*} \\[2mm] \rho w_1 v - \mu S v_{z^*} \\[2mm] \rho w_1 w - \mu S w_{z^*} \\[2mm] \rho C_p w_t T - S \lambda T_{z^*} \end{array} \right\}, \quad AD = \left\{ \begin{array}{l} -\rho c_c \\[2mm] F_x - P_x - P_{z^*} \dfrac{\partial z^*}{\partial x} + \mu c_u \\[2mm] F_y - P_y - P_{z^*} \dfrac{\partial z^*}{\partial y} + \mu c_v \\[2mm] F_z - P_{z^*} \dfrac{\partial z^*}{\partial z} + \mu c_w \\[2mm] Q_{vd} + c_t \lambda \end{array} \right\} \tag{8.17}$$

where

$$c_t = 2\left(\frac{\partial z^*}{\partial x} \frac{\partial^2 T}{\partial x \partial z^*} + \frac{\partial z^*}{\partial y} \frac{\partial^2 T}{\partial y \partial z^*} \right)$$

$$c_u = 2\left(\frac{\partial z^*}{\partial x} \frac{\partial^2 u}{\partial x \partial z^*} + \frac{\partial z^*}{\partial y} \frac{\partial^2 u}{\partial y \partial z^*} \right)$$

$$c_v = 2\left(\frac{\partial z^*}{\partial x} \frac{\partial^2 v}{\partial x \partial z^*} + \frac{\partial z^*}{\partial y} \frac{\partial^2 v}{\partial y \partial z^*} \right)$$

$$c_w = 2\left(\frac{\partial z^*}{\partial x} \frac{\partial^2 w}{\partial x \partial z^*} + \frac{\partial z^*}{\partial y} \frac{\partial^2 w}{\partial y \partial z^*} \right)$$

$$c_c = \frac{\partial u}{\partial z^*} \frac{\partial z^*}{\partial x} + \frac{\partial v}{\partial z^*} \frac{\partial z^*}{\partial y}$$

$$w_t = (u - v_0)\frac{\partial z^*}{\partial x} + v\frac{\partial z^*}{\partial y} + w\frac{\partial z^*}{\partial z} - \frac{\lambda}{\rho C_p}\left(\frac{\partial^2 z^*}{\partial x^2} + \frac{\partial^2 z^*}{\partial y^2} \right)$$

$$w_1 = (u - v_0)\frac{\partial z^*}{\partial x} + v\frac{\partial z^*}{\partial y} + w\frac{\partial z^*}{\partial z} - \frac{\mu}{\rho}\left(\frac{\partial^2 z^*}{\partial x^2} + \frac{\partial^2 z^*}{\partial y^2} \right)$$

$$S = \left(\frac{\partial z^*}{\partial x} \right)^2 + \left(\frac{\partial z^*}{\partial y} \right)^2 + \left(\frac{\partial z^*}{\partial z} \right)^2$$

In Eqs. (8.15) and (8.17), a variable with subscript x, y, z indicates the first-order partial derivatives of this variable with respect to x, y, and z, respectively, and Q_{vd} is the inner heat source item in the energy conservation equation, which indicates the droplet enthalpy transferred into the weld pool, and its magnitude and distribution are defined in Section 8.6.

8.2.4 Boundary Conditions

For the energy equation,

$$-\lambda \nabla T \cdot \vec{n}_b = q_s \tag{8.18}$$

At the symmetric plane $y = 0$, $q_s = 0$. The calculation of q_s on the top surface of the workpiece is introduced in Section 8.5. On other surfaces, $q_s = \alpha\,(T - T_f)$, where the heat transfer coefficient α is

$$\alpha = 24.1 \times 10^{-4} \varepsilon T^{1.61} \qquad \left(\text{w}/\text{m}^2 \circ \text{C}\right)$$

where ε is the emissivity of the body surface; for steel, $\varepsilon = 0.90$.[153]
For the momentum equation,

On the top surface of the workpiece,

$$-\mu\left(\frac{\partial u}{\partial z^*}\frac{\partial z^*}{\partial z} \cdot \vec{n}_{bx} \right) = \frac{\partial \gamma}{\partial T}\left(\frac{\partial T}{\partial x} \cdot \vec{t}_{bx} \right) \tag{8.19}$$

$$-\mu\left(\frac{\partial v}{\partial z^*}\frac{\partial z^*}{\partial z} \cdot \vec{n}_{by} \right) = \frac{\partial \gamma}{\partial T}\left(\frac{\partial T}{\partial y} \cdot \vec{t}_{by} \right) \tag{8.20}$$

$$\left(w \cdot \vec{n}_b \right) = 0 \tag{8.21}$$

where $\vec{t}_{bx}$ and $\vec{t}_{by}$ are tangential unit vectors on the workpiece surface along the x- and y-direction, respectively; $\vec{n}_{bx}$ and $\vec{n}_{by}$ are projections of normal unit vectors on the workpiece surface in the xoz plane and the yoz plane, respectively.

At the symmetric plane,

$$y = 0, \quad \frac{\partial u}{\partial y} = 0, \quad \frac{\partial w}{\partial y} = 0, \quad v = 0 \tag{8.22}$$

In the mushy zone and a solid ($T \le T_L$),

$$u = 0, \quad v = w = 0 \tag{8.23}$$

The boundary condition of the surface deformation equation, Eq. (8.11), is as follows:

$$\Phi(x, y) = 0, \qquad T \le T_L \tag{8.24}$$

where T_L is the liquidus temperature.

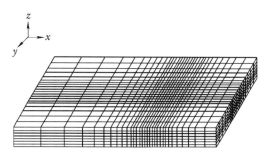

FIGURE 8.3
Schematic of grid system in the workpiece.

8.2.5 Discretization of Governing Equations

As shown in Figure 8.3, the nonuniform grid system is applied. Because there are relatively larger temperature gradients, larger surface deformation, and stronger fluid flow near the heat source, a finer grid is used around the heat source, while a coarser grid is used far away from the heat source. During the discretization of the momentum conservation equation and the continuity equation, the first-order derivatives of pressure and velocity in the momentum and continuity equations may cause considerable trouble in sawtooth-like distribution pressure and velocity fields.[154] Thus, a staggered grid system is utilized to overcome these troubles.

8.3 Distribution of Current Density on the Deformed MIG/MAG Weld Pool Surface

After the surface deformation of the MIG/MAG weld pool is calculated according to Eq. (8.11), the heat distribution mode of the arc on the deformed pool surface should be determined based on the surface deformation.

Since the 1980s, researchers have proposed quite a few numerical models to investigate heat transfer and fluid flow in tungsten inert gas (TIG) and MIG/MAG welding.[15–60,70–73] All these models used Gaussian function to describe the distribution mode of the welding current density, arc pressure, and heat flux density on the weld pool surface. For TIG welding, it is reasonable because there is relatively little surface deformation. However, in MIG/MAG welding severe surface deformation is generated in the weld pool by the greatly increased droplet impact and arc pressure. The pool surface beneath the arc is depressed to form a cavity, while there is a humping at the rear of the pool so that the pool surface is curved, which changes the arc behaviors significantly and leads to an alteration of distribution mode of both welding current density and heat flux density on the pool surface.[23] At the same time, because the arc pressure is the total pressure of arc plasma, the alteration of current density distribution will result in a change in arc pressure distribution. Then, the change of arc pressure distribution affects the pool surface deformation, which in turn affects the distribution of current density, arc pressure, and heat flux density. There is strong coupling among the pool surface deformation, current density distribution, arc pressure distribution, and heat flux density distribution in MIG/MAG welding. Therefore, when establishing the distribution mode of current density, arc pressure, and heat flux density, the pool surface deformation of the weld pool must be taken into consideration to predict the weld pool behaviors more accurately. In this section, the distribution mode of the welding current density will be established first based on the principles of arc physics, and then the distribution mode of arc pressure and heat flux density is determined.

The free arc is a gas conductor between two electrodes. Unlike a solid conductor with a fixed cross section, the gas conductor could enlarge and shrink its conductive cross section. But, when the current and other surrounding conditions (such as gas media, temperature, pressure, etc.) keep constant, a corresponding fixed cross section exists in every region of the arc, including the anode, cathode, and arc-column region, that satisfy the principle of minimum voltage. This principle means that the igniting arc will automatically choose a certain cross section, which is characterized by its diameter, to maintain a minimum electric field intensity in the arc or to keep a minimum voltage in a fixed length.[81,142] According to this principle, the current in the arc-column will flow into the weld pool through the shortest path. Figure 8.4 indicates the distribution of current density on section $y = 0$ of the workpiece.

In Figure 8.4, suppose L_1 and L_2 are the shortest distances from the wire tip to the weld pool surface, respectively. We have $O_1W = L_1$, $O_2W = L_2$. Point O' is the intersection between the wire centerline and the pool surface with

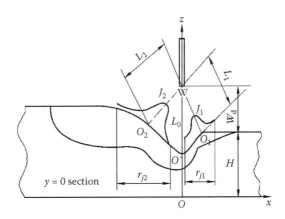

FIGURE 8.4
Schematic of current density distribution at the longitudinal cross section ($y = 0$).

maximum depression, and $O'W = L_0$. For a case of larger pool surface deformation, $L_0 > L_1$, $L_0 > L_2$. This makes more current flow into the weld pool
through the paths represented by L_1 and L_2, while there is less current at
locations far away from L_1 and L_2. Suppose that the local current densities
are Gaussian distributed. Taking O_1 and O_2 as the distribution origins on the
deformed weld pool surface, we obtain

At point O_1,

$$J(s) = J_{1m} \exp\left(-\frac{s^2}{2\sigma_{j1}^2}\right)$$

At point O_2,

$$J(s) = J_{2m} \exp\left(-\frac{s^2}{2\sigma_{j2}^2}\right)$$

where J_{1m} and J_{2m} are the maximum current density at points O_1 and O_2,
respectively; σ_{j1} and σ_{j2} are the corresponding current distribution parameters,
respectively; and s is the length of the curve starting from the origin, O_1 or O_2.

Tsai et al.[119] experimentally determined the effect of gas tungsten arc welding (GTAW) arc length on the current density distribution parameter. Their
test results showed that the distribution parameter σ_j is directly proportional
to the arc length L, that is, $\sigma_j/L = $ constant. Based on the experimental results,
we assume that there is the following formula:

$$\frac{\sigma_{j1}}{L_1} = \frac{\sigma_{j2}}{L_2} \tag{8.25}$$

Due to the fact that the greater is the distance L from the wire tip W to a
point on the weld pool surface, the less is the current density flowing through
this point, assume that J_m is inversely proportional to L,

$$J_{1m}L_1 = J_{2m}L_2 \tag{8.26}$$

The distance W_d from the wire tip W to the workpiece surface affects the
value of the current distribution parameter. A greater W_d causes a larger
deposition area of current on the weld pool surface and the corresponding
current distribution parameter. The current distribution area $r_{j1} + r_{j2}$ and W_d
meet the following relationship:

$$r_{j1} + r_{j2} = b(W_d) \tag{8.27}$$

where $b(W_d)$ is a function of W_d, which is selected based on the experimental and calculated results.

After solving the pool surface deformation equation, Eq. (8.11), points O_1 and O_2 can be determined according to the coordinates of each point on the weld pool surface and the coordinates $(0, 0, z_0)$ of the wire tip W. Similarly, we may determine points O_3 and O_4 on the transverse cross section $x = 0$. Thus, four special points are known:

At section $y = 0$,

$$O_1\left(x_1,0,H-\Phi\left(x_1,0\right)\right), \quad O_2\left(x_2,0,H-\Phi\left(x_2,0\right)\right)$$

At section $x = 0$,

$$O_3\left(0,y_1,H-\Phi\left(0,y_1\right)\right), \quad O_4\left(0,-y_1,H-\Phi\left(0,-y_1\right)\right)$$

The coordinates of point O', which is the lowest point on the deformed weld pool surface, are $\left(0,0,H-\Phi\left(0,0\right)\right)$.

As shown in Figure 8.4, the coordinates of O_2 and O' are $\left(x_2,0,H-\Phi\left(x_2,0\right)\right)$ and $\left(0,0,H-\Phi\left(0,0\right)\right)$, respectively. α is the angle between line $O'O_2$ and the x-axis. In the current density distribution curve, α' is the angle between connection lines of two points with a current density value of $0.05\,J_{1m}$ and the x-axis. Because α is closer to α', take $\alpha = \alpha'$. According to the geometric relationship shown in Figure 8.4, the following equation is available:

$$r_{j1} = \frac{2\sqrt{6}\sigma_{j1}x_1}{\sqrt{x_1^{2}+\left[\Phi\left(x_1,0\right)-\Phi\left(0,0\right)\right]^{2}}}$$

Similarly, we obtain

$$r_{j2} = \frac{2\sqrt{6}\sigma_{j2}\left|x_2\right|}{\sqrt{x_2^{2}+\left[\Phi\left(x_2,0\right)-\Phi\left(0,0\right)\right]^{2}}}$$

Substituting these two expressions into Eq. (8.27), we have

$$\frac{2\sqrt{6}\sigma_{j1}x_1}{\sqrt{x_1^{2}+\left[\Phi\left(x_1,0\right)-\Phi\left(0,0\right)\right]^{2}}} + \frac{2\sqrt{6}\sigma_{j2}\left|x_2\right|}{\sqrt{x_2^{2}+\left[\Phi\left(x_2,0\right)-\Phi\left(0,0\right)\right]^{2}}} = b(W_d) \qquad (8.28)$$

By simultaneous solution of Eq. (8.25) and Eq. (8.28), parameters σ_{j1} and σ_{j2} can be obtained.

To determine the overall current density distribution on the deformed weld pool surface, if points O_1, O_2, O_3, and O_4 are not in the same horizontal plane, take the average value of their z-coordinates as the z-coordinate value of a series of current density distribution origin points, namely,

$$z^* = H - \frac{\Phi(x_1,0) + \Phi(x_2,0) + \Phi(0,y_1) + \Phi(0,-y_1)}{4} \tag{8.29}$$

That is, on the deformed surface of the weld pool, link all points with value z^* of the z-coordinates, and a closed curve can be constituted. A series of the origins of local current density distribution are all located on this curve. Figure 8.5 is the projection of this closed curve in the plane of xoy, where the symbol "o" denotes the calculated points with the nearest distance from the weld pool surface to the wire tip W, and the solid line represents a double-elliptic curve approximating to most of the calculated points. The two axis lengths of the ellipse are x_1 and y_1 when $x \geq 0$ and x_2 and y_1 when $x \leq 0$. Thus, the calculated points fall mostly on the elliptic curve. For the sake of calculation convenience, the closed curve is taken as the double-elliptic curve for subsequent treatment.

As shown in Figure 8.6, the current density distribution is related to the distance L from the wire tip W to the weld pool surface. Any section through the z-axis (with the angle θ between this section and the x-axis, $\theta \neq 0$) and the elliptic curve has an intersection point. At this point, the current density distribution parameters $J_m(\theta)$ and $\sigma_j(\theta)$ can be estimated through linear interpolation by J_{1m}, J_{2m} and σ_{j1}, σ_{j2} (the distribution parameters of points O_1 and O_2 on the $y = 0$ section, respectively). That is,

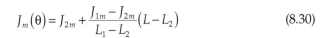

$$J_m(\theta) = J_{2m} + \frac{J_{1m} - J_{2m}}{L_1 - L_2}(L - L_2) \tag{8.30}$$

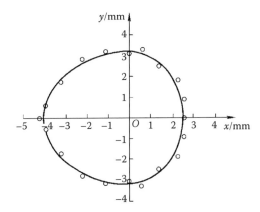

FIGURE 8.5
The closed curve composed of the origins of the current distribution.

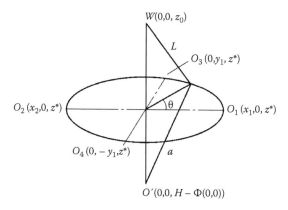

FIGURE 8.6
The geometric relationship between the origins of the welding current density distribution.

$$\sigma_j(\theta) = \sigma_{j1} + \frac{\sigma_{j2} - \sigma_{j1}}{L_2 - L_1}(L - L_1) \tag{8.31}$$

As the weld pool shape is symmetric with respect to $y = 0$, only the case of $y \geq 0$ ($\theta = 0$–180°) is discussed. According to the geometric relationship shown in Figure 8.6 and the double-elliptic curve equation formed by four points O_1, O_3, O_2, and O_4, we obtain

When $\theta = 0$–90°,

$$L(\theta) = \sqrt{x_1^2 \cos^2\theta + y_1^2 \sin^2\theta + (z_0 - z^*)^2} \tag{8.32}$$

$$a(\theta) = \sqrt{x_1^2 \cos^2\theta + y_1^2 \sin^2\theta + (z^* - H + \Phi(0,0))^2} \tag{8.33}$$

When $\theta = 90$–180°,

$$L(\theta) = \sqrt{x_2^2 \cos^2\theta + y_1^2 \sin^2\theta + (z_0 - z^*)^2} \tag{8.34}$$

$$a(\theta) = \sqrt{x_2^2 \cos^2\theta + y_1^2 \sin^2\theta + (z^* - H + \Phi(0,0))^2} \tag{8.35}$$

where $L(\theta)$ and $a(\theta)$ are, respectively, the distances from the wire tip W and the point O' with the maximum surface depression of weld pool to the origins of the current density distribution (all origins located in the closed double-elliptic curve).

Substituting Eqs. (8.32)–(8.35) into Eqs. (8.30) and (8.31), we have

When $\theta = 0$–$90°$

$$J_m(\theta) = J_{2m} + \frac{J_{1m} - J_{2m}}{L_1 - L_2}\left(\sqrt{x_1^2 \cos^2\theta + y_1^2 \sin^2\theta + (z_0 - z^*)^2} - L_2\right) \qquad (8.36)$$

$$\sigma_j(\theta) = \sigma_{j1} + \frac{\sigma_{j2} - \sigma_{j1}}{L_2 - L_1}\left(\sqrt{x_1^2 \cos^2\theta + y_1^2 \sin^2\theta + (z_0 - z^*)^2} - L_1\right) \qquad (8.37)$$

When $\theta = 90$–$180°$

$$J_m(\theta) = J_{2m} + \frac{J_{1m} - J_{2m}}{L_1 - L_2}\left(\sqrt{x_2^2 \cos^2\theta + y_1^2 \sin^2\theta + (z_0 - z^*)^2} - L_2\right) \qquad (8.38)$$

$$\sigma_j(\theta) = \sigma_{j1} + \frac{\sigma_{j2} - \sigma_{j1}}{L_2 - L_1}\left(\sqrt{x_2^2 \cos^2\theta + y_1^2 \sin^2\theta + (z_0 - z^*)^2} - L_1\right) \qquad (8.39)$$

On the deformed weld pool surface and at the closed curve composed by a series of origins of local current density distribution, the local current density distribution parameters $J_m(\theta)$ and $\sigma_j(\theta)$ of every point can be calculated by solving Eqs. (8.36)–(8.39). Therefore, the current density distribution at any point can be determined. As shown in Figure 8.7, O_s is the origin of the local current density distribution. By integrating the local current

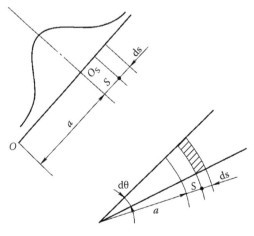

FIGURE 8.7
The integral area of the current density.

density throughout the whole distribution area, the following relationship is established:

$$\int_0^\pi \int_{-\infty}^\infty J_m(\theta)\exp\left(-\frac{s^2}{2\sigma_j^2(\theta)}\right)[s+a(\theta)]dsd\theta = \frac{I}{2} \tag{8.40}$$

Substituting Eqs. (8.33) and (8.35)–(8.38) into Eq. (8.40) and then making it be simultaneous with Eq. (8.26), a binary variable linear equation group about J_{1m} and J_{2m} can be obtained. With known J_{1m}, J_{2m} and σ_{j1}, σ_{j2}, $J_m(\theta)$ and $\sigma_j(\theta)$ for any angle θ in the closed curve of the local current density distribution on the weld pool surface can be calculated by solving Eqs. (8.36)–(8.39). Then, the current density value of any point on the deformed weld pool surface may be written as

$$J(\theta,s) = J_m(\theta)\exp\left(-\frac{s^2}{2\sigma_j^2(\theta)}\right) \tag{8.41}$$

where s is the radial distance from any point to the origin of the local current density distribution.

Equation (8.41) shows that the current density value of each point on the deformed weld pool surface is related to parameters θ and s. When the weld pool surface deformation is known, the current density distribution parameters $J_m(\theta)$ and $\sigma_j(\theta)$ at different angles θ are different. The relationships between the current density distribution parameters $J_m(\theta)$ and $\sigma_j(\theta)$ of each point and the weld pool deformation match with the fact: The shorter the distance is from a point on the weld pool surface to the wire tip, the greater is the current density flowing through this point. Once the weld pool deformation is determined, $J_m(\theta)$ and $\sigma_j(\theta)$ at different locations are subsequently determined. For estimating the current density at point $(x^*, y^*, H - \Phi\,(x^*, y^*))$, first calculate the weld pool surface deformation and find a point to which the point $(x^*, y^*, H - \Phi\,(x^*, y^*))$ corresponds at the closed double-elliptic curve. Then, the method is used to determine the parameters $J_m(\theta)$ and $\sigma_j(\theta)$. Next, the corresponding values of θ and s are discovered. Finally, the current density at the relevant point is calculated according to Eq. (8.41). θ and s can be determined by the following method:

First, the angle θ between the xoz plane and a plane passing through the point O' $(0, 0, H - \Phi\,(0, 0))$ and the point $(x^*, y^*, H - \Phi\,(x^*, y^*))$ as well as the z-axis is determined as

$$\theta = \arctan\frac{y^*}{x^*}$$

Then, find out the points at the closed double-elliptic curve in Figure 8.6 according to different values of θ, that is, the origins of the local current density distributions.

For $0° \le θ \le 90°$, the elliptic equation is defined as

$$\frac{x^2}{x_1^2} + \frac{y^2}{y_1^2} = 1$$

Since $x = r \cos θ$, $y = r \sin θ$,
then we get

$$\frac{r^2 \cos^2 θ}{x_1^2} + \frac{r^2 \sin^2 θ}{y_1^2} = 1$$

Its solution is as follows:

$$r = \sqrt{\frac{1}{\cos^2 θ / x_1^2 + \sin^2 θ / y_1^2}}$$

The origins at the double-elliptic curve of the local current density distribution have the following coordinates:

$$x_0 = \cos θ \cdot \sqrt{\frac{1}{\cos^2 θ / x_1^2 + \sin^2 θ / y_1^2}},$$

$$y_0 = \sin θ \cdot \sqrt{\frac{1}{\cos^2 θ / x_1^2 + \sin^2 θ / y_1^2}}$$

Thus,

$$s = \sqrt{\left(x_0 - x^*\right)^2 + \left(y_0 - y^*\right)^2 + \left(\Phi(x_0, y_0) - \Phi(x^*, y^*)\right)^2}.$$

And for $90° \le θ \le 180°$, the elliptic equation is $(x^2 / x_2^2) + (y^2 / y_1^2) = 1$; s can be derived by the same way.

For any point on the deformed weld pool surface, the method and steps for determining its current density are shown in Figure 8.8.

For a group of welding process conditions (such as shielding gas argon plus 2% O_2, low-carbon steel plate 6-mm thick, 1.2-mm wire diameter, $I = 240$ A welding current, $U_a = 25$ V arc voltage, $v_0 = 430$ mm/min welding speed, $W_e = 16$-mm wire extension, defined as MAG welding conditions A), numerical simulation is carried out.

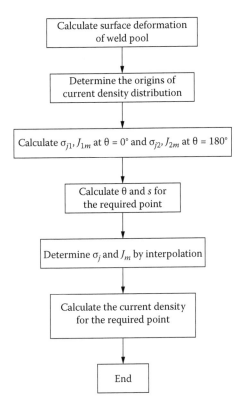

FIGURE 8.8
Calculation flowchart of current density on the weld pool surface.

Figure 8.9 shows the calculated weld pool shape. Obviously, as affected by the action of droplet impact and arc pressure, the weld pool surface is depressed beneath the arc, which results in a lower surface than that of the unmelted specimen. Under MAG welding conditions A, the maximum value of the pool surface depression is 1.89 mm. Because of this depression and addition of filler metal, the weld pool surface behind the arc rises above the unmelted specimen surface; thus, the reinforcement with a value of 2.59 mm forms after weld solidification.

Figure 8.10 indicates the calculated results of the current density distribution, and its corresponding pool surface shape is shown in Figure 8.9. It can be seen that on the longitudinal cross section ($y = 0$), the current density presents a bimodal distribution along the welding direction, and the peaks shift to the negative direction of the x-axis, which is symmetrical corresponding to the xoz plane in its arc space but is not symmetrical corresponding to the yoz plane. With the value of the y-coordinate increasing, two peaks gradually disappear.

Table 8.1 shows the variation of current density distribution parameters J_m and σ_j with angle θ at the different points on the weld pool surface.

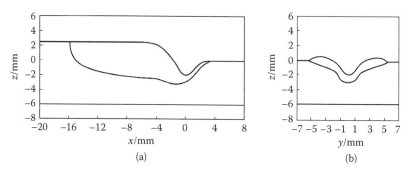

FIGURE 8.9
The calculated weld pool geometry of the MIG/MAG weld pool (MAG welding condition A):
(a) side view ($y = 0$); (b) front view ($x = 0$).

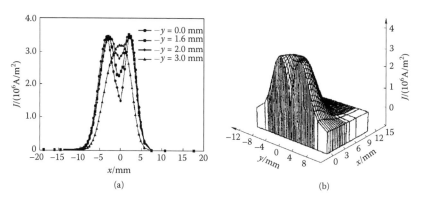

FIGURE 8.10
Predicted current density distribution (MAG welding condition A).

TABLE 8.1

Origin Coordinates and Distributed Parameters of
the Current Density Distribution of Different
Points on the Weld Pool Surface

θ (°)	(x, y, z^*)	J_m ($\times 10^6$ A/m²)	σ_j ($\times 10^{-3}$ m)
0	(2.50, 0.00, 5.46)	3.4890	2.7581
30	(2.28, 1.32, 5.46)	3.4776	2.8229
60	(1.49, 2.57, 5.46)	3.4627	2.9045
90	(0.00, 3.20, 5.46)	3.4540	2.9270
120	(−1.68, 2.91, 5.46)	3.4316	2.9392
150	(−3.24, 1.87, 5.46)	3.4183	2.9428
180	(−4.00, 0.00, 5.46)	3.4095	2.9540

Note: MAG welding condition A.

8.4 Calculation of Body Force and Arc Pressure

8.4.1 Body Force in the Weld Pool

The governing equations, Eqs. (8.1), (8.16), and (8.17), to describe the weld pool behaviors contain the body force terms F_x, F_y, and F_z. Inside the weld pool, there is buoyancy due to the nonuniform temperature distribution and the electromagnetic force due to the divergence of the current path. So, the body force in the weld pool includes electromagnetic force and buoyancy as follows:

$$\vec{F}_v = \left(\vec{J} \times \vec{B}\right) - \rho \vec{g} \beta \cdot \Delta T \tag{8.42}$$

where $\vec{J}$ is the current density, $\vec{B}$ is the magnetic induction, β is the linear expansion coefficient, g is acceleration due to gravity, and ΔT is the temperature difference. Due to the nonuniform distribution of current density, only the numerical method can be used to determine the body force of each node. As shown in Figure 8.11, assume that N_v is the node where electromagnetic force needs to be calculated with a coordinate (x_v, y_v, z_v). The current density flowing through the node N_v is j_v, which can be calculated by the current density distribution model. The connection vector between points W and N_v is $\vec{n}_v = \{x_v, y_v, z_v - z_0\}$, and components of j_v along the coordinate axis can be calculated through the direction cosine of $\vec{n}_v$ by the following formulas:

$$j_{vx} = j_v \cdot \frac{x_v}{\sqrt{x_v^2 + y_v^2 + z_v^2}} \tag{8.43}$$

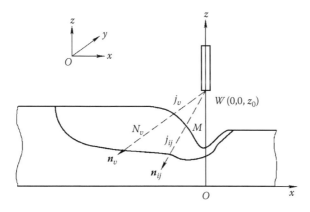

FIGURE 8.11
Schematic of the current line within the weld pool.

$$j_{vy} = j_v \cdot \frac{y_v}{\sqrt{x_v^2 + y_v^2 + z_v^2}} \tag{8.44}$$

$$j_{vz} = j_v \cdot \frac{z_v}{\sqrt{x_v^2 + y_v^2 + z_v^2}} \tag{8.45}$$

Suppose point $M(x_i, y_j, H - \Phi(x_i, y_j))$ is a node on the weld pool surface, and the current density flowing through this node is $\vec{j}_{ij}$, which is determined by the current density distribution model. The connection vector between points W and M is $\vec{n}_{ij} = \{x_i, y_j, H - \Phi(x_i, y_j) - z_0\}$. At the point M, taking an infinitesimal element $d\vec{S}_{ij}$, the magnetic induction intensity generated by the current $\vec{j}_{ij} \cdot d\vec{S}_{ij}$ at point N_v can be determined as follows:

The distance from N_v to the WM line is

$$a_{ij} = \frac{|\vec{n}_v \times \vec{n}_{ij}|}{|\vec{n}_{ij}|}$$

Magnetic induction B_{ij} generated at point N_v by current $\vec{j}_{ij} \cdot d\vec{S}_{ij}$ is

$$B_{ij} = \frac{\mu_m}{2\pi} \frac{\vec{j}_{ij} \cdot d\vec{S}_{ij}}{a_{ij}}$$

And B_{ij} has the same direction as the following vector:

$$-\vec{n}_v \times \vec{n}_{ij} = -\begin{vmatrix} \vec{i} & \vec{j} & \vec{k} \\ x_v & y_v & z_v - z_0 \\ x_i & y_j & H - \Phi(x_i, y_j) - z_0 \end{vmatrix} = A_x\vec{i} + A_y\vec{j} + A_z\vec{k}$$

where

$$A_x = -\begin{vmatrix} y_v & z_v - z_0 \\ y_j & H - \Phi(x_i, y_j) - z_0 \end{vmatrix}, \quad A_y = \begin{vmatrix} x_v & z_v - z_0 \\ x_i & H - \Phi(x_i, y_j) - z_0 \end{vmatrix},$$

$$A_z = -\begin{vmatrix} x_v & y_v \\ x_i & y_j \end{vmatrix}$$

According to the direction cosine of vector $(-\vec{n}_v \times \vec{n}_{ij})$, the components of magnetic induction in the x-, y-, and z-axes, which are generated at the point N_v by the current $\vec{j}_{ij} \cdot d\vec{S}_{ij}$, can be expressed as, respectively,

$$B_{ij}^x = B_{ij} \cdot \frac{A_x}{\sqrt{A_x^2 + A_y^2 + A_z^2}} = \frac{\mu_m}{2\pi} \frac{\vec{j}_{ij} \cdot d\vec{S}_{ij}}{a_{ij}} \cdot \frac{A_x}{\sqrt{A_x^2 + A_y^2 + A_z^2}} \qquad (8.46)$$

$$B_{ij}^y = B_{ij} \cdot \frac{A_y}{\sqrt{A_x^2 + A_y^2 + A_z^2}} = \frac{\mu_m}{2\pi} \frac{\vec{j}_{ij} \cdot d\vec{S}_{ij}}{a_{ij}} \cdot \frac{A_y}{\sqrt{A_x^2 + A_y^2 + A_z^2}} \qquad (8.47)$$

$$B_{ij}^z = B_{ij} \cdot \frac{A_z}{\sqrt{A_x^2 + A_y^2 + A_z^2}} = \frac{\mu_m}{2\pi} \frac{\vec{j}_{ij} \cdot d\vec{S}_{ij}}{a_{ij}} \cdot \frac{A_z}{\sqrt{A_x^2 + A_y^2 + A_z^2}} \qquad (8.48)$$

From Eqs. (8.43)–(8.48), the components of electromagnetic force in the x-, y-, and z-axes, which are generated in point N_v by the current $\vec{j}_{ij} \cdot d\vec{S}_{ij}$, can be written as

x-direction:

$$j_{vy}B_{ij}^z - j_{vz}B_{ij}^y$$

y-direction:

$$-j_{vx}B_{ij}^z - j_{vz}B_{ij}^x$$

z-direction:

$$j_{vx}B_{ij}^y - j_{vy}B_{ij}^x$$

After the calculations for each node and summation for all nodes on the weld pool surface, the body force of point N_v can be determined as

$$F_x = \sum_{i,j} (j_{vy}B_{ij}^z - j_{vz}B_{ij}^y) \qquad (8.49)$$

$$F_y = \sum_{i,j} (-j_{vx}B_{ij}^z - j_{vz}B_{ij}^x)$$ (8.50)

$$F_z = \sum_{i,j} (j_{vx}B_{ij}^y - j_{vy}B_{ij}^x) - \rho g\beta \cdot \Delta T$$ (8.51)

8.4.2 Distribution of Welding Arc Pressure on the Deformed Weld Pool Surface

Arc pressure is the total pressure of the plasma stream falling on the workpiece surface, which is related to the welding current and can be expressed as[81]

$$P_a = \frac{\mu_m IJ}{4\pi}$$ (8.52)

Substituting Eq. (8.41) into Eq. (8.52), the arc pressure distribution may be written as

$$P_a(\theta,s) = \frac{\mu_m IJ(\theta,s)}{4\pi} = \frac{\mu_m I}{4\pi} J_m(\theta) \exp\left(-\frac{s^2}{2\sigma_j^2(\theta)}\right)$$ (8.53)

Figure 8.12 shows the calculated results of the arc pressure distribution. Similar to the current density distributions, the arc pressure also presents as

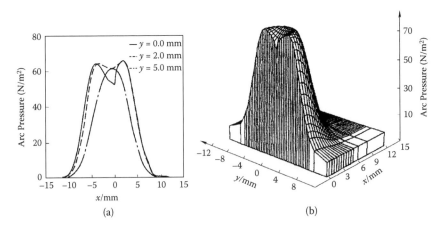

FIGURE 8.12
The calculated pressure distribution of the arc (MAG welding condition A): (a) longitudinal cross section; (b) three-dimensional distribution.

a bimodal distribution. The bimodal peak shifts to the negative direction of the x-axis, and the bimodal peaks gradually disappear with the increase of the y-coordinate.

8.5 Distribution of Arc Heat Flux on the Deformed Pool Surface

The distribution of the heat flux density is determined by that of current density. The derivation of the distribution equation of arc heat flux density on the weld pool surface is similar to that of current density distribution model.

Figure 8.13 illustrates the distribution of the arc heat flux density on the MIG/MAG weld pool surface. L_1 and L_2 are the shortest distances between the wire tip W to the intersecting line of the $y = 0$ plane and the top surface of the weld pool, that is, $O_1W = L_1$, $O_2W = L_2$. Taking points O_1 and O_2 as the origin points, the local heat flux density on the depressed pool surface is assumed to have a Gaussian distribution.

At point O_1,

$$q_1(s) = q_{1m} \exp\left(-\frac{s^2}{2\sigma_{q_1}^2}\right)$$

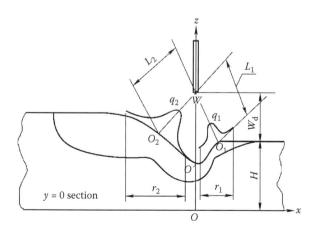

FIGURE 8.13
Schematic of heat flux density distribution on the MIG/MAG weld pool surface (on section $y = 0$).

At point O_2,

$$q_2(s) = q_{2m} \exp\left(-\frac{s^2}{2\sigma_{q_2}^2}\right)$$

where q_{1m} and q_{2m} are the maximum heat flux at points O_1 and O_2, respectively; σ_{q_1} and σ_{q_2} are the heat density distribution parameters, respectively; and s is the length of curve that starts from origin O_1 or O_2.

As discussed in Section 8.3, the longer is the distance L from the wire tip W to some point on the pool surface, the less is the current density at this point, so that the lower is the heat flux here. Assuming that q_m is proportional to L, that is,

$$q_{1m}L_1 = q_{2m}L_2 \tag{8.54}$$

According to Tsai's experimentally determined relationship between GTAW arc length and the arc heat flux distribution parameter,[119] σ_q is proportional to L, that is, σ_q/L = constant. Thus, we have

$$L_2 / \sigma_{q_2} = L_1 / \sigma_{q_1} \tag{8.55}$$

The distance W_d between the wire tip W and the workpiece surface affects the magnitude of the heat flux distribution parameters. The greater is the W_d, the larger is the area (on the weld pool surface) within which heat flux acts, so the greater is the heat flux distribution parameter. The distribution area of heat flux $r_1 + r_2$ and W_d satisfies the following equation:

$$r_1 + r_2 = a(W_d) \tag{8.56}$$

where $a(W_d)$ is a function of W_d, which is selected according to the experimental and calculated results.

Referring to the geometric relationship shown in Figure 8.13, O_2 and O' have the coordinates $(x_2, 0, H - \Phi(x_2, 0))$ and $(0, 0, H - \Phi(0, 0))$, respectively. α is the angle between line $O' O_2$ and the x-axis, and α' is the angle between the x-axis and the line connecting two points (on the heat flux density distribution curve) with the heat flux $0.05q_{1m}$. Since α is close to α', we assume that $\alpha = \alpha'$. After a similar derivation in Section 8.3, an equation can be obtained as follows:

$$\frac{2\sqrt{6}x_1\sigma_{q_1}}{\sqrt{x_1^2 + \left[\Phi(x_1,0) - \Phi(0,0)\right]^2}} + \frac{2\sqrt{6}|x_2|\sigma_{q_2}}{\sqrt{x_2^2 + \left[\Phi(x_2,0) - \Phi(0,0)\right]^2}} = a(W_d) \tag{8.57}$$

Through simultaneously solving Eq. (8.55) and Eq. (8.57), σ_{q_1} and σ_{q_2} can be obtained.

Similar to the derivation process of Eqs. (8.32)–(8.35), q_m and σ_q at other points on the weld pool surface can be obtained using q_{1m}, q_{2m}, σ_{q_1}, and σ_{q_2} at points O_1 and O_2 on section $y = 0$.

If $\theta = 0$–$90°$,

$$q_m(\theta) = q_{2m} + \frac{q_{1m} - q_{2m}}{L_1 - L_2}\left(\sqrt{x_1^2\cos^2\theta + y_1^2\sin^2\theta + (z_0 - z^*)^2} - L_2\right) \quad (8.58)$$

$$\sigma_q(\theta) = \sigma_{q_1} + \frac{\sigma_{q_2} - \sigma_{q_1}}{L_2 - L_1}\left(\sqrt{x_1^2\cos^2\theta + y_1^2\sin^2\theta + (z_0 - z^*)^2} - L_1\right) \quad (8.59)$$

If $\theta = 90$–$180°$,

$$q_m(\theta) = q_{2m} + \frac{q_{1m} - q_{2m}}{L_1 - L_2}\left(\sqrt{x_2^2\cos^2\theta + y_1^2\sin^2\theta + (z_0 - z^*)^2} - L_2\right) \quad (8.60)$$

$$\sigma_q(\theta) = \sigma_{q_1} + \frac{\sigma_{q_2} - \sigma_{q_1}}{L_2 - L_1}\left(\sqrt{x_2^2\cos^2\theta + y_1^2\sin^2\theta + (z_0 - z^*)^2} - L_1\right) \quad (8.61)$$

Integrating the arc heat flux over its action domain, we obtain

$$\int_0^\pi \int_{-\infty}^\infty q_m(\theta)\exp\left(-\frac{s^2}{2\sigma_q^2(\theta)}\right)[s + a(\theta)]dsd\theta = \frac{\eta I U_a}{2} \quad (8.62)$$

where $q_m(\theta)$ and $\sigma_q(\theta)$ can be obtained by solving Eqs. (8.58)–(8.61). Linking Eq. (8.62) and Eq. (8.54), a binary variable linear equation set about q_{1m}, q_{2m} can be obtained, so that q_{1m} and q_{2m} can be solved. Given the value of q_{1m} and q_{2m}, $q_m(\theta)$ and $\sigma_q(\theta)$ for any angle θ can be found according to Eqs. (8.58)–(8.61). The distribution of arc heat flux is determined as follows:

$$q(\theta, s) = q_m(\theta)\exp\left(-\frac{s^2}{2\sigma_q^2(\theta)}\right) \quad (8.63)$$

Equation (8.63) shows that the heat flux density on the deformed surface of the weld pool is related to θ and s, while the heat flux distribution parameters $q_m(\theta)$ and $\sigma_q(\theta)$ are related to the weld pool surface deformation and angle θ. It agrees with the facts that the shorter the distance is between a point at the pool surface and the wire tip, the greater the heat flux density is at this point.

After the pool surface deformation is known, based on the coordinate values of the wire tip W and the nodes on the weld pool surface, the points with the shortest distance to the wire tip W can be identified on the deformed pool surface. All these points on the pool surface consist of a closed curve with a double-elliptic shape. Then heat flux density distribution parameters q_{1m} and σ_{q_1} when $\theta = 0°$ as well as q_{2m} and σ_{q_2} when $\theta = 180°$ at cross section $y = 0$ can be determined. Heat flux density distribution parameters $q_m(\theta)$ and $\sigma_q(\theta)$ at any angle θ can be determined by using the interpolation method. θ and s can be obtained based on the coordinates of a checked point, and then the heat flux density at this point can be determined by Eq. (8.63). Figure 8.14 is the block diagram for calculating the heat flux at the pool surface.

Figure 8.15(a) shows the distribution of the arc heat flux density at different longitudinal cross sections with different y-coordinates. It can be seen that heat flux density appears as a bimodal distribution on the longitudinal cross section ($y = 0$). This bimodal distribution is not symmetrical about the y-axis; it is biased toward the negative direction of the x-axis, and two peaks are located at $x = -4$ mm and $x = 2.5$ mm. This is because the pool surface deformation behind the arc centerline is comparatively larger. As the longitudinal

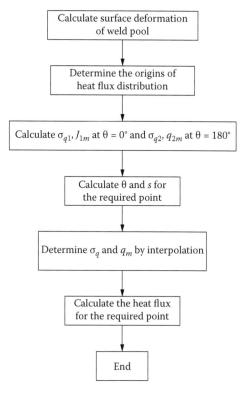

FIGURE 8.14
Block diagram for calculating heat flux density on the weld pool surface.

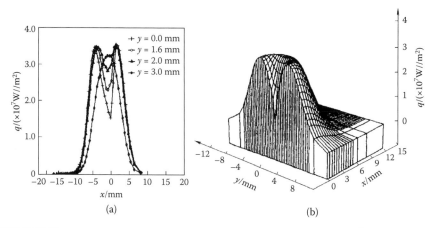

FIGURE 8.15
Calculated heat flux distribution of the arc (MAG welding condition): (a) side view; (b) three-dimensional view. (From Sun J S and Wu C S, *Int. J. Joining Mater.*, 11(4), 158–166, 1999.)

cross section is away from the weld centerline, that is, the y-coordinate value increases, bimodal distribution gradually disappears. When the value of the y-coordinate is greater than or equal to 3.0 mm, it becomes a single-peak distribution. As shown in Figure 8.9, around the arc centerline, the weld pool surface becomes depressed, and the distance from the wire tip to the weld pool surface is comparatively larger. But, for points (on the pool surface) away from the arc centerline, their distance to the wire tip is comparatively less. Therefore, there is less current flowing through the depressed pool surface, and more current flows through the points (on the weld pool surface) with some distance from the arc centerline. Movement of charged particles is the main means of heat transfer from the arc to the weld pool.[81] The higher is the current density at a point, the larger is the heat flux density at this point. The current density determines the heat flux density. Because the current density appears to have a bimodal distribution on the deformed pool surface, so does the distribution of the arc heat flux density.

Table 8.2 shows the heat flux density distribution parameters (q_m, σ_q) and the origin coordinates of the heat flux density on the weld pool surface under different values of θ.

8.6 The Distribution of Droplet Heat Content in the MIG/MAG Weld Pool

The difference in the droplet transfer mode is one of the most difficult questions in numerical simulation of MIG/MAG weld pool behaviors under

TABLE 8.2

Origin Point and Distribution Parameters q_m and σ_q
at Different Points

θ (°)	(x, y, z^*)	q_m ($\times 10^7$ W/m^2)	σ_q ($\times 10^{-3}$ m)
0	(2.50, 0.00, 5.46)	3.4679	2.9797
30	(2.28, 1.32, 5.46)	3.4511	2.9952
60	(1.49, 2.57, 5.46)	3.4165	3.0270
90	(0.00 3.20, 5.46)	3.3995	3.0426
120	(−1.68, 2.91, 5.46)	3.3576	3.0818
150	(−3.24, 1.87, 5.46)	3.2758	3.1566
180	(−4.00, 0.00, 5.46)	3.2344	3.1948

Note: MAG welding condition A.

different welding conditions.[155] The influence of droplet transfer on the welding process is reflected by the action of the droplet on the weld pool; the influence consists of energy, momentum, and mass transferred from droplets into the weld pool. The momentum and mass transfer lead to a large surface deformation and resultant weld reinforcement. The energy transfer is that droplets deliver extra thermal enthalpy into the weld pool. The distribution of droplet enthalpy inside the weld pool with a deformed surface is one of the key problems for accurately simulating weld pool behaviors in the MIG/MAG welding process. It is the focus of this section.

8.6.1 Momentum and Energy Analysis in the Droplet Transfer Process

During MIG/MAG welding, droplets obtain kinetic energy when they pass through the arc column. Then, they possess certain momentum. Their impact on the weld pool surface results in a severely deformed pool surface. Part of kinetic energy is consumed in the impact process, while the remaining part makes the droplets inject into the weld pool at a certain speed and transfer some heat to the weld pool. This section develops the droplet-enthalpy distribution model according to the energy balance in the process of droplet transfer.

Compared with the weld pool, droplets are clearly overheated. The enthalpy Q_{vd} that is brought into weld pool by droplets is treated as an internal heat source in the energy equations, Eqs. (8.1), (8.16), and (8.17)[59,62] and is given as follows:

$$Q_{vd} = \frac{4 v m_D \cdot \Delta H_d}{\pi d_w^2 S_m} \tag{8.64}$$

where v is the droplet transfer frequency, m_D is the droplet mass, d_w is the wire diameter, S_m is the wire melting speed, and ΔH_d is the heat content difference between the droplets and the weld pool.

Under the action of arc pressure P_a, droplet impact P_d, gravitational force of liquid metal $\rho g \Phi$, and surface tension γ, the weld pool surface will be deformed as shown in Figure 8.16. In the coordinate system $O_1 - xy\Phi$, the deformation can be described by Eq. (8.11), which is rewritten as follows:

$$P_a - \rho g \Phi + P_d + C_3 = -\gamma \frac{\left(1+\Phi_y^2\right)\Phi_{xx} - 2\Phi_x\Phi_y\Phi_{xy} + \left(1+\Phi_x^2\right)\Phi_{yy}}{\left(1+\Phi_x^2+\Phi_y^2\right)^{3/2}} \quad (8.65)$$

As shown in Figure 8.16, shape functions $\Phi_1(x, y)$ can be obtained by solving the deformation equation, Eq. (8.65), when the four forces mentioned are in a dynamic balance. In deformation Eq. (8.65), if we let $P_d = 0$, it is the case of TIG welding, and there is no droplet impact. In this case, there also exists a slightly deformed pool surface, that is, $\Phi_2(x, y)$ (which is shown as a dashed line in Figure 8.16). In Figure 8.16, $\Phi_1(0, 0)$ and $\Phi_2(0, 0)$ are, respectively, the maximum surface depression of the weld pool in the two cases mentioned. It is evident that the weld pool surface shape change from $\Phi_2(x, y)$ to $\Phi_1(x, y)$ is caused by the droplet impact P_d.

Suppose that the weld pool surface shape is $\Phi(x, y)$. Then, the maximum depression is $\Phi(0, 0)$, while the maximum hump height H_{max} appears in the rear of the weld pool. In coordinate system $O_1 - xy\Phi$, the energy required to produce the depression and hump can be expressed as

$$E_D = \int_0^{\Phi(0, 0)} \rho g z_1 \cdot A_D\left(z_1\right) dz_1 \quad (8.66)$$

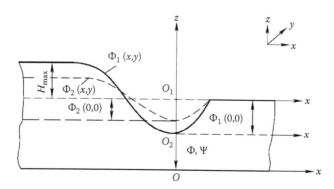

FIGURE 8.16
Schematic of surface deformation of the weld pool under the action of different forces.

$$E_{\mathrm{H}} = \int\limits_{0}^{H_{\max}} \rho g z_2 \cdot A_{\mathrm{H}}\left(z_2\right) dz_2 \qquad (8.67)$$

where $A_{\mathrm{D}}(z_1)$ and $A_{\mathrm{H}}(z_2)$ are the horizontal areas of the depressed region and humped region, respectively.

Therefore, when a deformation is produced on the weld pool surface as described by the shape function $\Phi(x, y)$, the corresponding energy E required is the sum of E_{D} and E_{H}. That is,

$$E = E_{\mathrm{D}} + E_{\mathrm{H}} = \int\limits_{0}^{\Phi(0,\,0)} \rho g z_1 \cdot A_{\mathrm{D}}\left(z_1\right) dz_1 + \int\limits_{0}^{H_{\max}} \rho g z_2 \cdot A_{\mathrm{H}}\left(z_2\right) dz_2 \qquad (8.68)$$

When the configuration function of the weld pool surface $\Phi(x, y)$ is known, both $A_{\mathrm{D}}(h_1)$ and $A_{\mathrm{H}}(z_2)$ are determined. The corresponding energy E will be obtained by Eq. (8.68).

When the configuration functions of the weld pool surface are $\Phi_1(x, y)$ and $\Phi_2(x, y)$, respectively, the energy E_1 and E_2 required for deformation can be calculated by Eq. (8.68). The droplet impact P_{d} not only deforms the weld pool surface but also it makes the droplet enter the weld pool. After pool surface deformation is produced by droplet impact and other factors, the remaining energy drives the droplet to enter into the pool. By analyzing the remaining energy and impact of droplets, the distribution region of droplet enthalpy inside the weld pool can be determined after surface deformation. According to energy balance, $\Delta E = E_1 - E_2$ represents the energy exhausted during the process that the droplet impact P_{d} causes a surface deformation. If the energy that a droplet possesses at the moment that it just makes contact with the undeformed pool surface is written as $(1/2)m_{\mathrm{D}}v_{\mathrm{D}}^2$, $(m_{\mathrm{D}}v_{\mathrm{D}}^2/2) - \Delta E$ is the remaining kinetic energy at the moment the droplet just enters the deformed pool surface. Suppose the droplet has a velocity of v_{R} at that moment, then we have

$$\frac{1}{2}m_{\mathrm{D}}v_{\mathrm{R}}^2 = \frac{1}{2}m_{\mathrm{D}}v_{\mathrm{D}}^2 - \Delta E \qquad (8.69)$$

After manipulation,

$$v_{\mathrm{R}} = \sqrt{\frac{m_{\mathrm{D}}v_{\mathrm{D}}^2 - 2 \cdot \Delta E}{m_{\mathrm{D}}}} \qquad (8.70)$$

By using the numerical method, the instantaneous velocity v_{R} of the droplet entering into the pool can be derived from this formula.

8.6.2 The Distribution Region of Droplet Heat Content in the MIG/MAG Weld Pool

If a droplet enters into the weld pool with a speed of v_R, its remaining impact force is vm_Dv_R. This remaining impact combining with the gravity of liquid metal acts on the deformed pool surface and makes a further depression of the weld pool surface, which will directly affect the distribution of droplet enthalpy in the weld pool. Next, we derive the basic equations describing the relation between the forces (surface pressure) and the pool surface (interface) shape.

Taking the lowest point O_2 on the deformed pool surface as the origin of the coordinates, the interaction of droplet impact vm_Dv_R, gravity of the liquid metal, and surface tension γ will lead to a further deformation expressed by the shape function $\Psi(x, y)$ in the coordinate system $O_2 - xy\Psi$, as shown in Figure 8.17. The droplet impact on unit area is $vm_Dv_R/\pi r_D^2$. Still using Eq. (8.11), we get

$$\frac{vm_Dv_R}{\pi r_D^2} - \rho g\Psi + C_4 = -\gamma\frac{\left(1+\Psi_y^2\right)\Psi_{xx} - 2\Psi_x\Psi_y\Psi_{xy} + \left(1+\Psi_x^2\right)\Psi_{yy}}{\left(1+\Psi_x^2+\Psi_y^2\right)^{3/2}} \tag{8.71}$$

where m_D, v_R, and v are droplet mass, droplet speed, and droplet transfer frequency, respectively. Their values are determined by experimental results.[80,147] Constant C_4 can be determined by the following formula according to conservation of mass:

$$\iint \Psi(x,y)dxdy = \frac{vm_D}{\rho} \tag{8.72}$$

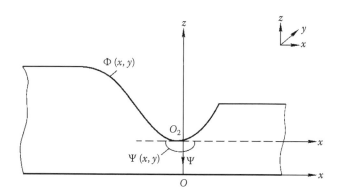

FIGURE 8.17
The coordinates in determining the distribution region of droplet enthalpy.

Shape function $\Psi(x, y)$ can be obtained by solving Eq. (8.71). Imagine that there should be a further cavity at the depressed pool surface, and this cavity is determined by shape function $\Psi(x, y)$. But, in fact this cavity does not exist because of the filler droplets. Thus, the droplet enthalpy should be distributed within a virtual depression area (cavity) that is determined by shape function $\Psi(x, y)$. For the convenience of calculation, the volume surrounded by the cavity defined by $\Psi(x, y)$ and the horizontal plane passing the lowest depression point O_2 is considered as the distribution volume of droplet enthalpy.

8.6.3 The Calculated Results of Droplet Enthalpy Distribution within the Weld Pool

The calculated distribution region of droplet enthalpy is shown in Figure 8.18. According to the selected process parameters, it is a spray transfer.[156] Figures 8.18(a) and 8.18(b) show the shape of the droplet enthalpy distribution region at a longitudinal cross section ($y = 0$) and a transverse cross section ($x = 0$), respectively. The droplet enthalpy is distributed in a volume beneath the point with the largest depression, and the volume is similar to a cone in geometry. The volume is symmetrical about the *xoz* plane and asymmetrical about the *yoz* plane. It is biased toward a negative *x*-axis. For the MAG welding condition A, the maximum length, width, and depth of the cone are 2.51, 1.20, and 0.66 mm, respectively. Table 8.3 shows the influences of welding parameters on the magnitude and distribution region of droplet enthalpy. As the welding current increases, the maximum length, width, and depth of the distribution region of droplet enthalpy increase. This is because with increasing welding current, droplets obtain more kinetic energy when they pass through the arc column. At the same time, the momentum of the droplet increases, and the droplet impact increases. It can also be seen from

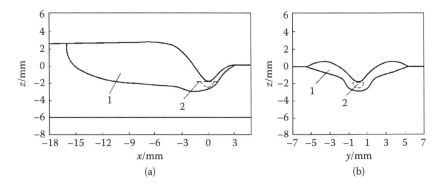

FIGURE 8.18
The distribution region of droplet enthalpy in the weld pool (1, weld pool; 2, distribution region of droplet enthalpy) (MAG welding condition A): (a) side view; (b) front view.

TABLE 8.3

Influence of Welding Parameters on Droplet Enthalpy and Its Distribution Region

Welding Current (A)	Wire Feeding Rate (mm/s)	Arc Voltage (V)	Welding Speed (mm/min)	Distribution Region of Droplet Enthalpy					Magnitude of Enthalpy (J/mm^3)
				Maximum Length (mm)	Maximum Width (mm)	Maximum Depth (mm)	Volume (mm^3)		
150	39.45	25	430	1.12	0.53	0.27	0.061		2.342
180	57.20	25	430	1.52	0.67	0.48	0.169		2.472
210	77.06	25	430	1.88	0.96	0.53	0.329		2.602
240	99.04	25	430	2.51	1.20	0.66	0.681		2.732
270	123.15	25	430	2.93	1.31	0.69	0.887		2.863
300	143.47	25	430	3.21	1.49	0.73	1.174		2.993

Note: MAG welding condition A with a changeable welding current.

Table 8.3 that the magnitude of droplet enthalpy increases as the welding current increases.

8.7 Analysis Results of MIG/MAG Weld Pool Behaviors

8.7.1 Program and Parameters Used in Simulation

The numerical analysis program for MIG/MAG weld pool behaviors is composed of a main program and three subroutines. They are the weld pool surface deformation subroutine (SDeform), the temperature field calculation subroutine (Temp), and the fluid flow field subroutine (Flow).

Figure 8.19 is the main program block diagram for numerical analysis of MIG/MAG weld pool behaviors. The convergence criteria for the fluid flow field and the temperature field are as follows:

$$\left| T_{n+1} - T_n \right| / T_n \le 10^{-3}$$

$$\left| u_{n+1} - u_n \right| / \left| u_n \right| \le 10^{-3}$$

$$\left| v_{n+1} - v_n \right| / \left| v_n \right| \le 10^{-3} \qquad (8.73)$$

$$\left| w_{n+1} - w_n \right| / \left| w_n \right| \le 10^{-3}$$

If the convergent conditions are not satisfied, recalculate thermal parameters according to the available temperature field and again call for SDeform, Flow, and Temp subroutines. Repeat this iterative process until the convergence conditions are satisfied.

8.7.1.1 Thermal Physical Properties of Materials

Numerical calculation is carried out by using a mathematical model to describe the weld pool behaviors in MIG/MAG welding. The workpiece material is low-carbon steel Q195, and its size is 150 mm (length) by 80 mm (width) by 6 mm (thickness). Table 8.4 and Eqs. (8.74)–(8.77) show the material thermal physical properties.

$$C_p = \begin{cases} 0.51376 - 3.3504 \times 10^{-4}T + 6.8929 \times 10^{-7}T^2 & T \le 973K \\ -10.539 + 1.17 \times 10^{-2}T & 973K \le T \le 1023K \\ 11.873 - 1.0208 \times 10^{-2}T \quad (kJ \cdot kg^{-1} \cdot K^{-1}) & 1023K \le T \le 1100K \\ 0.644 & 1100K \le T \le 1379K \\ 0.35434 + 2.1 \times 10^{-4}T & 1379K \le T \end{cases}$$

$$(8.74)$$

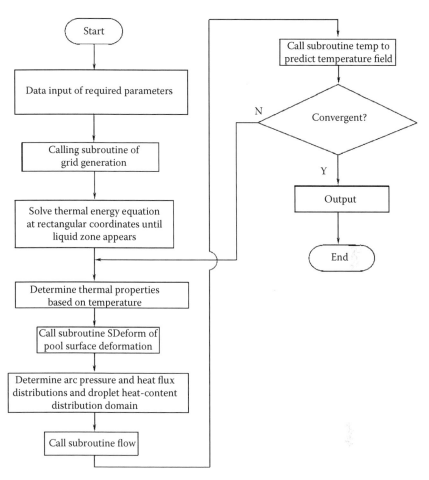

FIGURE 8.19
Block diagram of the main program.

TABLE 8.4

Thermophysical Parameters of Low-Carbon Steel

Thermophysical Parameters	Value
Density ρ	7,860 (kg·m^{-3})
Liquidus temperature T_L	1,793 (K)
Solidus temperature T_s	1,700 (K)
Expansion coefficient of liquid metal β	10^{-4} (K^{-1})
Surface tension temperature coefficient $\partial\gamma/\partial T$	-0.35×10^{-3} (N·m^{-1}·K^{-1})
Surface tension coefficient γ	1.0 (N·m^{-1})
Vacuum permeabilityμ_m	1.26×10^{-6} (N·m^{-1})
Latent heat of fusion L_m	273,790 (J·kg^{-1})

$$\lambda = \begin{cases} 60.719 - 0.027857T & & T \leq 851\text{K} \\ 78.542 - 0.0488T & & 851\text{K} \leq T \leq 1082\text{K} \\ 15.192 + 0.0097T & (\text{w} \cdot \text{m}^{-1} \cdot \text{K}^{-1}) & 1082\text{K} \leq T \leq 1768\text{K} \\ 349.99 - 0.1797T & & 1768\text{K} \leq T \leq 1798\text{K} \end{cases} \quad (8.75)$$

$$\mu = \begin{cases} 119.003 - 0.061T & & 1823\text{K} \leq T \leq 1853\text{K} \\ 10.603 - 0.025T & (\text{p}_a \cdot \text{s}) \times 10^{-3} & 1853\text{K} \leq T \leq 1873\text{K} \\ 36.263 - 0.0162T & & 1873\text{K} \leq T \leq 1973\text{K} \end{cases} \quad (8.76)$$

$$\alpha = 24.1 \times 10^{-4} \cdot \varepsilon \cdot T^{1.61} \quad (\text{w} \cdot \text{m}^{-2} \cdot \text{K}^{-1}) \qquad \varepsilon = 0.9 \qquad (8.77)$$

8.7.1.2 Determining the Droplet Transfer Parameters

In MIG/MAG welding, the droplet transfer mode is classified as globular transfer, projected transfer, spray transfer, and rotating spray transfer. The welding current is a principal parameter that has a dominating effect on different droplet transfer modes. For example, when the welding current is low, globular transfer occurs; when the current is higher, spray transfer occurs. A study showed that if low-carbon steel wire and argon plus 2% O_2 shielding gas are used, the droplet transfer mode changes gradually from globular transfer to projected transfer and finally spray transfer as the welding current increases.[81] Because the droplet transfer mode affects droplet parameters, it is critical to establish the correlation of the welding parameters and the droplet parameters in different conditions for numerical analysis of weld pool behaviors during MIG/MAG welding. This section determines the relationships of the welding parameters and the droplet parameters according to some experiments.

A low-carbon steel wire with 1.2-mm diameter, direct current electrode positive (DCEP) polarity, and argon plus 2% O_2 shielding gas were used to conduct experiments for measuring the droplet parameters at different welding current levels.[80,147] By analyzing the measured data, the relationship between droplet radius (mm) and welding current I (A) is obtained as follows:

$$r_D = 1/(9.4722 \times 10^{-5} I^2 - 1.4827 \times 10^{-2} I + 1.5412) \qquad (8.78)$$

The related coefficient is −0.969.

The statistical equation for the droplet speed v_D (m/s) reaching the weld pool surface versus the welding current can be expressed as follows:

$$v_D = 9.3829 \times 10^{-5} I^2 - 1.6377 \times 10^{-2} I + 1.0626 \qquad (8.79)$$

The related coefficient is −0.940.

The relationship between the droplet heat content H_{dr} (J/g) and the welding current I (A) can be described by the following equation:

$$H_{dr} = 0.5519I + 1715.1801 \tag{8.80}$$

The related coefficient is −0.998.

The average heat content of the weld pools in MIG welding of low-carbon steel is as follows: H_{pv} = 1.45–1.61 (KJ/g).[157] So, the heat content delivered into the weld pool by droplet transfer can be expressed as $\Delta H_d = H_{dr} - H_{pv}$.

The relationship between the melting rate of wires S_m (m/s) versus I (A), U_a (V), L_e (mm), and d_w (mm) is determined by the following equation[158]:

$$S_m = \left(8.9997 \times 10^{-10} I - 2.4298 \times 10^{-11} IU_a - 2.5198 \times 10^{-8}\right) / d_w^2$$
$$+ 1.5254 \times 10^{-3} I^2 \cdot L_e / d_w^4 \tag{8.81}$$

The related coefficient is −0.980.

From the related coefficients mentioned, it can be seen that the given regression equations are in good agreement with the experimental data.

In numerical analysis, if the welding parameters I, U_a, L_e, and d_w are given, the droplet parameters r_D, S_m, v_D, and H_{dr} can be calculated by these equations, and other droplet parameters needed are derived from the following formulas:

$$m_D = \frac{4}{3} \pi r_D^3 \rho \tag{8.82}$$

$$v = \frac{\rho \pi d_w^2 S_m}{4 m_D} \tag{8.83}$$

8.7.2 Influences of Welding Parameters on Weld Pool Surface Deformation

The effects of welding current on the weld pool geometry are shown in Figures 8.20 and 8.21, where the (a), (b), and (c) portions are the calculated results when the welding current is 150, 240, and 300 A, respectively. The experiments showed that the droplet transfer mode changes from globular transfer to spray transfer as welding current increases. For a 1.2-mm diameter low-carbon steel wire, the transition current is about 220 ± 10 A,[156] so Figure 8.20(a) and Figure 8.21(a) are the geometry of the weld pool in a globular transfer mode, while Figures 8.20(b), 8.20(c), 8.21(b), and 8.21(c) are the geometry of the weld pool in a spray transfer mode. The calculated results show that the weld pool surface beneath the arc is depressed by both the

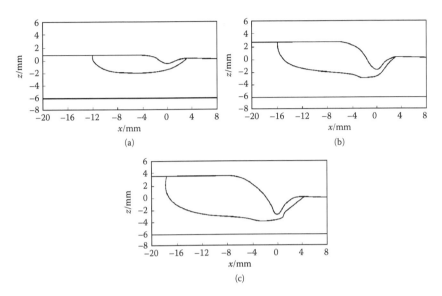

FIGURE 8.20
The predicted weld pool geometry at the longitudinal cross section (MAG welding condition A with a varying current): (a) $I = 150$ A; (b) $I = 240$ A; (c) $I = 300$ A.

droplet impact and the arc pressure, and it is lower than the unmelted top surface of the workpiece. Due to the deformation and filler metal, the rear part of the weld pool surface is humped, which is higher than the unmelted top surface of workpiece. After solidification, this humping part forms the weld reinforcement. When the welding current is 150 A, the metal transfer mode is globular transfer. The depressed cavity of the weld pool is shallow and narrow with a maximum depression of 0.66 mm and a maximum hump height of 0.77 mm. When the welding current is 240 A, the transfer mode falls in the zone of spray transfer, and the surface deformation of the weld pool becomes deep and wide with a maximum depression of 1.89 mm and a maximum hump height of 2.59 mm. When the welding current increases to 300 A, the surface deformation increases further; the maximum depression increases to 2.85 mm and maximum hump height increases to 3.39 mm. Therefore, as the welding current increases, the droplet transfer mode will be changed, and the maximum depression and hump height of the weld pool surface increase correspondingly. The main reason is that the droplet impact in spray transfer is much greater than that in globular transfer. And in the spray transfer mode, the heat input and wire melting rate as well as the arc pressure and droplet impact all become larger as the welding current increases. The increasing of all these variables results in an increase of both surface depression and hump height, so that the surface deformation becomes more severe. Figures 8.22(a) and 8.22(b) show the calculated results of surface

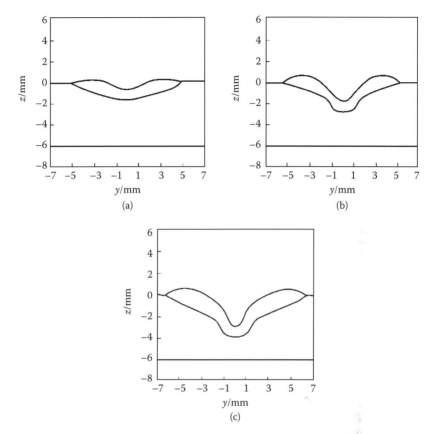

FIGURE 8.21
The predicted weld pool geometry at the transverse cross section (MAG welding condition A with a changeable current): (a) $I = 150$ A; (b) $I = 240$ A; (c) $I = 300$ A.

deformation and weld reinforcement when the droplet transfer modes are globular transfer ($I = 150$ A) and spray transfer ($I = 240$ A), respectively.

Table 8.5 shows the influence of the distance W_h between the nozzle and workpiece surface (contact tube-to-workpiece [CTW] distance) on the geometry of the weld pool. Variation of the CTW distance changes the magnitude of the arc pressure and affects the distribution of the welding current, arc pressure, and heat flux on the weld pool surface. Therefore, the weld pool geometry inevitably will be affected by the CTW distance. When the CTW distance increases, the concave deformation of the pool surface becomes broad and shallow, the length of the weld pool increases, but the height of the weld reinforcement decreases slightly. This is because as the CTW distance increases, the distribution regions of current density, arc pressure, and heat flow expand while their maximum values decrease.

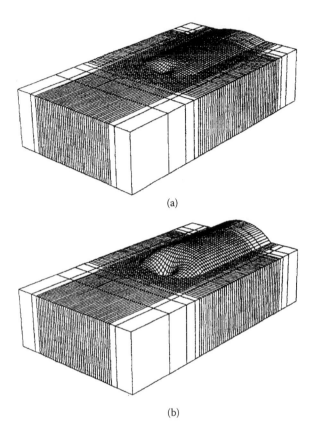

(a)

(b)

FIGURE 8.22
The weld reinforcement and three-dimensional shape of the weld pool (MAG welding condition A with a changeable current): (a) $I = 150$ A; (b) $I = 240$ A.

TABLE 8.5

Influences of CTW Distance W_h on Weld Pool Geometry

W_h (mm)	Weld Pool Length (mm)	Reinforcement (mm)	Maximum Depression (mm)	Weld Pool Width (mm)	Weld Pool Depth (mm)
15	17.19	2.78	2.49	11.28	3.92
20	19.01	2.59	1.90	12.00	2.90
25	20.03	2.51	1.12	14.21	2.01

Note: MAG welding condition A with varying W_h.

8.7.3 Influences of Welding Parameters on the Heat Flux Distribution

Figures 8.23(a), 8.23(b), and 8.23(c) indicate the heat flux distribution along the x-direction (welding direction) under welding currents of 150, 240, and 300 A, respectively. It can be seen that regardless of whether the welding current value is 150 A (globular transfer) or 240 A and 300 A (spray transfer),

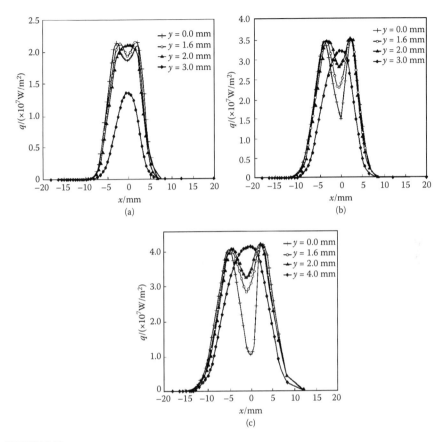

FIGURE 8.23
The predicted arc heat flux distribution on the weld pool surface (MAG welding condition A with varying welding current): (a) $I = 150$ A; (b) $I = 240$ A; (c) $I = 300$ A. (From Sun J S and Wu C S, *Int. J. Joining Mater.*, 11(4), 158–166, 1999.)

the heat flow along the x-direction near the center of the arc (with small y-coordinate value) shows a bimodal distribution, and the two peaks shift to the negative x-axis direction. As the y-coordinate value (the distance away from the weld centerline) increases, the peaks disappear, and a single-peak distribution reappears. The region of bimodal distribution is different with different welding currents. When the welding current is 150 A, the distribution becomes a single peak when the y-coordinate is greater than or equal to 2.0 mm; for $I = 240$ A or $I = 300$ A, it becomes a single-peak distribution when the y-coordinate values are greater than or equal to 3.0 or 4.0 mm. On the other hand, the x-coordinate value at which heat flux reaches its peak value at the $y = 0$ section is also related to the value of the welding current. The peaks of the x-coordinate values are −3.0 and 2.0 mm for $I = 150$ A; −4.0 and 2.5 mm for $I = 240$A; and −5.0 and 3.0 mm for $I = 300$ A. The distances between two peaks at the x-direction are 5.0, 6.5, and 8.0 mm when the welding current is 150, 240, and 300 A, respectively. When the welding current increases, the distance between the two peaks gradually increases along the x-direction.

Figure 8.23 shows that welding current also affects the ratio HF of the maximum heat flux value to the heat flux value at the arc centerline on the transverse cross section $y = 0$. Obviously, the larger the ratio, the more remarkable the bimodal distribution of heat flux is. It can be seen from Figure 8.23 that HF at $y = 0$ cross section is about 1.35 when the welding current is 150 A, while the HF increases to about 2.51 and 4.00 when the welding current is 240 and 300 A, respectively.

The influence of welding currents on the distribution of heat flux density is due to the fact that the surface shape of the weld pool is greatly affected by the welding current. It can be seen from Figure 8.20 and Figure 8.21 that the maximum depression of the weld pool surface increases as the welding current rises. This maximum depression is 0.66, 1.89, and 2.85 mm when the welding current takes values of 150, 240, and 300 A, respectively. The surface depression at the center of the weld pool makes the distance from the wire end to the weld pool surface become larger, whereas the distance between points further away from the arc centerline to the wire end become smaller. Thus, more current flows into those surface points a little bit farther away from the arc centerline on the depressed weld pool surface. The heat flux shows a bimodal distribution, as shown in Figure 8.23. The larger the welding current, the greater the maximum depression of the weld pool surface and the smaller the current density is near the centerline at the weld pool surface. The bimodal distribution is more notable, that is, HF has the maximum value at the transverse cross section $y = 0$.

8.7.4 The Calculated Temperature Field

Figure 8.24 shows the calculated isotherms with a temperature interval of 250 K at the longitudinal and transverse cross sections of the weld pool in a

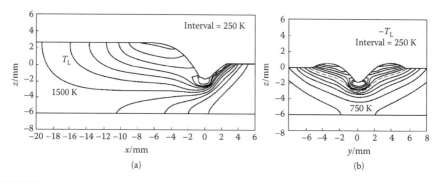

FIGURE 8.24
Calculated temperature distribution in the workpiece (T_1 = 1,793 K) (MAG welding condition A): (a) $y = 0$ side view; (b) $x = 0$ front view.

spray transfer mode. The isothermal with a value of 1,500 K is the melt line. It can be seen that the developed model demonstrates the finger-like weld penetration in spray transfer.

8.7.5 The Calculated Fluid Fields in the Weld Pool

Figure 8.25 shows the calculation results for the fluid flow fields in a case of spray transfer. In Figure 8.25, (a) is the predicted fluid flow field at the longitudinal cross section ($y = 0$), while (b), (c), (d), and (e) are the calculated fluid flow fields at the transverse cross sections with $x = 0$ mm, $x = -1.5$ mm, $x = -3.0$ mm, and $x = -5.0$, respectively. Figure 8.26 shows a schematic of the fluid flow field in the weld pool. From these calculated results, it can be seen that there are two circulating loops inside the weld pool. One is in the middle region of the weld pool, and another is in the rear part of the weld pool. Around the arc centerline, the liquid metal flows to the bottom of the weld pool. The combination effect of the electromagnetic force and the droplet impact results in the circulation loop formed in the middle part of the weld pool. Near the arc centerline z-axis, as both electromagnetic force and droplet impact act downward, they push the hot liquid metal flowing to the bottom of weld pool, which brings more heat into the bottom of the weld pool.

The circulation loop at the pool rear is formed due to the surface tension temperature gradient at the pool surface. Usually, the surface tension temperature gradient is negative, so that for the region closer to the arc centerline, the higher temperature of liquid metal causes a lower surface tension. Near the pool boundary, the liquid metal temperature is lower, and the surface tension is higher. Thus, the liquid metal at the weld pool surface flows outward from the center region to the boundary of the weld pool. Figure 8.25 indicates that, in a spray transfer case, the weld pool forms a finger-like penetration. In spay transfer, the larger welding current leads to the larger droplet enthalpy and a larger distribution region of enthalpy (as shown in Table 8.3). A larger welding current produces a significantly enhanced arc pressure, greater droplet velocity, and bigger droplet impact force. All these factors make a stronger downward flow of molten metal, which delivers more heat to the weld pool bottom. Therefore, finger-like penetration occurs.

Figure 8.27 shows the calculated results of the fluid field in the weld pool with a globular transfer case. Figures 8.27(a), 8.27(b), and 8.27(c) indicate the predicted fluid flow fields at the longitudinal cross section ($y = 0$) and the transverse cross sections ($x = 0$ mm, $x = 5.0$ mm), respectively. There is only one circulation loop in the weld pool under the globular transfer mode. This circulation brings heat to the pool rear rather than the pool bottom. Figure 8.28 shows the schematic of the fluid flow field in the weld pool. Comparing Figure 8.28 with Figure 8.26, it can be seen that the fluid flow fields in globular transfer and spray transfer are obviously different.

As can be seen from Figure 8.28, the flow direction of the liquid metal is also outward. During the globular transfer, the welding current is relatively

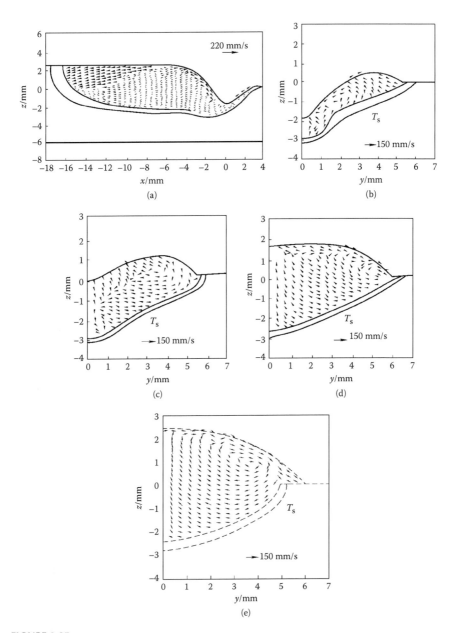

FIGURE 8.25
Calculated flow fields in the weld pool: (a) side view, $y = 0$; (b) front view, $x = 0$ mm; (c) front view, $x = -1.5$ mm; (d) front view, $x = -3$ mm; (e) front view, $x = -5$ mm (MAG welding condition A). (From Sun J S and Wu C S, *Int. J. Joining Mater.*, 11(4), 158–166, 1999.)

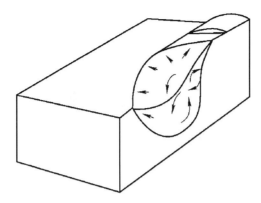

FIGURE 8.26
Schematic of fluid flow in the weld pool (MAG welding condition A).

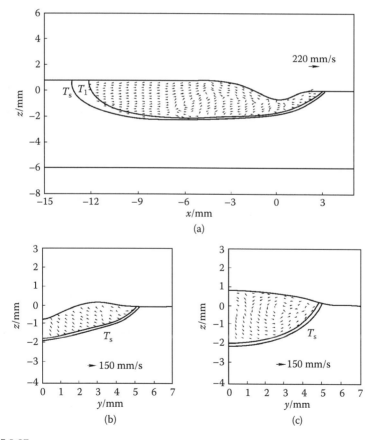

FIGURE 8.27
The calculated flow field in the weld pool (MAG welding condition A, $I = 150$ A): (a) longitudinal cross section at $y = 0$; (b) transverse cross section at $x = 0$ mm; (c) transverse cross section at $x = -5$ mm.

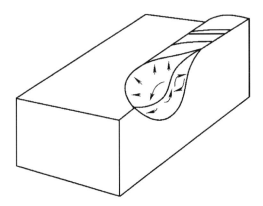

FIGURE 8.28
Schematic of fluid flow in the weld pool (MAG welding condition A, I = 150 A).

low; the arc pressure and the droplet impact are also lower. Meanwhile, the relatively low welding current also leads to a small droplet enthalpy and a small distribution area in the weld pool (Table 8.3). Therefore, under the condition of globular transfer, an approximate hemicycle transverse cross section of weld is formed, and no finger-like penetration forms.

In summary, the variations in welding parameters change the type of droplet transfer, which influences the mode of fluid flow inside the weld pool to a large extent. The mode of fluid flow has a great effect on the temperature field of the weld pool, which determines the geometrical shape of the weld pool and the weld bead.

In most cases, the value of the surface tension temperature coefficient $\partial\gamma/\partial T$ is negative. When the coefficient $\partial\gamma/\partial T$ is changed from negative to positive by applying a metallurgical method, it inevitably affects the fluid flow mode inside the weld pool. Figure 8.29 demonstrates the effect of the variation of $\partial\gamma/\partial T$ on the flow field inside the weld pool. Figure 8.30 represents the schematic of fluid flow inside the weld pool. It can be seen that when $\partial\gamma/\partial T$ is changed from negative to positive, the mode of fluid flow inside the weld pool gets a significant change, especially in the tail of the weld pool, where the vertical component of fluid velocity $\vec{w}$ changes its direction (from downward to upward).

The mode of fluid flow inside the weld pool has an important influence on the welding metallurgical process, and it is also closely related to the formation of porosity and impurity inside the weld pool. The temperature distribution inside the weld pool is not uniform. In the front of the weld pool, the heat input is more than the heat dissipation, and the temperature beneath the arc is the highest on the weld pool surface. So, in this area silicon, manganese, and other alloy elements will be reduced out of their compounds. More gas, such as nitrogen, hydrogen, oxygen, and the like, will also be dissolved in liquid metal. Meanwhile, at the tail of the weld pool, since the temperature

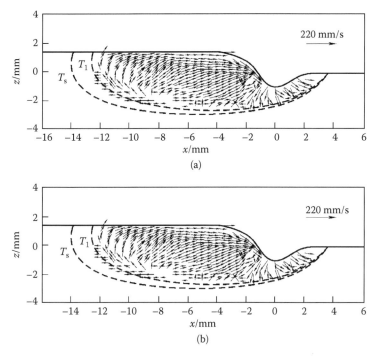

FIGURE 8.29
The calculated results indicating the sign of $\partial\gamma/\partial T$ on the flow fields inside the weld pool (MAG welding condition A, $I = 180$ A): (a) value of $\partial\gamma/\partial T$ is negative; (b) value of $\partial\gamma/\partial T$ is positive.

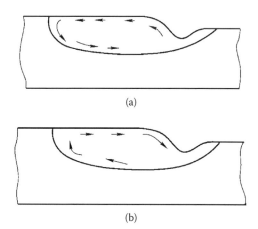

FIGURE 8.30
The effect of $\partial\gamma/\partial T$ on the fluid flow fields inside the weld pool (MAG welding condition A, $I = 180$ A): (a) value of $\partial\gamma/\partial T$ is negative; (b) value of $\partial\gamma/\partial T$ is positive.

at the tail gradually decreases, part of silicon, manganese, and other alloy elements will be reoxidized to form their oxides. Moreover, some gases (such as nitrogen, hydrogen, oxygen, etc.) dissolved at the higher temperature and gases produced from the welding metallurgy reaction (such as carbon monoxide) will be in the supersaturating states, creating conditions for the formation of porosity and other weld defects. If these oxides or supersaturated gases cannot escape from the weld pool, inclusions or porosity will be formed, which will further reduce the mechanical properties and compactness of the weld metal. In particular, a significant reduction in toughness may occur. On the other hand, because the heat input cannot keep up with the heat dissipation at the tail of the weld pool, liquid metal solidifies, which is extremely unfavorable for the transgression of oxides and saturated gas. Thus, the behavior at the weld pool tail has a significant influence on the transgression of oxides and saturated gas.

Heiple's experimental studies[159] showed that the surface tension temperature coefficient $\partial\gamma/\partial T$ of liquid steel changes from negative to positive with the addition of a small amount of surface-active elements, such as sulfur, oxygen, selenium, and tellurium. Obviously, if the addition of a small amount of surface-active elements makes the sign of $\partial\gamma/\partial T$ become positive, the value of $\vec{w}$ (vertical component of fluid velocity) at the weld pool tail is positive (i.e., flow velocity component w with upward direction), which is very beneficial for the transgression of oxides and saturated gas, which results in less gas and fewer inclusions inside the deposited metal.

8.8 Experimental Verification of the Model

Welding experiments must be made to verify whether the developed models of weld pool behaviors reflect the nature of physical phenomena in MIG/MAG welding and whether the calculated results of weld pool behaviors are in accordance with the actual welding situations.

8.8.1 Test Method and Test Materials

Specimen material: Low-carbon steel, Q195

Specimen size: 150 × 80 × 6 mm (6-mm thick)

Wire: H08Mn2Si, diameter φ1.2 mm

Shielding gas: Argon plus 2% O_2, 18.L/min gas flow rate

To verify the reliability of the model in a larger range of processing parameters, the main welding parameters (e.g., welding current I, welding speed v_0, and CTW distance W_h) are changed in a wide range. For example, the welding current *I* is chosen from 150 to 300 A, and the variation range of welding speed v_0 is 300–550 mm/min; three values, such as 15, 20 (normal), and 25 mm, are adopted for CTW distance.

For the verification of fluid flow fields inside the weld pool, there are still many difficulties in the experimental technique at present. However, because fluid flow fields are correlated with the temperature fields, the verification of the temperature field is also an indirect verification of the calculated results of the fluid flow fields inside the weld pool. In actual operation, the verification of the calculated temperature profiles is made by comparing the measured temperature distribution in base metal and the size and shape of the weld bead with predicted results. During the welding process, the temperature variations of some points in the specimen are measured to obtain the temperature distribution curves on the specimen. Thermocouples with the *x-y* function recorder are used to detect the temperature. After welding, the welds are sectioned at different positions along the welding bead to make macrophotographs. Then, geometrical information about the weld depth, width, and penetration and the weld reinforcement are obtained using the reading microscope. Many experimental tests are performed, and the average of several repeated measurements are taken as the final results.

8.8.2 Experimental Verification of Weld Geometry

Figure 8.31 shows the relationship between the measured and the calculated values of the penetration depth, bead width, weld reinforcement height and area, and penetrated area under different levels of welding current *I*, while welding speed v_0 and CTW distance are fixed (v_0 = 430 mm/min and W_h = 20 mm). It can be seen that the calculated and measured values are in good agreement. The range of experimental and calculated welding current is from 150 to 300 A, so two common types of droplet transfer, such as globular transfer and spray transfer, are included. Therefore, the established model is applicable not only to the spray transfer but also to the globular transfer mode.

When welding current *I* and welding speed v_0 are fixed but CTW distance W_h is different, the calculated and measured values of the weld dimensions are shown in Table 8.6. It can be seen that the relative error of the calculated and measured results is less than 10%.

According to the calculated and measured results, Figure 8.32 shows the influences of changes in welding parameters (welding current *I*, welding speed v_0, CTW distance W_h) on the transverse cross-sectional shape of welds. Figures 8.32(a), 8.32(b), and 8.32(c) demonstrate the variation of geometry shape of the transverse cross section when the welding current gradually increases, Figures 8.32(c), 8.32(d), and 8.32(e) illustrate the comparison between the predicted and measured geometry shape of the weld pool on the cross section at different welding speeds, while Figures 8.32(c) and 8.32(f) show those when CTW changes. The calculated and measured data on both the fusion line and the contour of the weld reinforcement are in good agreement.

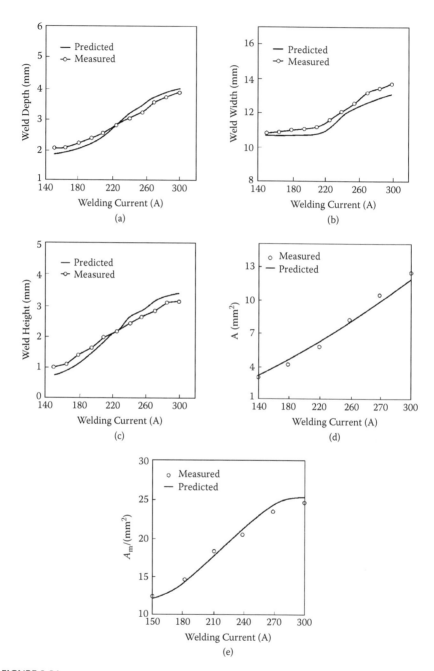

FIGURE 8.31

Comparison of calculated and measured results of weld geometry at different welding currents (MAG welding condition A, W_h = 20 mm): (a) penetration; (b) bead width; (c) reinforcement height; (d) the cross-sectional area of the reinforcement; (e) the cross-sectional area of the melting part of the base metal.

TABLE 8.6

The Effect of CTW W_h on Weld Dimensions

W_h (mm)	Depth (mm)			Width (mm)			Reinforcement (mm)		
	Predicted	Measured	Error(%)	Predicted	Measured	Error(%)	Predicted	Measured	Error(%)
15	3.92	3.78	3.70	11.28	11.46	1.57	2.78	2.72	2.20
20	2.90	3.09	6.15	12.00	12.34	2.76	2.59	2.67	3.00
25	2.01	2.21	9.05	14.21	13.98	1.65	2.51	2.62	4.20

Note: MAG welding condition A with a changing W_h.

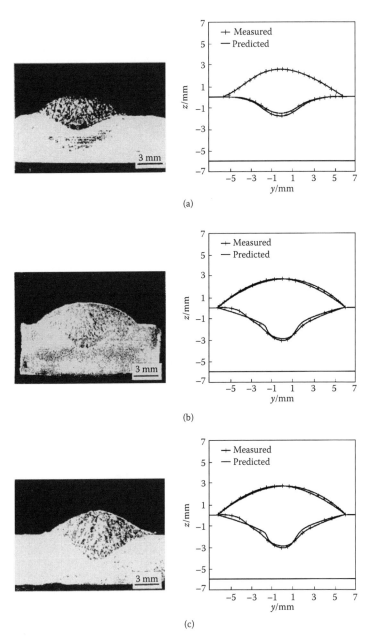

FIGURE 8.32
The comparison of calculated and experimental weld geometry (macrophotography on left and transverse cross sections on right): (a) $I = 200$ A, $v_0 = 430$ mm/min, $W_h = 20$ mm; (b) $I = 240$ A, $v_0 = 430$ mm/min, $W_h = 20$ mm; (c) $I = 270$ A, $v_0 = 430$ mm/min, $W_h = 20$ mm.

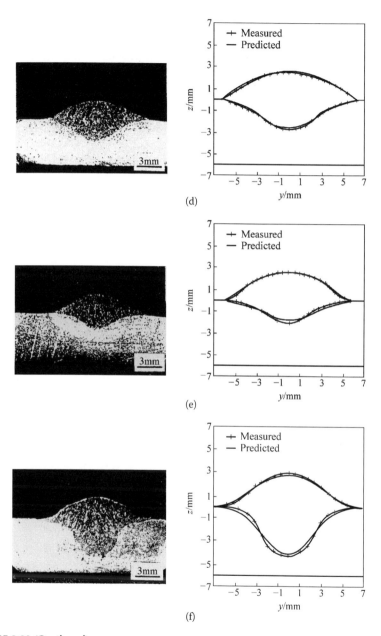

(d)

(e)

(f)

FIGURE 8.32 (Continued)
The comparison of calculated and experimental weld geometry (macrophotography on left and transverse cross sections on right): (d) $I = 270$ A, $v_0 = 460$ mm/min, $W_h = 20$ mm; (e) $I = 270$ A, $v_0 = 510$ mm/min, $W_h = 20$ mm; (f) $I = 270$ A, $v_0 = 430$ mm/min, $W_h = 15$ mm).

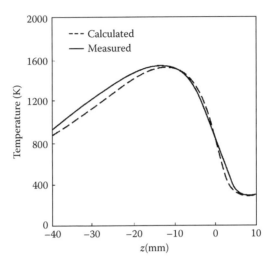

FIGURE 8.33
Comparison of calculated and measured temperature distribution along the weld centerline on the bottom surface (MAG welding condition A). (From Wu C S and Sun J S, *Proc. Instn. Mech. Eng.*, Part B: *J. Eng. Manuf.*, 212: 525–531, 1998.)

8.8.3 The Experimental Verification of Temperature Distribution in the Specimen

Figure 8.33 illustrates the calculated and experimental results of temperature distribution along the weld centerline ($y = 0$ mm) on the bottom surface of the workpiece. The solid line is the measured temperature, while the dashed line represents the calculated temperature. It can be seen that the calculated and measured temperature values are basically consistent with each other.

9

Weld Pool and Keyhole Behaviors in Plasma Arc Welding

9.1 Introduction

The density of energy deposited on a workpiece is the prime determinant in the efficiency with which melting in fusion welding processes takes place. If the energy density coming from a heat source is high enough, like the cases in plasma arc welding (PAW), laser beam welding (LBW), and electron beam welding (EBW), the rate at which it is deposited greatly exceeds the rate at which it is lost by conduction into the workpiece. In this case, the material at the point of deposition rises in temperature not just to the melting point but also well above that. In fact, the temperature can rise to the boiling point, converting liquid metal to vapor, superheating the vapor. When this occurs, the energy source is said to be operating in the *keyhole mode*. This mode begins to occur at energy densities above around 10^9 W/m^2.

Once vapor is formed by such high energy densities, it expands, is released upward away from the surface, and it produces a reaction force that presses the melt downward and sideways. With a combination of arc pressure, if any, the result is a depression that permits additional electrons and ions (from a plasma arc), photons (from a laser beam), or electrons (from an electric beam) to impinge on fresh material, which is then heated in the same way. The depression becomes larger and transforms to a keyhole, the entire central core of which consists of plasma or vapor surrounded on all sides by an envelope of liquid metal. For sufficient energy input, this keyhole will penetrate entirely through the thickness of the workpiece, even if this is several centimeters. In this way, the faces of the two joint elements can be melted, the molten material flows around the keyhole and joins together at the weld pool, metallurgical continuity is obtained, and solidification can occur to produce a weld.

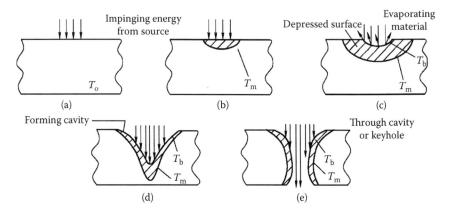

FIGURE 9.1
Schematic showing the extremely rapid sequence of steps leading to the formation of the keyhole. T_m, melting point; T_b, boiling point.

Figure 9.1 schematically shows the extremely rapid sequence of steps leading to the formation of a keyhole when the energy density deposited is high enough: (a) initial deposition of impinging energy on the surface; (b) surface heating and inward propagation of heat to cause melting anywhere $T > T_m$; (c) continued inward propagation of heat to raise a larger volume of material to above T_m, and elevation of temperature near the surface to above T_b, causing vaporization and a downward force on the liquid; (d) continued deposition of heat, increased vaporization and greater depression of liquid, and increased growth of melted volume; and (e) eventual penetration of vapor cavity through the thickness to produce a keyhole of vapor or plasma surrounded by molten material.

This chapter focuses on the keyhole and weld pool behaviors in PAW. The fluid flow and heat transfer phenomena are very complicated when the weld pool is coupled with a keyhole. From the viewpoint of modeling, there are two main methods for numerical simulation of the PAW process: (1) Consider the physical mechanism of keyhole formation and establish a unified model to integrate the keyhole and weld pool behaviors together. Due to the complexity of the keyhole formation process, some simplifying assumptions have to be made, and the keyhole shape is prescribed as a cylinder[99,105] or a cone.[102,160] With such a prescribed keyhole, the weld pool shape and size and its surrounding temperature profiles are computed. (2) Do not consider the formation process of the keyhole; put emphasis on the effect of the keyhole on the volumetric distribution of heat density inside the weld pool and develop suitable heat source modes.[100,109] In this chapter, the second method is used first to conduct numerical simulation of the temperature field in PAW, and then a simple approach describing the keyhole shape is introduced.

9.2 Finite Element Analysis of Quasi-Steady State Temperature Field in PAW

The arc used in PAW is constricted by a small nozzle and has a much higher gas velocity (300–2,000 m/s)[160] and heat input intensity (10^9–10^{10} w/m²)[81] than that in gas tungsten arc welding (GTAW). As the plasma arc impinges on the area where two workpieces are to be joined, it can melt material and create a molten liquid pool. Because of its high velocity and the associated momentum, the arc can penetrate through the molten pool and form a keyhole in the weld pool. Moving the welding torch and the associated keyhole will cause the flow of the molten metal surrounding the keyhole to the rear region, where it resolidifies to form a weld bead. The keyhole mode of welding is the primary attribute of high-power-density welding processes (PAW, LBW, and EBW) that makes them penetrate thicker pieces with a single pass. Compared to LBW and EBW, keyhole PAW is more cost-effective and more tolerant of joint preparation. However, in keyhole PAW, the quality of the weld depends on keyhole stability, which itself depends on a large number of factors, especially the physical characteristics of the material to be welded and the welding process parameters to be used. Thus, keyhole PAW is susceptible to the variation of welding process parameters, which makes it have a narrower range of applicable process parameters for good weld quality so that keyhole PAW is still limited to wide application in industry.[161] The temperature profile around the weld pool has great influence on the formation and stability of the keyhole. Through numerical simulation of the temperature field and the weld pool behavior in keyhole PAW, the process parameters can be optimized for obtaining a high-quality weld structure.[162,163] Therefore, it is of great significance to model and simulate the temperature distribution and weld pool geometry in the keyhole PAW process.

Because of the complexity of the phenomena associated with the formation of a keyhole, only a limited number of theoretical studies treating the PAW process have been reported, each of varying degrees of approximation and each focusing on different aspects of the problem. General Gaussian or double-ellipsoid heat source models[75,76,122,164–166] used widely in simulation of arc welding processes are not suitable for keyhole PAW.[77] As a first step in a series of studies, this section focuses on developing a suitable heat source model for finite element analysis of the temperature profile in keyhole PAW.

9.2.1 Volumetric Heat Source Mode in Keyhole PAW

As mentioned, the key problem in the finite element analysis of keyhole PAW is how to model the welding heat source. Most researchers employed a Gaussian distribution of heat flux (W/m²) deposited on the surface of the workpiece. Although such a surface mode of a heat source may be used for the shallow penetration arc welding processes like GTAW, it does not reflect the digging action of the arc, so it is not suitable for modeling the welding processes with deeper penetration like gas metal arc welding (GMAW). Goldak et al.[122] proposed a double-ellipsoid heat source model that has the capability to analyze the thermal history of deep penetration welds like those of GMAW. However, the double-ellipsoid distribution of heat intensity (W/m^3) is still not applicable to the high-density-welding processes with a high ratio of the weld penetration to width such as keyhole PAW.

Figure 9.2 shows the schematic of the weld pool and keyhole, while Figure 9.3 shows a transverse cross section of a keyhole PAW weld. Keyhole PAW produces a weld with a high depth-to-width ratio. The transverse cross section of the weld is of the "bugle-like" configuration. To consider the strong digging action of the plasma jet and the resulting bugle-like weld configuration in keyhole PAW, new types of welding heat source models must be proposed.

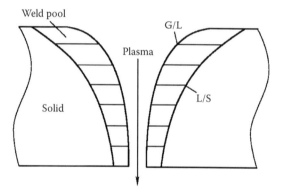

FIGURE 9.2
Schematic of the weld pool and keyhole.

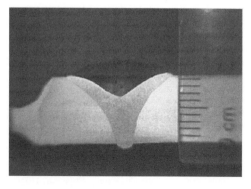

FIGURE 9.3
Transverse cross section of the PAW weld.

The software package SYSWELD is equipped with the following four kinds of volumetric heat source modes: double-ellipsoid heat source, volumetric heat source mode exerted along the plate thickness, three-dimensional conical (TDC) heat source, and modified three-dimensional conical (MTDC) heat source.

9.2.1.1 Double-Ellipsoid Heat Source

As shown in Figure 9.4, a double-ellipsoid (DE) heat source is introduced in Section 2.6. The welding direction is along the y-axis. For the heat density in a half-ellipsoid with a positive y-coordinate,

$$q_V(x,y,z) = \frac{6\sqrt{3}f_fQ}{a_h b_f c_h \pi \sqrt{\pi}} \times \exp\left(-\frac{3x^2}{a_h^2}\right)\exp\left(-\frac{3z^2}{c_h^2}\right)\exp\left(-\frac{3y^2}{b_f^2}\right) \qquad (9.1)$$

For the heat density in a half-ellipsoid with a negative y-coordinate,

$$q_V(x,y,z) = \frac{6\sqrt{3}f_rQ}{a_h b_r c_h \pi \sqrt{\pi}} \times \exp\left(-\frac{3x^2}{a_h^2}\right)\exp\left(-\frac{3z^2}{c_h^2}\right)\exp\left(-\frac{3y^2}{b_r^2}\right) \qquad (9.2)$$

where (a_h, b_f, b_r, c_h) are the distribution parameters, (f_f, f_r) are the fractions of the heat deposited in the front and rear semiellipsoids ($f_f + f_r = 2$), and $Q = \eta I U_a$ is the effective energy deposited (η is the weld thermal efficiency, I is the welding current, U_a is the arc voltage).

9.2.1.2 Volumetric Heat Source Mode Exerted along the Plate Thickness

The volumetric heat source mode exerted along the plate thickness is based on two assumptions: (1) The heat intensity deposited region is maximum at the top surface of the workpiece and is minimum at the bottom surface of the

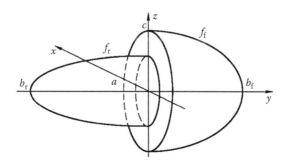

FIGURE 9.4
Double-ellipsoid (DE) heat source mode.

workpiece; (2) along the thickness of the workpiece, the diameter of the heat density distribution region is linearly decreased. But, the heat density at the central axis (z-direction) is kept constant. At any plane perpendicular to the z-axis, the heat intensity is distributed in a Gaussian form. Thus, in fact, it is the repeated addition of a series of Gaussian heat sources with different distribution parameters and the same central maximum values of heat density along the workpiece thickness. In this way, the heating action of the plasma jet through the workpiece in keyhole PAW is considered.

At any plane perpendicular to the z-axis, the heat flux distribution may be written as

$$q_V(r,z) = q_{V0} \exp\left(-\frac{3r^2}{r_0^2}\right) \tag{9.3}$$

where q_{v0} is the maximum value of heat intensity, r_0 is the distribution parameter, and r is the radial coordinate. The key problem is how to determine the parameters q_{v0} and r_0.

According to energy balance on the workpiece,

$$
\begin{aligned}
\eta U_a I &= \int_0^H \int_0^{2\pi} \int_0^{r_0} q_V(r,z)\, r\, dr\, d\theta\, dh = \int_0^H \int_0^{2\pi} \int_0^{r_0} q_{V0} \exp\left(-\frac{3r^2}{r_0^2}\right) r\, dr\, d\theta\, dh \\
&= -\frac{\pi q_{V0}}{3} \int_0^H r_0^2 \int_0^{r_0} \exp\left(-\frac{3r^2}{r_0^2}\right) d\left(-\frac{3r^2}{r_0^2}\right) dh \\
&= \frac{\pi q_{V0}(1-e^{-3})}{3} \int_0^H r_0^2 \, dh
\end{aligned}
\tag{9.4}
$$

where $\eta U_a I$ is the effective energy deposited, and H is the thickness of the workpiece. Different modes of volumetric heat source correspond to different values of q_{V0} and r_0.

9.2.1.3 Three-Dimensional Conical Heat Source

As shown in Figure 9.5, the height of the conical heat source is $H = z_e - z_i$; the z-coordinates of the top and bottom surfaces are z_e and z_i, respectively; and the diameters at the top and bottom are r_e and r_i, respectively. The distribution parameter r_0 can be expressed as

$$r_0(z) = r_e - (r_e - r_i)\frac{z_e - z}{z_e - z_i} \tag{9.5}$$

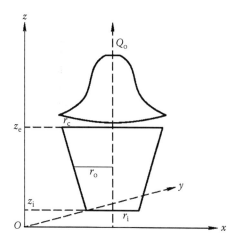

FIGURE 9.5
Schematic of three-dimensional conical (TDC) heat source.

It may written as

$$r_0^2 = \left[r_i + (r_e - r_i)\frac{h}{H} \right]^2, h = z - z_i \tag{9.6}$$

Substituting Eq. (9.6) into Eq. (9.4) to calculate the integral term,

$$\int_0^H r_0^2 dh = \int_0^H \left[r_i + (r_e - r_i)\frac{h}{H} \right]^2 dh = \frac{H}{3}\left(r_e^2 + r_e r_i + r_i^2 \right) \tag{9.7}$$

Substituting Eq. (9.7) into Eq. (9.4), we have

$$\eta I U_a = \frac{\pi q_{V0} H(1 - e^{-3})}{9}\left(r_e^2 + r_e r_i + r_i^2 \right)$$

After manipulation,

$$q_{V0} = \frac{9\eta I U_a e^3}{\pi(e^3 - 1)} \times \frac{1}{H\left(r_e^2 + r_e r_i + r_i^2 \right)} \tag{9.8}$$

Substituting Eq. (9.8) into Eq. (9.3), we obtain

$$q_V(r, z) = \frac{9\eta I U_a e^3}{\pi(e^3 - 1)} \times \frac{1}{(z_e - z_i)\left(r_e^2 + r_e r_i + r_i^2 \right)} \exp\left(-\frac{3r^2}{r_0^2} \right) \tag{9.9}$$

where

$$r_0 = r_0(z) = r_i + (r_e - r_i)\frac{z - z_i}{z_e - z_i}$$

9.2.1.4 Modified Three-Dimensional Conical Heat Source

In keyhole PAW, the bugle-like weld configuration results. To reflect this feature, an MTDC heat source is proposed (Figure 9.6). For an MTDC heat source, the distribution parameter r_0 no longer declines as linearly as for TDC, but in a curvilinear way.

As shown in Figure 9.6, the height of the MTDC heat source is $H = z_e - z_i$; the z-coordinates of the top and bottom surfaces are z_e and z_i, respectively; and the diameters at the top and bottom are r_e and r_i, respectively. Let r_0 represent the distribution parameter at z. Then, r_0 can be expressed as

$$r_0 (z) = a_3 \ln z + b_3 \tag{9.10}$$

since

$$r_i = a_3 \ln z_i + b_3$$

$$r_e = a_3 \ln z_e + b_3$$

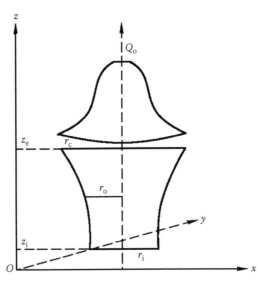

FIGURE 9.6
Modified three-dimensional conical (MTDC) heat source.

We have

$$a_3 = \frac{r_e - r_i}{\ln z_e - \ln z_i} \tag{9.11}$$

$$b_3 = \frac{r_i \ln z_e - r_e \ln z_i}{\ln z_e - \ln z_i} \tag{9.12}$$

Substituting Eqs. (9.11) and (9.12) into Eq. (9.10), we obtain

$$r_0(z) = \frac{(r_e - r_i) \times \ln(z)}{\ln(z_e) - \ln(z_i)} + \frac{r_i \times \ln(z_e) - r_e \times \ln(z_i)}{\ln(z_e) - \ln(z_i)} \tag{9.13}$$

Let $h = z - z_i$. Since

$$\int_0^H r_0^2 dh = \int_0^H (a_3 \ln z + b_3)^2 dh = \int_0^H [a_3 \ln(h + z_i) + b_3]^2 dh$$

$$= a_3^2 \int_0^H \ln^2(h + z_i) \, dh + 2a_3 b_3 \int_0^H \ln(h + z_i) dh + b_3^2 H \tag{9.14}$$

Based on a mathematics handbook,

$$\int \ln u \, du = u \ln u - u + C$$

$$\int \ln^2 u \, du = u \ln^2 u - 2 \int \ln u \, du = u \ln^2 u - 2(u \ln u - u) + C$$

We get

$$\int_0^H r_0^2 dh = a_3^2 \int_0^H \ln^2(h + z_i) dh + 2a_3 b_3 \int_0^H \ln(h + z_i) dh + b_3^2 H$$

$$= a^2 \left[(h + z_i) \ln^2(h + z_i) - 2(h + z_i) \ln(h + z_i) + 2(h + z_i) \right] \Big|_0^H$$

$$+ 2a_3 b_3 \left[(h + z_i) \ln(h + z_i) - (h + z_i) \right] \Big|_0^H + b_3^2 H$$

$$= a_3^2 \left[(H + z_i) \ln^2(H + z_i) - z_i \ln^2 z_i \right]$$

$$- 2a_3 (a_3 - b_3) \left[(H + z_i) \ln(H + z_i) - z_i \ln z_i - H \right] + b_3^2 H$$

Let

$$A_V = \int_0^H r_0^2 dh = a_3^2[(H+z_i)\ln^2(H+z_i) - z_i \ln^2 z_i]$$
$$-2a_3(a_3 - b_3)[(H+z_i)\ln(H+z_i) - z_i \ln z_i - H] + b_3^2 H \qquad (9.15)$$

Substituting Eq. (9.15) into Eq. (9.4), we get

$$\eta I U_a = \frac{\pi q_{V0}(1 - e^{-3})}{3} \int_0^H r_0^2 dh = \frac{\pi q_{V0}(1 - e^{-3})}{3} A_V$$

After manipulation,

$$q_{V0} = \frac{3\eta U_a I e^3}{A_V(e^3 - 1)} \qquad (9.16)$$

Substituting Eq. (9.16) into Eq. (9.3), we obtain

$$q_V(r,z) = \frac{3\eta U_a I e^3}{A_V \pi (e^3 - 1)} \exp\left(-\frac{3r^2}{r_0^2}\right) \qquad (9.17)$$

where $r_0 = r_0(z)$ is defined by Eq. (9.13).

9.2.2 Quasi-Steady State PAW Heat Source

The high plasma gas velocity and the associated momentum force the plasma jet to penetrate the base metal, forming a complete penetration keyhole in the weld pool. Thus, the plasma jet emerges from the underbead at the bottom of the workpiece. The size and shape of the keyhole produced by PAW depend mainly on the pressure of the impinging gas. Figure 9.2 is the approximate schematic of the weld pool and keyhole in PAW. The key issue for modeling the PAW process is how to develop a suitable heat source mode that takes into account the "digging action" of plasma and its distribution along the workpiece thickness.

The existence of the keyhole, on one hand, makes the heat intensity from the plasma arc distribute through the workpiece thickness, but on the other hand, it causes some vaporization, which results in some heat losses. Therefore, the net heat input is not totally deposited on the workpiece. To consider this point, part of the heat density at the upper section of the keyhole is excluded. As shown in Figure 9.7, the heat intensity is distributed within a domain bounded by the curves $f_0(z)$ and $f_1(z)$ at the upper section

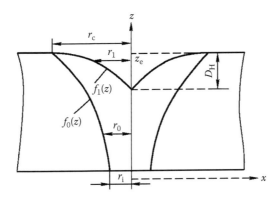

FIGURE 9.7
The heat source mode of QPAW.

of the keyhole, while at the lower section of the keyhole, the heat intensity is distributed as a modified conical heat source. This model of heat source is proposed for the quasi-steady state PAW process, so it is referred to as QPAW for short.[167] The heat density distribution of QPAW is described by Eqs. (9.15)–(9.17), while $f_0(z)$ and $f_1(z)$ are expressed by

$$f_0(z) = r_0(z) = \frac{(r_e - r_i) \times \ln(z)}{\ln(z_e) - \ln(z_i)} + \frac{r_i \times \ln(z_e) - r_e \times \ln(z_i)}{\ln(z_e) - \ln(z_i)} \qquad (9.18)$$

$$f_1(z) = \frac{r_e}{\sqrt{D_H}}\sqrt{z + D_H - z_e}, \quad (z_e - D_H \le z \le z_e) \qquad (9.19)$$

where D_H is the height of the excluded part corresponding to the surface depression, as shown in Figure 9.7. Only the domain between $f_0(z)$ and $f_1(z)$ is related to the volumetric heat source, within which there is volumetric distribution of heat density. The distribution mode of heat density follows Eq. (9.17), but $r_0(z)$ is defined as

$$r_0 = r_0(z) = f_0(z), \quad z_i \le z \le (z_e - D_H);$$

$$r_0 = f_0(z) \text{ but } f_1(z) \le r \le f_0(z), \quad (z_e - D_H \le z \le z_e) \qquad (9.20)$$

9.2.3 Mesh Generation

The workpiece material is stainless steel SS304 with dimensions $200 \times 80 \times 9.5$ (mm³). The half-workpiece is divided into eight-node hexahedrons.

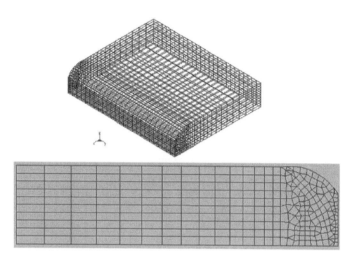

FIGURE 9.8
Mesh generation corresponding to the QPAW heat source mode.

As mentioned, part of heat density at the upper section of the keyhole is excluded because of evaporation loss and the keyhole effect. To reflect this characteristic and match the distribution mode of heat source QPAW, some elements in the upper section of the keyhole are treated as "dead." The discrete grids are shown in Figure 9.8. In total, the half-workpiece is discretized into 14,000 elements and 16,968 nodes.

9.2.4 FEM Analysis Results

The software package SYSWELD was used to conduct an FEM analysis of the temperature field in keyhole PAW. The PAW process parameters were as follows: $I = 250$ A, $U_a = 31.7$ V, $v_0 = 120$ mm/min, 7.1-L/min plasma gas flow rate, 11.8 L/min shielding gas flow rate, 5-mm nozzle-workpiece distance. Note this process condition as PAW test A. Under the conditions of PAW test A, the experiment demonstrated that it took about 4.5 s to achieve the quasi-steady state of the welding process, and an open keyhole through the workpiece thickness was formed. Only the weld pool and its surrounding temperature field in the quasi-steady state were computed.

First, two kinds of volumetric heat source modes (i.e., TDC and MTDC) were used to predict the PAW weld pool shape and size and temperature profile. The results are shown in Figures 9.9 and 9.10. It can be seen that the heat source mode had a great effect on the computed results.

Figure 9.11 shows the comparison of the experimental results with the predicted ones based on different heat source modes. Table 9.1 gives the data for the weld width. From the point of view of agreement, the calculation precision of TDC is poorer. It can be seen that TDC is not suitable for determining

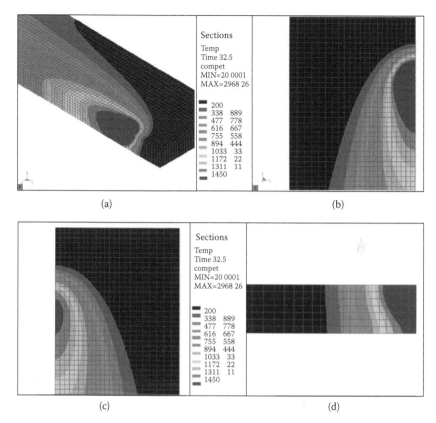

FIGURE 9.9
The weld pool shape and temperature field based on the TDC heat source mode (test A): (a) three-dimensional drawing; (b) top surface; (c) bottom surface; (d) transverse cross section.

keyhole PAW weld dimensions. Although the precision of MTDC is improved compared to TDC, especially the weld width at both the top and bottom surfaces, the calculation precision for the location and locus of the melt line in the weld cross section is still lower.

Second, the developed heat source mode for quasi-steady state keyhole PAW (i.e., QPAW) was employed to calculate the weld geometry. Figure 9.12 shows the comparison of experimental results with the predicted ones based on the QPAW mode for welding condition test A. It can be seen that the predicted weld geometry agreed well with experimental measurements. Since the QPAW mode depicts the character of the keyhole PAW process by considering the bugle-like configuration of the keyhole and the decay of heat intensity distribution of the plasma arc along the direction of the workpiece thickness, the calculation precision of the weld geometry at the transverse cross section is quite satisfactory.

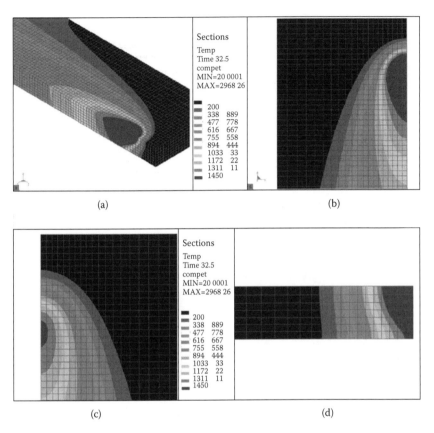

FIGURE 9.10
The weld pool shape and temperature field based on the MTDC heat source mode (test A): (a) three-dimensional drawing; (b) top surface; (c) bottom surface; (d) transverse cross section.

9.3 Numerical Simulation of Transient Development of PAW Temperature Profiles

Starting from the arc ignition, the PAW thermal process is transient until the quasi-steady state is achieved. Thus, it is necessary to develop a transient model.

9.3.1 Transient PAW Heat Source

During the period from arc ignition to quasi-steady state, both the weld pool and the keyhole experience gradual expansion. According to the PAW process characteristics and experimental observations, a heat source mode for the transient thermal process in keyhole PAW (i.e., TPAW) is developed.

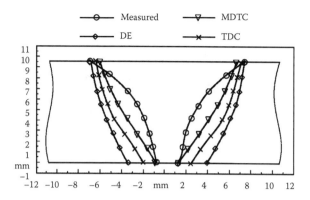

FIGURE 9.11
Comparison of predicted PAW weld cross section with experimental result under test A.

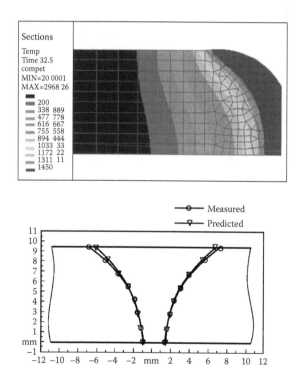

FIGURE 9.12
Transverse cross section of PAW weld (Test A). (From WUCS, Wang H G and Zhang Y M, *Weld. J.*, 85(12): 2845–2915, 2006.)

TABLE 9.1

Comparison of Predicted and Measured Weld Width under Test A

Heat Source Mode	Top-Side Weld Width (mm)		Back-Side Weld Width (mm)	
	Predicted	Measured	Predicted	Measured
3-D conical	13.60	14.35	4.25	2.11
Modified 3-D conical	12.90	14.35	2.30	2.11

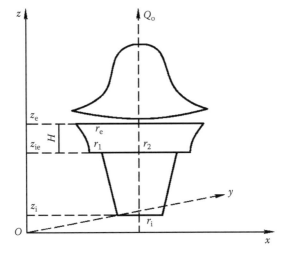

FIGURE 9.13
Schematic of TPAW mode.

As shown in Figure 9.13, the heat source mode TPAW consists of two parts based on the tapering of the radius r_0 (the domain radius within which the heat density is distributed). In the upper part, r_0 is tapered nonlinearly; it decays linearly in the lower part. This means that the upper part is the MTDC heat source mode, while the lower part is the TDC mode. The combined heat source mode is employed to describe the transient temperature field. By using a trial-and-error method, we found that the boundary between the upper and lower parts is located at the maximum depression point of the weld bead surface after welding. The TPAW mode may be expressed as

$$q_V(r,z) = \frac{9\eta IU_a e^3}{\pi(e^3 - 1)} \frac{\chi_i}{B_v} \exp\left(-\frac{3r^2}{r_0^2}\right) \tag{9.21}$$

For the upper and lower parts of the heat source, B_v takes different expressions, which are computed from Eq. (9.15) or Eq. (9.7) according to the corresponding parameters (r_e, r_1, z_e, z_{ie}) and (r_2, r_i, z_{ie}, z_i). χ_i $(i=1,2)$ corresponds to

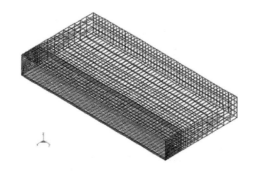

FIGURE 9.14
Mesh generation.

the fraction of heat power in the upper and lower parts of the heat source
$(\chi_1 + \chi_2 = 1)$.

$$r = \sqrt{x^2 + y^2} = \sqrt{(x - x_0)^2 + (y - y_0 - v_0 t)^2} \tag{9.22}$$

$$z_i \leq z \leq z_{ie}, \quad r_0(z) = r_e - (r_e - r_i)\frac{z_e - z}{z_e - z_i} \tag{9.23}$$

$$z_{ie} \leq z \leq z_e, \quad r_0(z) = \frac{(r_e - r_i) \times \ln(z)}{\ln(z_e) - \ln(z_i)} + \frac{r_i \times \ln(z_e) - r_e \times \ln(z_i)}{\ln(z_e) - \ln(z_i)} \tag{9.24}$$

9.3.2 FEM Analysis Results of the Transient Temperature Field in PAW

For FEM analysis of the results of the transient temperature field in PAW,
Figure 9.14 shows the grid system used in the computation. The modified
value of thermal conductivity λ and the combined heat source mode TPAW
are employed to consider the effects of plasma force action and fluid flow in
the weld pool indirectly.

As mentioned, under the conditions of PAW test A, it took 4.5 s to achieve
quasi-steady state and to form an open keyhole. Figure 9.15 is the calculated
evolution process of the weld pool and temperature profiles during this
transit period.

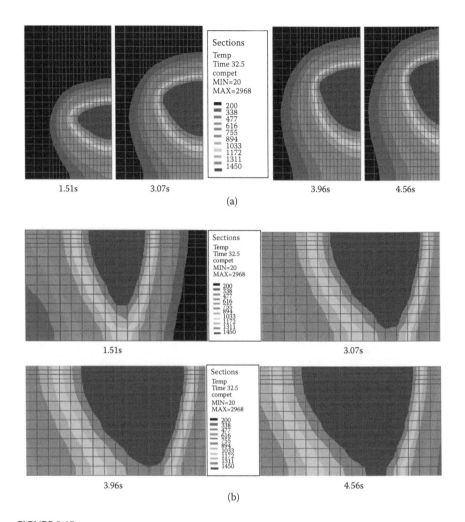

FIGURE 9.15
The computed results of the transient PAW temperature field (test A): (a) top surface; (b) longitudinal cross section; (c) (next page) transverse cross section. (From Wu C S, Wang H G, and Zhang M X, *Acta Metall. Sinica*, 42(3): 311–316, 2006.)

To demonstrate the variation of the weld pool with time, the fusion line was extracted from Figure 9.15. Figure 9.16 illustrates the evolution of the weld pool shape and size computed by the TPAW mode. It can be seen that, during the PAW process, the penetration rises more quickly to show the heat density distribution along the thickness of the workpiece. The transmit period from arc ignition to quasi-steady state is computed as 4.56 s, which matches well with the measured value of 4.50 s.

To observe the weld geometry evolution of keyhole PAW during the period from the arc ignition to the quasi-steady state, a special program

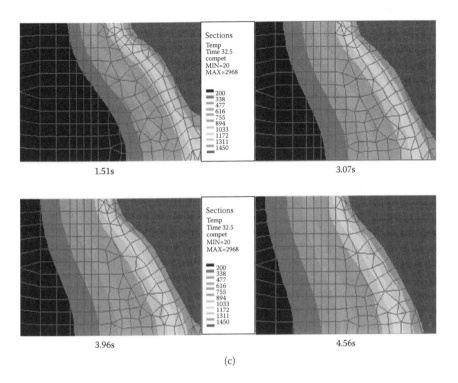

(c)

FIGURE 9.15 (Continued)
The computed results of the transient PAW temperature field (test A): (a) top surface; (b) longitudinal cross section; (c) transverse cross section. (From Wu C S, Wang H G, and Zhang M X, *Acta Metall. Sinica*, 42(3): 311–316, 2006.)

was developed to control the welding experiments. For each case, quite a few experiments were carried out under the same conditions but with different welding times. For example, after the arc ignition, welding was started and then stopped at 3.0 s. Then, the weld geometry and penetration at this moment remained. Another welding was started and then stopped at 4.5 s. In this way, the weld geometry and penetration at different moments could be obtained after processing of weld samples so that indirect information on keyhole formation and evolution could be extracted. Figure 9.17 shows the transverse cross section of the PAW weld at six different instants.

Figure 9.18 compares the predicted and measured transverse cross sections of the PAW weld. For the initial period (less than 3.0 s), there is a difference between the calculated and measured weld dimensions. But, as time passes, the difference narrows, and both are in good agreement at 4.0 s.

The above-mentioned work is only a first stage in modeling the keyhole PAW thermal process because it is just based on the macroscopic heat transfer principle. It should be pointed out that further investigation needs to be carried out for accurate simulation of the keyhole PAW process.

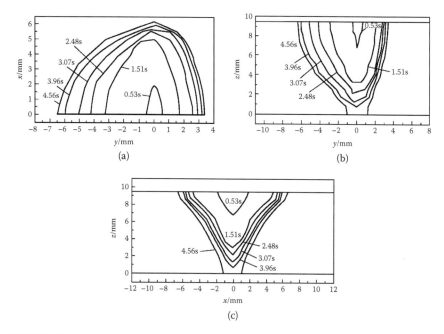

FIGURE 9.16

Variation of weld pool with time (PAW test A): (a) top surface; (b) longitudinal cross section; (c) transverse cross section. (From Wu C S, Wang H G, and Zhang M X, *Acta Metall. Sinica*, 42(3): 311–316, 2006.)

9.4 Numerical Analysis of Double-Sided PAW plus Tungsten Inert Gas Welding

As can be seen in Figure 9.19, a regular PAW system uses an electrical connection (ground cable) between the workpiece and the power supply to allow the welding current to complete the loop. The electric arc is established between the workpiece and the torch, and there is almost no current flowing along the thickness direction, so the plasma arc has limited penetration. To increase the welding productivity of thick plates, Zhang of the University of Kentucky invented and developed a new arc welding process, double-sided arc welding (DSAW).[168,169] As shown in Figure 9.20, in a DSAW system the workpiece is disconnected from the power supply, and a second torch (tungsten inert gas, TIG) is placed on the opposite side of the workpiece to complete the current loop. As a result, electric arcs are simultaneously established between the workpiece and each of two torches. It is found that this configuration has greatly improved penetration capability. For example, stainless steel plates up to 12.7-mm (1/2-inch) thick can be welded in a single pass without bevels under 70-A welding current.[168] Wu et al. conducted joint

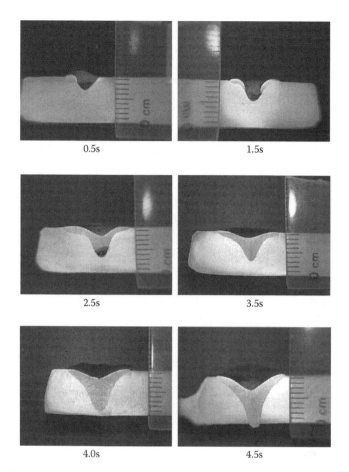

FIGURE 9.17
Macrophotographs of PAW welds.

research with Zhang and simulated the thermal process associated with DSAW numerically. This section introduces the relevant results.[169,170]

9.4.1 Formulation

As shown in Figure 9.20, the workpiece is at a horizontal position; the PAW torch is positioned above and the TIG torch is positioned below the workpiece. Two arcs move at constant speed u_0 along the x-axis. The moving coordinate system o-xyz travels with the same speed as the arcs, and its origin is located at the intersection of the electrode centerline and the bottom surface of the workpiece.

Figure 9.21 is the schematic of a longitudinal cross section of the weld pool during the keyhole formation process in DSAW. The top-side pool surface is

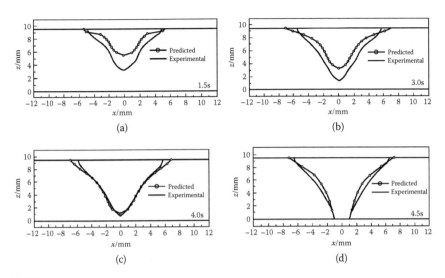

FIGURE 9.18
Comparison of predicted and measured transverse cross sections of PAW weld at different times: (a) 1.5 s; (b) 3.0 s; (c) 4.0 s; (d) 4.5 s (test A). (From Wu C S, Wang H G, and Zhang M X, *Acta Metall. Sinica*, 42(3): 311–316, 2006.)

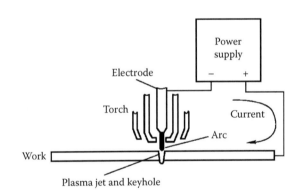

FIGURE 9.19
Schematic of regular PAW system.

deformed under action of plasma arc pressure P_p, gravity $\rho g \varphi$, and surface tension γ, which is governed by

$$P_p - \rho g \varphi + C_4 = -\gamma \frac{\left(1+\varphi_y^2\right)\varphi_{xx} - 2\varphi_x\varphi_y\varphi_{xy} + \left(1+\varphi_x^2\right)\varphi_{yy}}{\left(1+\varphi_x^2+\varphi_y^2\right)^{\frac{3}{2}}} \qquad (9.25)$$

where C_4 is a constant. The plasma arc pressure P_p depends on the distribution of the current density.

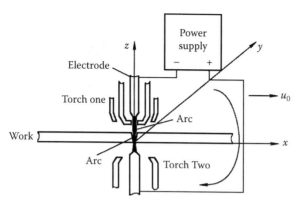

FIGURE 9.20
Schematic of DSAW system.

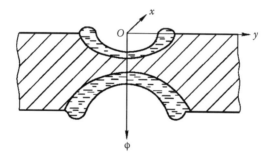

FIGURE 9.21
Schematic of weld pool before the open keyhole is formed in DSAW.

With a coordinate system in Figure 9.21, the back-side pool surface deformation ψ is expressed as

$$-P_T + \rho g(\psi - H) + C_5 = -\gamma \frac{\left(1 + \psi_y^2\right)\psi_{xx} - 2\psi_x \psi_y \psi_{xy} + \left(1 + \psi_x^2\right)\psi_{yy}}{\left(1 + \psi_x^2 + \psi_y^2\right)^{\frac{3}{2}}} \tag{9.26}$$

where P_T is the TIG arc pressure, C_5 is a constant, and H is the workpiece thickness.

When an open keyhole is established through the thickness of the workpiece, the upper and bottom weld pools merge together to form one molten pool, which is schematically illustrated in Figure 9.22. The pool surface is governed by

$$P_p - P_T + \rho g \Psi + C_6 = -\gamma \frac{\left(1 + \Psi_y^2\right)\Psi_{xx} - 2\Psi_x \Psi_y \Psi_{xy} + \left(1 + \Psi_x^2\right)\Psi_{yy}}{\left(1 + \Psi_x^2 + \Psi_y^2\right)^{\frac{3}{2}}} \tag{9.27}$$

where C_6 is a constant.

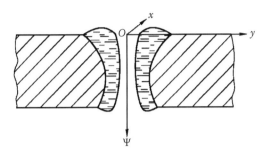

FIGURE 9.22
Schematic of weld pool with an open keyhole in DSAW.

The boundary conditions associated with Eqs. (9.25)–(9.27) are as follows:

$$\varphi, \psi, \Psi(x,y) = 0, \qquad T \le T_{\mathrm{L}} \tag{9.28}$$

where T_{L} is the liquidus temperature. The physical meaning of Eq. (9.28) is that there is no surface deformation outside the molten region.

Take the weld thermal process in DSAW as the heat conduction problem. Using enhanced thermal conductivity, account for the effect of fluid flow indirectly. At each time step, calculate the temperature field first and determine the weld pool shape and size. Then, compute the surface deformation for both sides and judge if a single pool is formed. Determine the heat density distribution according to the weld pool and keyhole shape. Iteration continues until convergence.

9.4.1 Computed Results

The finite difference method was used to solve the equations numerically in the surface-fitted coordinate system utilizing uneven computational grids as determined based on the temperature gradient. In the computation, the thickness, width, and length of the workpiece were 9.5, 80, and 150 mm, respectively. The DSAW conditions were as follows: 4.8 mm electrode diameter, 6-mm PAW torch nozzle-to-workpiece distance, 10-mm TIG arc length, 23.19.kJ/cm heat input.

Figures 9.23 and 9.24 illustrate the results of computed temperature profiles in longitudinal and transverse directions, respectively, during the dynamic process of keyhole formation. The vertical dotted line of $x = 0$ indicates the starting point of welding. It can be seen that the full penetration and the fully penetrated keyhole, for the given current (67 A), welding speed (1.3 mm/s), and workpiece (9.5-mm thick stainless steel), were established in approximately 1.35 and 1.65 s after the arc was ignited, respectively. As can be observed in the figures, when full penetration was established, the sum of the depths of the partial cavities on the two sides of the workpiece was

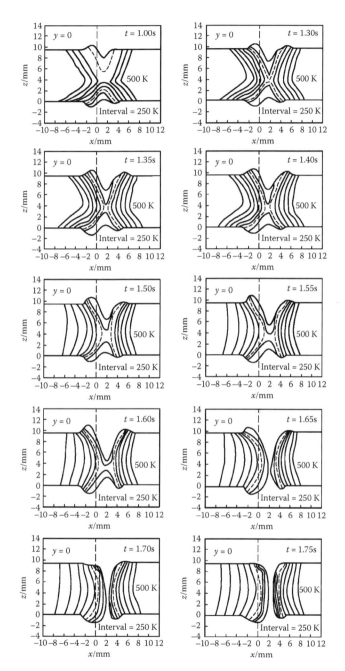

FIGURE 9.23

Dynamic development of temperature field at a longitudinal cross section in DSAW (from Wu C S, Sun J S, and Zhang Y M, *Model. Simul. Mater. Sci. Eng.*, 12: 423–442, 2004). The vertical dotted line at $x = 0$ indicates the location where the arc was ignited. Solid line, isothermal; dotted line, solid-liquid boundary. Isothermal interval = 250 K.

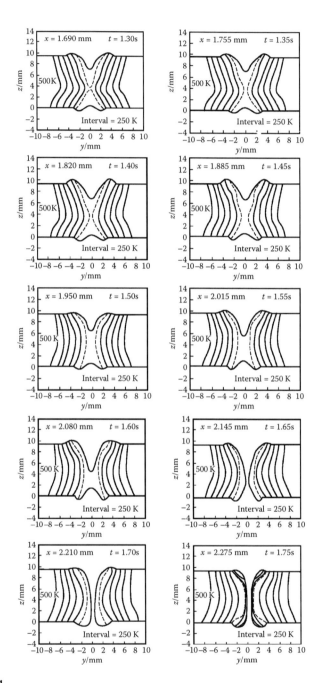

FIGURE 9.24
Dynamic development of temperature field at a transverse cross section in DSAW (from Wu C S, Sun J S, and Zhang Y M, *Model. Simul. Mater. Sci. Eng.*, 12: 423–442, 2004). The vertical dotted-line at $x = 0$ indicates the location where the arc was ignited. Solid line, isothermal; dotted line, solid-liquid boundary. Isothermal interval = 250 K.

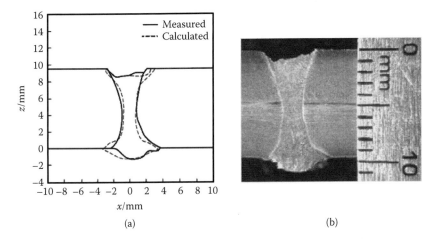

FIGURE 9.25
Comparison of predicted and measured weld dimension: (a) comparison; (b) macrophotograph. (From Wu C S, Sun J S, and Zhang Y M, *Model. Simul. Mater. Sci. Eng.*,12: 423–442, 2004.)

only approximately 3 mm, less than one-third of the workpiece thickness. The remaining 6.5 mm (more than two-thirds) of the workpiece thickness was rapidly penetrated in only approximately 0.3 s to establish the fully penetrated keyhole. It appears that the establishment of full penetration accelerates the establishment of the keyhole.

Figure 9.25 compares the computed and measured weld shape and size at a transverse cross section. Both are in good agreement.

9.5 Description of Keyhole Shape

The keyhole dynamics is a very complicated physical phenomenon still unsolved satisfactorily. At the starting stage of keyhole PAW, both the torch and the workpiece keep stationary until the full penetration is achieved and an open keyhole is formed. An open keyhole is actually established instantly. First, we do not consider the transient process of keyhole formation but focus on the following points: What process parameters make an open keyhole form? What shape and size do an open keyhole have under certain conditions? We discuss these next and take a simple case of a stationary axisymmetric keyhole as an example.

9.5.1 Axisymmetric Keyhole

As a first step, the keyhole shape is assumed to be axisymmetric. The inkeyhole plasma jet is assumed to be a one-dimensional incompressible

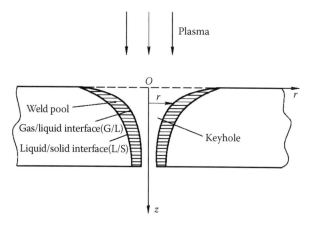

FIGURE 9.26
Schematic of axisymmetric keyhole.

ideal gas flow. As shown in Figure 9.26, the coordinate system (r, z) is established.

For a thin control volume, the conservations of mass and momentum hold. For mass conservation,

$$\rho_p u_p \pi r_p^2 = \text{Constant}$$

where ρ_p is the plasma density, u_p is the plasma velocity, and r_p is the keyhole radius.

Differentiating this equation with respect to the axial coordinate z and manipulating, we obtain

$$\rho_p \frac{du_p}{dz} \pi r_p^2 + \rho_p u_p \pi 2 r_p \frac{dr_p}{dz} = 0$$

After manipulation,

$$r_p \frac{du_p}{dz} + 2 u_p \frac{dr_p}{dz} = 0 \tag{9.29}$$

For momentum conservation,

$$\rho_p \pi r_p^2 dz \left(u_p \frac{du_p}{dz} \right) = \pi r_p^2 \left[p_p - \left(p_p + \frac{dp_p}{dz} \right) \right]$$

where p_p is the pressure.

Let $\dot{m} = \rho_p (\pi r_p^2 u_p)$; $\dot{m}$ is the flow rate of plasma mass (kg/s),

$$\dot{m} \frac{du_p}{dz} + \pi r_p^2 \frac{dp_p}{dz} = 0 \tag{9.30}$$

Applying the Young-Laplace equation to the interface between the plasma and liquid metal, the location of the keyhole wall is governed by

$$p_P - p_L = \gamma\left(\frac{1}{R_1} + \frac{1}{R_2}\right)$$

Substituting the curvatures,

$$p_P - \rho g z = -\gamma\left\{\frac{\dfrac{d^2 r_P}{dz^2}}{\left[1+\left(\dfrac{dr_P}{dz}\right)^2\right]^{3/2}} + \frac{\dfrac{dr_P}{dz}}{r_P\left[1+\left(\dfrac{dr_P}{dz}\right)^2\right]^{1/2}}\right\} \tag{9.31}$$

where ρ is the liquid density, g is the acceleration of gravity, and γ is the surface tension.

Equations (9.29)–(9.31) constitute a group of governing equations with three unknowns (r_P, u_P, p_P). After solving the governing equations, the variables (r_P, u_P, p_P) are known. When the value r_P at every axial point z is known, the keyhole geometry is determined. To solve Eqs. (9.29)–(9.31) simultaneously, discretization and iteration have to be carried out, and suitable boundary conditions are set.

9.5.2 Three-Dimensional Keyhole

When the plasma arc travels, the keyhole shape is no long axisymmetric but biased toward the pool rear. As shown in Figure 9.27, the coordinate system (x, y, z) is set up. Assume that at any plane perpendicular to the z-axis, the

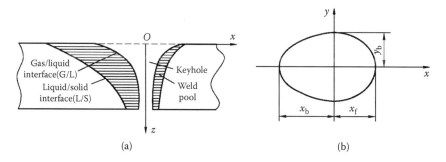

(a) (b)

FIGURE 9.27
Schematic of three-dimensional keyhole: (a) longitudinal cross section, (b) keyhole shape at a plane perpendicular to the z-axis.

keyhole cross section is an oblate ellipse. Three parameters (x_f, x_b, y_b) are used to describe the ellipse geometry. If the keyhole shape function $z = \varphi(x,y)$ is known, the parameters (x_f, x_b, y_b) at each plane are known. The area of the oblate ellipse at any plane perpendicular to the z-axis may be written as

$$A_P = \pi\varphi(0,y_b)\frac{\varphi(x_f,0)+\varphi(x_b,0)}{2} \tag{9.32}$$

Based on the conservation of mass,

$$\rho_P A_P u_P = \text{Constant}$$

Differentiate it with respect to z,

$$u_P\frac{dA_P}{dz} + A_P\frac{du_P}{dz} = 0 \tag{9.33}$$

According to conservation of momentum,

$$\rho_P A_P dz\left(u_P\frac{du_P}{dz}\right) = A_P\left[p_P - \left(p_P + \frac{dp_P}{dz}\right)\right]$$

Let $\rho_P A_P u_P = \dot{m}$; $\dot{m}$ is the flow rate of plasma mass (kg/s),

$$\dot{m}\frac{du_P}{dz} + A_P\frac{dp_P}{dz} = 0 \tag{9.34}$$

At the keyhole wall (plasma-molten metal interface), we have

$$p_P - \rho g z = -\gamma\frac{\left(1+\varphi_x^2\right)\varphi_{yy} - 2\varphi_x\varphi_y\varphi_{xy} + \left(1+\varphi_y^2\right)\varphi_{xx}}{\left(1+\varphi_x^2+\varphi_y^2\right)^{3/2}} \tag{9.35}$$

where φ is the keyhole shape function, and the subscript of φ is its partial derivative.

Equations (9.32)–(9.35) constitute a full description of the keyhole shape. Their solution procedure may be written as follows:

1. Guess a value for p_p, solve Eq. (9.35), and obtain the preliminary keyhole shape $z = \varphi(x,y)$.
2. Get A_P from Eq. (9.32).

3. Get u_p from Eq. (9.33).
4. Get p_p from Eq. (9.34).
5. Return to step 1; according the new value of p_p, re-solve Eq. (9.35) and obtain the new keyhole shape $z = \varphi(x,y)$. Repeat these steps and conduct iteration until convergence.

9.5.3 Computation of the Weld Pool and Temperature Field in PAW

As soon as the keyhole shape and size are determined, we can develop a suitable and adaptive mode of the welding heat source for the PAW process.

1. Take the keyhole as the domain within which the inner volumetric heat source deposits.
2. Designate the adaptive volumetric heat source mode according to the keyhole shape and size.
3. Take the keyhole wall as the inner boundary with temperature of $T_b = 2,800$ K.

As long as an adaptive heat source mode is developed, the temperature field can be computed as a heat conduction problem.

10

Vision-Based Sensing of Weld Pool Geometric Parameters

10.1 Introduction

The weld pool encodes abundant and direct information about the welding process and quality. It is of great theoretical and practical significance to capture weld pool images from the top side of the workpiece and extract geometric parameters of the weld pool. It embodies two aspects: (1) The measured data of weld pool geometry, including shape and size, are useful for verifying the mathematical models of weld pool behaviors. Numerical simulation is a powerful tool for transforming welding research from a "qualitative" to a "quantitative" sense and from "art" to "science." But, the mathematical models and simulation results must be verified before engineering applications. If the data for weld pool geometry under different welding conditions are measured, then they can be used to verify the mathematical models of welding processes. (2) Based on the weld pool geometric parameters detected from the top side of the workpiece, the correlation of the weld pool surface geometry to the weld penetration can be described, laying a good foundation for welding penetration control.

10.2 Vision-Based Experiment System for Detecting Tungsten Inert Gas Weld Pool Geometry

Based on low cost and automatic detection methodology, the common commercial CCD (charge-coupled device) camera was used to capture weld pool images from the top side of the workpiece in a constant current tungsten inert gas (TIG) welding process.[171] Clear and high-resolution images are captured by the system. Special image-processing software has been developed, and the edges of the whole weld pool are extracted.

10.2.1 Components of the Experimental System

Figure 10.1 shows the block diagram of the experimental system. The major functional elements of the experimental system include a process control computer, a precision welding power supply, a welding table, a control unit, a CCD camera with special filter, an image grabber, and a monitor. According to the function of the elements, the system is divided into two parts: the welding system and the image-sensing system.

The welding power supply is a pulsed current TIG power supply (Panasonic YC-300TSPVTA), but its welding current control section is modified by using a microprocessor-based unit. A welding control unit is developed to control the welding process parameters. The main elements of the control unit are a single-chip microcomputer type 8031, program memory, data memory, and digital-to-analog (D/A) and analog-to-digital (A/D) converter. Its main functions include receiving the instructions from the process control computer, controlling the welding current, and sending signals of image sampling to the sensing system.

The process control computer is the center of the whole system. It maintains dual-direction communicating with the welding control unit and the image grabber and controls the welding current through the welding control unit. The welding control unit sends the image-sampling signal to the process control computer at a preset interval. When the process control computer receives such a sampling signal, it captures the weld pool image immediately via the image grabber, conducts image processing, and extracts the geometric parameters of the weld pool. According to the current sampled online and certain control strategy, the process

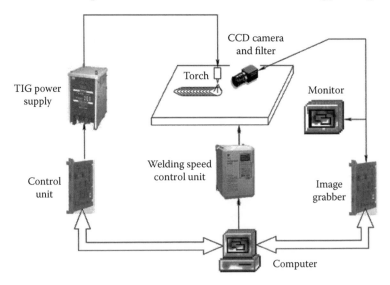

FIGURE 10.1
Schematic of the experimental system.

control computer outputs the welding current data to the welding control unit by a serial port. Then, the welding control unit adjusts the welding current data and sends them to the TIG welding power supply to control the welding current. Figure 10.2 shows a program flowchart of the process control computer.

During the welding process, the welding torch is stationary while the workpiece assembled on the welding table is moving. The process control computer controls the speed of the welding table (i.e., the welding speed).

The image-sensing system consists of a CCD camera, narrowband filter, image grabber, and control computer. It conducts arc light filtering, image capturing, image digitalizing, image processing, and geometric parameter extracting. Figure 10.3 shows a block diagram of the image-sensing system.

The narrowband pass filter is composed of one narrowband filter (610 nm central wavelength, 10 nm half bandwidth, and 27% transparency) and two neutral filters (10% transparency).[171] Its function is to filter the arc light, minimize the arc light interference, and capture clear images of the weld

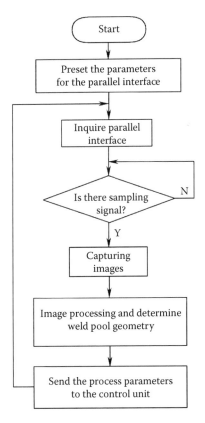

FIGURE 10.2
The program flowchart of the process control computer.

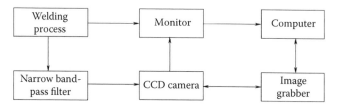

FIGURE 10.3
Block diagram of the image-sensing system.

TABLE 10.1

Main Specification of MTV-1802CB Camera

Resolution	600 lines
Sensitivity	0.02 lux
Power	12-V DC
Electronic shutter	8 levels(1/60–1/10,000)
Gain	Hand-adjusted permit

TABLE 10.2

Main Specification of CA-MPE-1000 Image Grabber

Resolution	768 × 576
Sampling accuracy	1/256
Conversion time	1/30 s
Synchronizing mode	Internal, external synchronizing
Scan mode	Alternative-line scanning
Geometric linearity	No distortion

pool. Table 10.1 gives the main specification for the CCD camera (type MTV-1802CB). The image grabber is a black-and-white image grabber (CA-MPE-1000). Its main specifications are shown in Table 10.2.

10.2.2 Calibration of Weld Pool Images

The vision-sensing system should be calibrated to get the information on shape and size of the weld pool from the captured images. The captured weld pool image is transformed to a digital matrix in the computer memory. One element (pixel) in the matrix corresponds to a point in the image. The real dimension represented by one pixel must be determined by calibration. Here, concentric rectangles are taken as the calibration object (shown in Figure 10.4). The distance between two adjacent rectangles is 1 mm. The coefficients between the image coordinate and the space coordinate along the horizontal direction (perpendicular to the weld centerline) and along

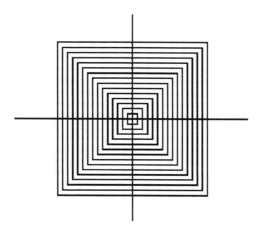

FIGURE 10.4
Reference object for calibration.

the vertical direction (parallel to the weld centerline) should be calculated. Software with a user-friendly interface carries out the calibration. The coefficients are as follows:

Along the horizontal direction: 0.0430 mm/pixel.

Along the vertical direction: 0.0752 mm/pixel.

10.2.3 Analysis of Weld Pool Image Characteristics

The aim of image processing is to extract useful information from the captured images. Generally, image processing should include image filtering, image enhancing, edge sharpening, and extracting. For an object image, we have to know the characteristics of the object image so that a good algorithm for image processing can be developed. Thus, the object image should be analyzed before image processing to develop or select a good algorithm for image processing.

Generally, the edges of an object are defined as the points where the gray values in the digital matrix vary most sharply. In the images captured by the system, the pool surface is a mirror-like reflection while the solid surface of the workpiece diffusely reflects arc light, so there is a bright region inside the weld pool. Due to the interaction of brightness of adjacent domains, the weld pool edges are located between the valley and peak of gray levels. According to this characteristic, the software developed searches for the weld pool edge and calculates the geometric parameter of the weld pool.

Figure 10.5 shows an image captured by the system. The *x*-axis is parallel to the welding line, and the *y*-axis is vertical to the welding line. The bright zone at the upper part in the image is the arc light. The high-brightness area in the middle of the image is the mirror image of the arc in

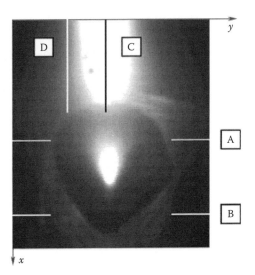

FIGURE 10.5
TIG weld pool image.

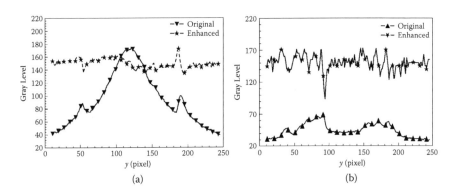

FIGURE 10.6
The distribution of original and enhanced gray levels along the y-direction: (a) line A; (b) line B.

the weld pool. The whole contour of the weld pool is clear in Figure 10.5. Here, two lines (A and B) along the y-axis and two lines (C and D) along the x-axis are made.

Figure 10.6(a) shows the distribution of original and enhanced gray levels along line A. From the original gray-level curve, it can be seen that the gray level is high in the middle part of the weld pool because of the reflective rays of the arc light. From the pool center to two sides, the gray level decreases first, changes sharply, and increases at the interface of molten and solid metal, then continues decreasing. The characteristic that the distribution of the gray level has a valley and peak around the pool edges is clear because of the sudden change of gray level around the pool edges. It provides effective information for identifying weld pool edges; that is, the point where the gray value varies most sharply between the valley and peak of grey levels is the edge point. To find the edge point, the original gray level should be enhanced. It can be seen from the curve of the enhanced gray level that the point with the lowest gray level in the left part of the image is the point at which the gray value varies most sharply in the curve of the original gray level (i.e., the left edge point of the weld pool). Also, the point with the highest gray level in the right part of the image is the right edge point of the weld pool.

Figure 10.6(b) shows the distribution of original and enhanced gray levels along line B. From the original gray-level curve, it can be seen that the characteristic of the valley and peak is not obvious, and it is difficult to find the edge of the weld pool from the variation of the gray level because many valley and peak points appear in the left part of the image.

Based on the analysis of many weld pool images, it was found that the gray-level distribution along the y-direction has an obvious characteristic of a valley and peak in the middle part of the pool image, whereas this characteristic is not obvious at the rear or front part of the pool image.

Figure 10.7(a) shows the distribution of the original and enhanced gray levels along line C. From the original gray-level curve, it can be seen that the gray level decreases rapidly from the center (arc light) to the front part of the weld pool. The point that decreases most sharply is regarded as the weld pool edge. At the rear of the weld pool, the gray level also changes rapidly, and valley and peak points appear, which can be treated as the characteristic of the weld pool edge. Figure 10.7(b) shows the distribution of original and enhanced gray levels along line D. The distribution of gray levels does not have such a characteristic because it is located near the side of the pool image.

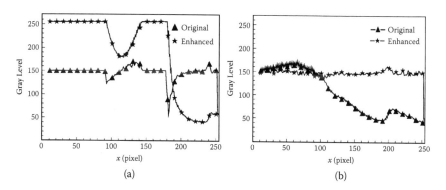

FIGURE 10.7
The distribution of original and enhanced gray levels along the x-direction: (a) line C; (b) line D.

Based on this analysis, it is concluded that the rear, middle, and front parts of weld pool images have different gray value characteristics so that the same algorithm is not available for all parts. Thus, the weld pool edges at the middle part of the image are first searched along the y-direction. Then, the pool edges at the rear and front parts of the image are searched along the x-direction.

10.2.4 Image Processing

The algorithms for noise elimination and image enhancement are combined to sharpen the points with a large gray-level variation. Here, $g(i, j)$ and $f(i, j)$ represent the gray values of point (i, j) in the image after and before processing, respectively. The processing is carried out based on the following equations:

Along the y-direction (perpendicular to the weld centerline):

$$g(x,y) = 150 + 1.25 \times [f(x,y+2) + f(x,y+1) - f(x,y) - f(x,y-1)] \quad (10.1)$$

Along the x-direction (parallel to the weld centerline):

$$g(x,y) = 150 + 1.25 \times [f(x+2,y) + f(x+1,y) - f(x,y) - f(x-1,y)] \quad (10.2)$$

Based on the different characteristics between the edge point and interference point, the edge of the weld pool can be searched according to the enhanced and original gray-level distributions of the pool image.

To record the searched weld pool edge points, a structure array and two integer arrays are defined to record the middle, front, and rear edge points of weld pool, respectively.

1. To search for the starting point, first search for the center point according to the feature that the gray level is high in the middle part of the weld pool. Then, enhance the line where the center point is located. If the weld pool edge feature in this line is not obvious, then enhance the lines above or below this line alternatively until the line in which the feature of the weld pool edge is obvious. Take this line as the start line. Find the weld pool edge points in this line and take the left and right edge points as the start point.

2. From the start line to the upper part of the image, enhance each line along the y-direction and search the left and right edge points of the weld pool one by one. The searched left and right edge points are stored in the defined arrays.

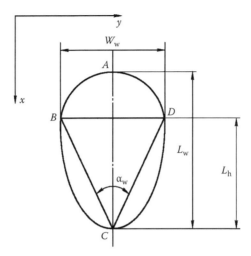

FIGURE 10.8
Definition of surface geometrical parameters of the weld pool.

3. From the start line to the lower part of the image, enhance each line along the *y*-direction and search the left and right edge points of the weld pool one by one. The searched left and right edge points are stored in the defined arrays.

10.2.5 Definition of Top-Side Surface Geometric Parameters of the Weld Pool

To provide data for verifying the mathematical models of the welding process and establish the correlation of the top-side pool surface geometry to the back-side weld width, top-side surface geometric parameters of the weld pool should be defined and determined based on the processed images of the weld pool.

In Figure 10.8, the surface geometrical parameters of the weld pool, such as the rear length of the weld pool L_w, the maximum width of the weld pool W_w, the rear angle of the weld pool α_w, the area of the weld pool S_w, that are surrounded by curve ABCD are defined.

10.3 Measuring Experiments of TIG Welding on Mild Steel Plates

The image captured by the camera is an analog image. After digitization, it becomes a digital image stored in the computer memory as a matrix, as shown in Eq. (10.3). Each element in the digital image is displayed as a dot on the computer screen (i.e., a pixel).

$$f(x_i, y_j) = \begin{bmatrix} f(0,0) & f(0,1) & \cdots & f(0,N-1) \\ f(1,0) & f(1,1) & \cdots & f(1,N-1) \\ \cdots & \cdots & \cdots & \cdots \\ f(M-1,0) & f(M-1,1) & \cdots & f(M-1,N-1) \end{bmatrix} \qquad (10.3)$$

After calibration, the transformation coefficients between the image coordinates and the real coordinates are obtained as mentioned. The relationship between the pixel coordinate $f(x_i, y_j)$ in the image coordinate and the corresponding point coordinate $g(x_g, y_g)$ in the space coordinate is as follows:

$$x_g = (255 - x_i) \times 0.00752 \ (\text{mm}) \qquad (10.4a)$$

$$y_g = y_j \times 0.0043 \ (\text{mm}) \qquad (10.4b)$$

10.3.1 Measuring Results with Varying Welding Current

Table 10.3 gives the process parameters in bead-on-plate welding of a mild steel workpiece with dimensions $250 \times 60 \times 2$ (mm³). Figure 10.9 shows the original images of TIG weld pool at top-side surface under different levels of welding current. Figure 10.10 shows the corresponding processed images. Figure 10.11 shows the real shape and size of weld pools. Table 10.4 lists the measured weld pool geometric parameters.

It can be seen that weld pool width, pool length, rear length of pool, and pool area all increase with the welding current, whereas the variation of the pool rear angle is not obvious. The maximum width of the pool is dominated by the distribution of arc heat flux. Since the temperature field moves at a certain speed in the quasi-steady state, the temperature gradient is lower

TABLE 10.3

TIG Welding Process Parameters

Test Number	Welding Current (A)	Welding Speed (mm/min)	Arc Length (mm)	Flow Rate of Ar (L/min)
①	100	180	6	10
②	105	180	6	10
③	110	180	6	10
④	115	180	6	10

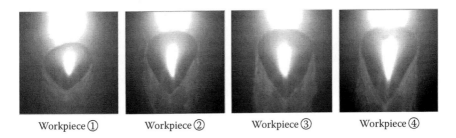

Workpiece ①   Workpiece ②   Workpiece ③   Workpiece ④

FIGURE 10.9
Original top-side images of the weld pool.

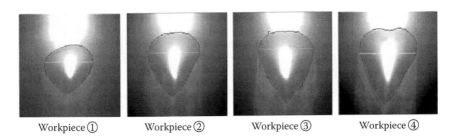

Workpiece ①   Workpiece ②   Workpiece ③   Workpiece ④

FIGURE 10.10
Processed top-side images of the weld pool.

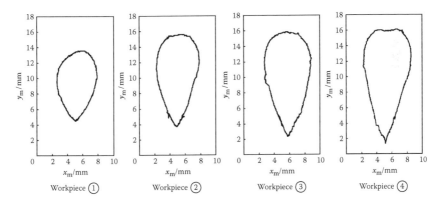

FIGURE 10.11
Real size of weld pools under different welding currents.

behind the arc centerline and is higher ahead of the arc centerline. Thus, the rear length of the weld pool is larger than the front length, and its variation magnitude with welding current is also larger. The variation of the front length of the weld pool is not obvious with increasing welding current. In total, the pool length increases with welding current, and the rising trend is clear. The weld pool area is dominated by the heat input to the workpiece

TABLE 10.4

Measured Weld Pool Geometric Parameters

Welding Current (A)	Welding Speed (mm/ min)	Weld Pool Width (mm)	Weld Pool Length (mm)	Rear Length of Weld Pool (mm)	Weld Pool Area (mm²)	Rear Angle of Weld Pool (°)
100	180	4.95	8.80	5.94	52.29	42.67
105	180	5.51	11.82	7.98	80.62	34.92
110	180	6.02	13.31	9.92	101.96	31.43
115	180	6.34	14.73	10.94	112.54	29.87

TABLE 10.5

TIG Welding Process Parameters

Test Number	Welding Current (A)	Welding Speed (mm/min)	Arc Length (mm)	Flow Rate of Ar (L/min)
⑤	100	150	6	10
⑥	100	160	6	10
⑦	100	180	6	10

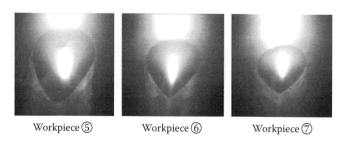

Workpiece ⑤   Workpiece ⑥   Workpiece ⑦

FIGURE 10.12
Original top-side image of the weld pool.

and the heat conduction speed in the workpiece. The pool area expands with increasing welding current. The pool rear angle is related to the rear length and maximum width of the weld pool. Both the rear length and the maximum width of the weld pool increase with increasing welding current, but the former increases with a larger magnitude. Thus, the pool rear angle decreases with increasing welding current.

10.3.2 Measuring Results under Different Welding Speeds

The same welding conditions were used as mentioned, but the welding speed changes as shown in Table 10.5.

Figures 10.12 and 10.13 show the original and the processed top-side images of the TIG weld pool under different welding speeds, respectively.

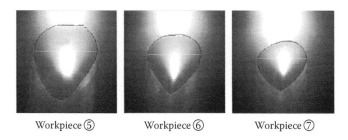

Workpiece ⑤   Workpiece ⑥   Workpiece ⑦

FIGURE 10.13
Processed top-side image of the weld pool.

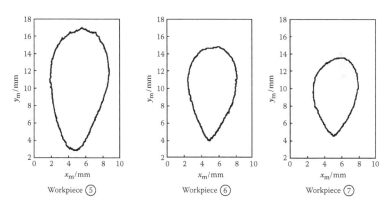

Workpiece ⑤   Workpiece ⑥   Workpiece ⑦

FIGURE 10.14
The real shape and size of the weld pool under different welding speeds.

TABLE 10.6

Weld Pool Geometric Parameters under Different Welding Speeds

Welding Current (A)	Travel Speed (mm/ min)	Weld Pool Width (mm)	Weld Pool Length (mm)	Rear Length of Weld Pool (mm)	Weld Pool Area (mm²)	Rear Angle of Weld Pool (°)
100	150	6.79	13.61	9.66	109.45	38.42
100	160	5.59	10.91	7.62	79.57	45.14
100	180	4.95	8.80	5.94	52.29	42.67

Figure 10.14 shows the real shape and size of the weld pool. Table 10.6 lists the measured geometric parameters of the weld pool. It can be seen that the weld pool width, pool length, rear length of the pool, and the pool area all decrease with increasing welding speed, whereas the variation of the pool rear angle is not obvious.

10.4 Measuring Experiments of TIG Welding on Stainless Steel Plates

Bead-on-plate welding was performed with constant current from the start to the end of the weld. The workpiece material was stainless steel (0Cr18Ni9Ti) with the dimensions 200 × 50 mm and 3-mm thickness. Argon was used as the shielding gas with a flow rate of 10 L/min. The diameter of the tungsten electrode was 3.2 mm with a 90° angle. The arc length was held at 6 mm. The ranges of welding current and welding speed were 90–130 A and 100–150 mm/min, respectively.

Figures 10.15 to 10.19 show the measured weld pool geometry under the different levels of welding current and travel speed. In these figures, the (a) portions are the raw images of the weld pool, captured by the CCD camera directly; the (b) portions are the weld pool images after processing by the algorithm; and the (c) portions are the weld pool real size after transformation from image to practical dimension.

10.5 LaserStrobe-Based Sensing System

The mentioned vision system can capture the entire weld pool images in continuous current TIG welding. It is a low-cost and automatic vision sensing system. However, its reliability is not so high if it is used in GMAW. During the GMAW process, there are very strong interferences from arc light, so a more reliable and advanced sensing method is needed. An advanced experimental

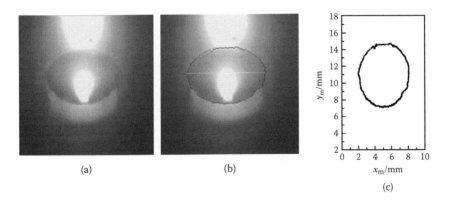

(a) (b) (c)

FIGURE 10.15
The weld pool geometry at 90-A welding current and 103-mm/min travel speed (from Wu C S, Gao J Q, Liu X F, et al., *Proc. Instn. Mech. Eng.*, Part B, *J. Eng. Manuf.*, 217: 879–882, 2003): (a) raw image; (b) processed image; (c) weld pool of real size.

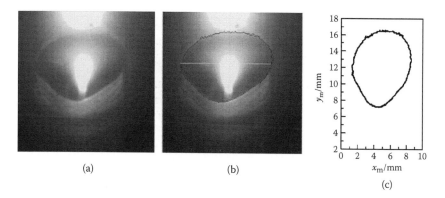

FIGURE 10.16
The weld pool geometry at 100-A welding current and 113-mm/min travel speed (From Wu C S, Gao J Q, Liu X F, et al., Proc. Instn. Mech. Eng., Part B, *J. Eng. Manuf.*, 217: 879–882, 2003): (a) raw image; (b) processed image; (c) weld pool of real size.

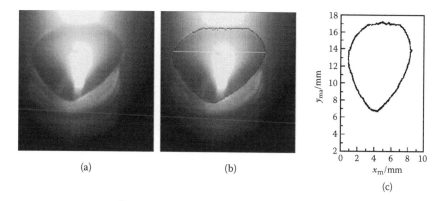

FIGURE 10.17
The weld pool geometry at 110-A welding current and 125-mm/min travel speed (From Wu C S, Gao J Q, Liu X F, et al., Proc. Instn. Mech. Eng., Part B, *J. Eng. Manuf.*, 217: 879–882, 2003): (a) raw image; (b) processed image; (c) weld pool of real size.

system assisted by LaserStrobe is able to capture the weld pool images in the GMAW process. Figure 10.20 shows the block diagram of the experimental system, which mainly consisted of LaserStrobe and a welding machine.

The LaserStrobe sensing system is produced by Control Vision Incorporated in the United States. The LaserStrobe system integrates high-frequency pulse laser technology, a high-speed electronic shutter camera, composite filter technology, and image process technology. It can effectively overcome the interferences from strong arc light and provides clear weld pool images.

The LaserStrobe system includes the camera head, controller unit, and two laser units. It utilizes the laser emitted from the laser unit to overcome the interference from the arc light. The average power of the laser is only 7 mW, but

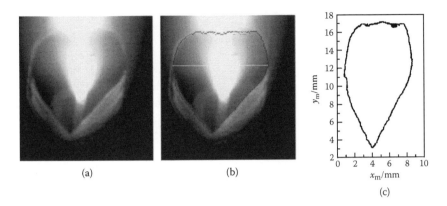

FIGURE 10.18
The weld pool geometry at 120-A welding current and 135-mm/min travel speed (From Wu C S, Gao J Q, Liu X F, et al., *Proc. Instn. Mech. Eng.*, Part B, *J. Eng. Manuf.*, 217: 879–882, 2003): (a) raw image; (b) processed image; (c) weld pool of real size.

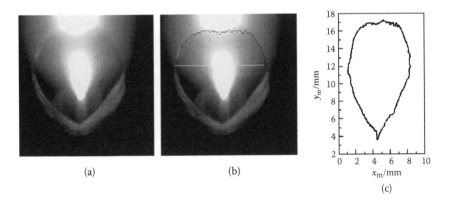

FIGURE 10.19
The weld pool geometry at 125-A welding current and 149-mm/min travel speed (From Wu C S, Gao J Q, Liu X F, et al., *Proc. Instn. Mech. Eng.*, Part B, *J. Eng. Manuf.*, 217: 879–882, 2003): (a) raw image; (b) processed image; (c) weld pool of real size.

its peak power during the pulse period (only 5 ns) can reach a value of 80 kW. The shutter of the camera is synchronized with the laser pulse. During the pulse duration, the intensity of the laser illumination is much stronger than that of the GMAW arc and molten metal. Thus, the arc light interference can be effectively minimized when the camera captures the weld pool image. In addition, there is a narrowband filter compatible with the laser wavelength, so the image signal-to-noise ratio is further improved. Consequently, clear images of the weld pool can be acquired during the GMAW process. After the process control computer processes the weld pool images, the weld pool

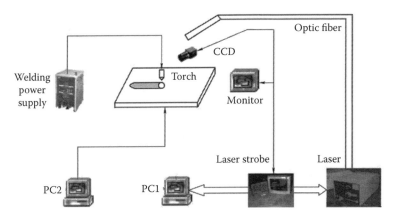

FIGURE 10.20
Block diagram of the experimental system.

TABLE 10.7

Focus Tube Parameters

Length of Focus Tube (cm)	Camera	Maximum Stand-off Distance (cm)	Typical Field of View (mm)	Minimum Stand-off Distance (cm)	Typical Field of View (mm)
8.3	P60	28	45	14.5	18
11.8	P60	14.5	18	11	11.5
15.2	P60	11	11.5	9.5	8

shape and size can be reliably detected and determined. Thus, it provides abundant information for welding process analysis and control.

During the very short pulse duration, the intensity of the laser is much stronger than that of the GMAW arc. The pool surface is a mirror-like reflection, while the solid surface of the workpiece diffusely reflects arc light, so high-contrast and high-quality weld pool images can be captured under the illumination of the laser beam, and the edges of weld pool can be reliably and quickly extracted.

The focus tube is a key part of the camera head and is attached to the front end of the camera head. Short- (8.3-cm), medium- (11.8-cm), and long- (15.2-cm) focus tubes are available, with each length providing a different level of magnification and having a different field of view. Table 10.7 provides the parameters of the focus tube. The focus tube selection is determined by the welding process and the welding parameters. Here, a short-focus tube was selected.

The method in Section 10.2.2 is used to calibrate the LaserStrobe experiment system. The transfer coefficients are as follows:[172]

Horizontal direction (vertical to welding direction): 0.0521 mm/pixel

Vertical direction (parallel to welding direction): 0.1316 mm/pixel

10.6 Measuring Results in GMAW

10.6.1 Tests in CO_2 Welding

GMAW with CO_2 shielding gas was performed with DCEP (direct current electrode positive) under different welding conditions. For each condition, bead-on-plate welding was performed. The workpiece material was mild steel (Q235) with 200×50 mm and 6-mm thickness. The flow rate of the shielding gas was 10 L/min. The diameter of the welding wire was 1.2 mm. The distance from the nozzle to the plate was 20 mm. Figures 10.21 and 10.22 show part of the measuring results.

10.6.2 GMAW Tests with Mixed Shielding Gas

GMAW welding experiments with DCEP and mixed shielding gas were performed under different welding conditions.[172] For each condition, bead-on-plate welding was performed. The workpiece material was mild

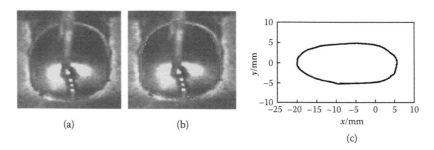

(a)   (b)   (c)

FIGURE 10.21
The weld pool geometry of CO_2 shield gas arc welding (from Wu C S, Gao J Q, and Zhang M, Proc. Instn. Mech. Eng., Part B, *J. Eng. Manuf*, 218: 813–818, 2004) ($I = 110$ A, $U_a = 20$ V, $v_0 = 120$ mm/min): (a) raw image; (b) processed image; (c) weld pool of real size.

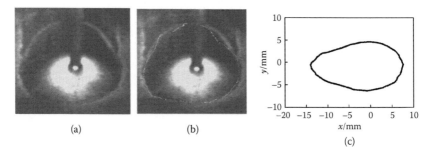

(a)   (b)   (c)

FIGURE 10.22
The weld pool geometry of CO_2 shield gas arc welding ($I = 120$ A, $U_a = 20$ V, $v_0 = 135$ mm/min): (a) raw image; (b) processed image; (c) weld pool of real size.

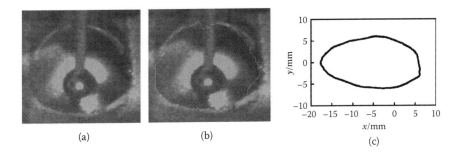

(a) (b) (c)

FIGURE 10.23
The weld pool geometry of GMAW with mixed shielding gas ($I = 110$ A, $U_a = 22$ V, $v_0 = 100$ mm/min): (a) raw image; (b) processed image; (c) weld pool of real size.

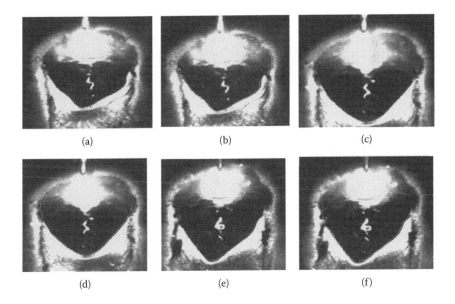

(a) (b) (c)

(d) (e) (f)

FIGURE 10.24
The weld pool geometry of GMAW with mixed shielding gas ($I = 250$ A, $U_a = 31$ V, $v_0 = 400$ mm/min): (a) $t = 2.5$ s; (b) $t = 2.5$ s; (c) $t = 3.5$ s; (d) $t = 3.5$ s; (e) $t = 10$ s; (f) $t = 10$ s; (a), (c), (e) raw images; (b), (d), (f) processed images.

steel (Q235) 200×50 mm and 6-mm thickness. The shielding gas was 80% argon plus 20% CO_2, and the flow rate was 12 L/min. The diameter of the welding wire was 1.2 mm. The distance from the nozzle to the plate was 20 mm. Figure 10.23 shows the measuring results.

GMAW welding experiments with 90% argon plus 10% CO_2 shielding gas with a flow rate of 12.5 L/min were carried out. The distance from the nozzle to the plate was 25 mm. The droplet transfer mode was by spray transfer. Other conditions were the same as mentioned. Figure 10.24 shows the measuring results for the weld pool.

11

Numerical Analysis of the Transport Mechanism in Welding Arcs

11.1 Introduction

The welding arc exerts current density, heat density, and pressure distributions on the weld pool surface, so it has important effects on the weld thermal efficiency, melting efficiency, weld pool behaviors, and weld thermal processes. Thus, understanding and comprehensive grasp of the physical transport mechanisms in the welding arc are the basis and prerequisites for numerical analysis of the weld thermal process and molten pool behaviors.

11.2 Model of the Tungsten Inert Gas Welding Arc

The behavior of an arc is governed by a coupled set of physical laws, that is, Ohm's law, Maxwell's equation, and the conservation equations of mass, momentum, energy, and electric charge. The interactions of electric, magnetic, thermal, and fluid flow fields and the strongly nonlinear thermodynamic and transport properties of the plasma make the analysis difficult. In spite of extensive efforts in arc physics and arc technology,[173-192] the understanding of arc behavior remains incomplete. In this section, a mathematical model for the tungsten inert gas (TIG) welding arc is introduced.

Figure 11.1 shows the computational domain used to model the TIG welding arc. The presence of an electric field between the cathode (a tungsten rod) and the anode (metal workpiece) causes the passage of an electric current through the ionized plasma region, which in turn gives rise to a self-induced magnetic field. The magnetic field interacts with the current, transferring momentum to the gas, which is accelerated toward the anode in the form of a characteristic cathode jet. Due to the electrical resistance of the plasma, the energy produced by the current keeps the plasma in the ionized state and provides the heating mechanism for the welding process. For modeling this

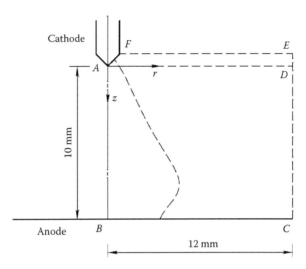

FIGURE 11.1
Calculation domain of the arc model.

complex welding arc, some assumptions must be made. The arc model in this section is based on the following assumptions:

1. The arc plasma is in local thermodynamic equilibrium (LTE), which means that the electron and heavy particle temperatures are equal.
2. The arc is steady and radially symmetrical.
3. The plasma is optically thin, so radiation may be accounted for using an optically thin radiation loss per unit volume.
4. The arc plasma consists of pure argon at atmospheric pressure.
5. The arc plasma shows a laminar flow.

Under these assumptions, the equations governing the arc column may be written as follows:

Current continuity in terms of electric potential is given by

$$\frac{\partial}{\partial z}\left(\sigma_e \frac{\partial \phi}{\partial z}\right) + \frac{1}{r}\frac{\partial}{\partial r}\left(\sigma_e r \frac{\partial \phi}{\partial r}\right) = 0 \tag{11.1}$$

where σ_e is the electric conductivity, and ϕ is the electric potential.
The current density is given by

$$J_r = -\sigma_e \frac{\partial \phi}{\partial r}, \qquad J_z = -\sigma_e \frac{\partial \phi}{\partial z} \tag{11.2}$$

Since the current distribution is axisymmetric, the self-induced magnetic field is given by the following relation from Ampere's law:

$$B_\theta = \frac{\mu_m}{r} \int_0^r J_z r \, dr \tag{11.3}$$

Let $u = u(r, z)$ and $v = v(r, z)$ represent the fluid velocity components in the z and r directions, respectively, as shown in Figure 11.1. Then, the continuity and momentum equations describing fluid flow in the arc may be written as

$$\frac{1}{r}\frac{\partial}{\partial r}(rv) + \frac{\partial u}{\partial z} = 0 \tag{11.4}$$

$$\rho\left(v\frac{\partial v}{\partial r} + u\frac{\partial v}{\partial z}\right) = F_r - \frac{\partial p}{\partial r} + \mu\left(\frac{\partial^2 v}{\partial r^2} + \frac{\partial v}{r\partial r} - \frac{v}{r^2} + \frac{\partial^2 v}{\partial z^2}\right) \tag{11.5}$$

$$\rho\left(v\frac{\partial u}{\partial r} + u\frac{\partial u}{\partial z}\right) = F_z - \frac{\partial p}{\partial z} + \mu\left(\frac{\partial^2 u}{\partial r^2} + \frac{\partial u}{r\partial r} + \frac{\partial^2 u}{\partial z^2}\right) \tag{11.6}$$

$$\rho C_P\left(v\frac{\partial T}{\partial r} + u\frac{\partial T}{\partial z}\right) = \frac{\partial}{r\partial r}\left(\lambda r\frac{\partial T}{\partial r}\right) + \frac{\partial}{\partial z}\left(\lambda\frac{\partial T}{\partial z}\right) + \frac{J_z^2 + J_r^2}{\sigma_e}$$

$$+ \frac{5}{2}\frac{k_e}{e}\left(J_z\frac{\partial T}{\partial z} + J_r\frac{\partial T}{\partial r}\right) - S_R \tag{11.7}$$

where ρ is density; μ is viscosity; p is pressure; F_r and F_z are the components of body force in the radial (r) and axial (z) directions, respectively; C_p is specific heat; λ is thermal conductivity; T is temperature; k_e is the Boltzmann constant; e is the electron charge; and S_R is the radiation loss term of argon plasma. The source terms on the right-hand side of Eq. (11.7) represent the joule heating, transport of enthalpy due to electron drift, and radiation losses.

In the arc column, the Lorentz force is the body force; its components are given as

$$F_r = -B_\theta J_z, \qquad F_z = -B_\theta J_r \tag{11.8}$$

For convenience for a numerical solution, we transform Eqs. (11.4)–(11.7) into the governing equations based on the method of stream function-vorticity. Define the vorticity ϖ

$$\varpi = \frac{\partial v}{\partial z} - \frac{\partial u}{\partial r} \tag{11.9}$$

and the stream function Ψ

$$u = \frac{1}{r}\frac{\partial\Psi}{\partial r} \tag{11.10}$$

$$v = -\frac{1}{r}\frac{\partial\Psi}{\partial z} \tag{11.11}$$

Substituting Eqs. (11.9)–(11.11) into Eqs. (11.4)–(11.7), the governing equation takes the following form:

$$\rho\left(\frac{1}{r}\frac{\partial\Psi}{\partial r}\frac{\partial\varpi}{\partial z} - \frac{1}{r}\frac{\partial\Psi}{\partial z}\frac{\partial\varpi}{\partial r} + \frac{1}{r}\frac{\partial\Psi}{\partial z}\frac{\varpi}{r}\right) = \mu\left(\frac{\partial^2\varpi}{\partial r^2} + \frac{\partial\varpi}{r\partial r} - \frac{\varpi^2}{r} + \frac{\partial^2\varpi}{\partial z^2}\right) - \left(\frac{\partial F_z}{\partial r} - \frac{\partial F_r}{\partial z}\right) \tag{11.12}$$

$$\frac{\partial}{\partial z}\left(\frac{1}{r}\frac{\partial\Psi}{\partial z}\right) + \frac{\partial}{\partial r}\left(\frac{1}{r}\frac{\partial\Psi}{\partial r}\right) + \varpi = 0 \tag{11.13}$$

$$\rho C_p\left(\frac{1}{r}\frac{\partial\Psi}{\partial r}\frac{\partial T}{\partial z} - \frac{1}{r}\frac{\partial\Psi}{\partial z}\frac{\partial T}{\partial r}\right) = \left[\frac{1}{r}\frac{\partial}{\partial r}\left(\lambda\frac{\partial T}{\partial r}\right) + \frac{\partial}{\partial z}\left(\lambda\frac{\partial T}{\partial z}\right)\right]$$

$$+ \frac{J_z^2 + J_r^2}{\sigma_e} + \frac{5}{2}\frac{k_e}{e}\left(J_z\frac{\partial T}{\partial z} + J_r\frac{\partial T}{\partial r}\right) - S_R \tag{11.14}$$

11.2.1 Boundary Conditions

A sketch of the computational domain is shown in Figure 11.1 and the corresponding boundary conditions are given in Table 11.1. As the calculation domain for u, v, and T, the area ABCEF is chosen, but the domain for ϕ is chosen as the smaller area ABCD since the exact boundary condition for ϕ along the line FA is unknown. Due to symmetry, only half of the flow

TABLE 11.1

Boundary Conditions

	AB	BC	CE	EF	FA	DA
ϕ	$\dfrac{\partial\phi}{\partial r} = 0$	$\phi = \text{Constant}$	$\dfrac{\partial\phi}{\partial r} = 0$			Eq. (11.15)
u	$\dfrac{\partial u}{\partial r} = 0$	$u = 0$	$\dfrac{\partial u}{\partial r} = 0$	$\dfrac{\partial\rho u}{\partial r} = 0$	$u = 0$	
v	$v = 0$	$v = 0$	$\dfrac{\partial v}{\partial r} = 0$	$v = 0$	$v = 0$	
T	$\dfrac{\partial T}{\partial r} = 0$	Eq. (11.20)	$\dfrac{\partial T}{\partial r} = 0$	1,000 K	3,000 K	

domain is considered for the calculation. Along the centerline AB, symmetry conditions are used. Zero velocities are specified along the solid boundaries BC and FA. Along the far-field boundary CE, zero radial gradients for all variables are specified. A constant electrical potential is specified along the anode surface BC because the anode is assumed to be a perfect conductor relative to the plasma. Along the boundary EF, the temperature is taken as 1,000 K, and the radial velocity component is neglected. At the cathode surface FA, the temperature is assumed to be 3,000 K. The sharpening angle of the electrode is 75°.

The current density distribution along the line DA is assumed to be of the form

$$J_c = J_0 \exp(-b_a r) \tag{11.15}$$

where b_a is a constant. The maximum current density is determined as follows:

$$J_0 = \frac{I}{2\pi r_a^2} \tag{11.16}$$

The constant b_a is determined from

$$I = 2\pi \int_0^{R_c} J_0 \exp(-b_a r) r \, dr \tag{11.17}$$

where $r_a = 0.51$ mm, and $R_c = 3$ mm. These two parameters correspond to the maximum temperature region and the current uniform distribution region in the cathode spot.[173]

11.2.2 Heat Transfer to the Anode Surface

A special treatment is required for the anode surface. Three principal modes of heat transfer contribute to the anode: (1) heat flux due to the electron flow, (2) convection from the plasma, and (3) radiation from the plasma. Heat loss due to vaporization of the anode material has been neglected.

The heat flux to the anode due to the flow of electrons may be expressed as

$$Q_e(r) = \frac{5}{2} k_e (C_\alpha T_b - T_a) \frac{J_a}{e} + J_a V_\varphi + J_a \phi_f \tag{11.18}$$

where r is the radial distance, k_e is the Boltzmann constant, C_α is the ratio of electron temperature to plasma temperature, T_b is the electron temperature at the free-fall edge, T_a is the anode temperature, J_a is the current density at the anode, V_f is the work function of the anode material, and ϕ_f is the anode fall. The first term in Eq. (11.18) is the transport of enthalpy due to the random thermal energy of the electrons, while the second term represents the heat given up by the electrons as they enter the lattice of the anode and release energy proportional to the anode work function. The third term is the energy acquired by the electrons in traversing the anode fall. In principle, this energy is required to sustain the production of positive ions in the vicinity of the anode and to ensure current conservation.

The convection contribution may be written as[188]

$$Q_c(r) = \frac{0.515}{Pr_a}\left(\frac{\mu_{ab}\rho_{ab}}{\mu_a\rho_a}\right)^{0.11}\left[\mu_{ab}\rho_{ab}\frac{v}{r}\right][h_{ab} - h_a] \tag{11.19}$$

where Pr_a is the plasma Prandtl number at the anode temperature; μ_a and μ_{ab} are the viscosity at the anode temperature and at the free-fall edge, respectively; ρ_a and ρ_{ab} are the density at the anode temperature and at the free-fall edge, respectively; v is the velocity component along the anode surface; h_{ab} and h_a are the plasma enthalpy at the anode temperature and at the free-fall edge, respectively.

Ushio et al. found the contribution of plasma radiation Q_r to the anode is less than 5% of the total heat input.[193] So, Q_r is taken as 3% here.

The heat flux to the anode surface can then be written as

$$Q_a(r) = Q_e(r) + Q_c(r) + Q_r \tag{11.20}$$

11.2.3 Calculation Results

The governing equations and boundary conditions are solved using a finite difference approach based on control volume.[194] To enhance the calculation accuracy, several variable meshes are used. The size of meshes near the cathode and the axis is finer to account for the steep gradients. Calculations of fluid flow and heat transfer in the arc plasma are carried out. The thermodynamic and transport properties required for the solution are taken from Reference 195.

Figures 11.2 and 11.3 show the centerline distribution of the electric field strength, the electrical potential, and the current density. The current density is very high near the cathode tip but decreases rapidly with increasing distance from the cathode. Since the electrical conductivity at the centerline is almost constant, the curves of the electric field strength and of the current density are similar. A steep gradient of electrical potential near the cathode tip results in the high current density, which generates a strong magnetic force. The constant density lines are plotted in Figure 11.4. The predicted magnetic field is given in Figure 11.5. The divergence of the current in the vicinity of the cathode, which gives rise to the cathode jet, is clearly demonstrated.

The strong magnetic force acting on the welding arc generates a steep pressure gradient near the electrode tip, which in return accelerates the argon plasma toward the baseplate. Figure 11.6 shows the axial variation of plasma

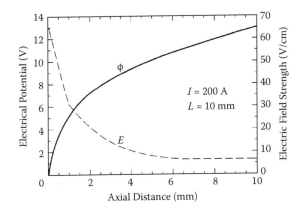

FIGURE 11.2
Electric field strength and potential in the arc axis. (From Wu C S, Ushio M, and Tanaka M, *Comput. Mater. Sci.*, 7: 308–314, 1997.)

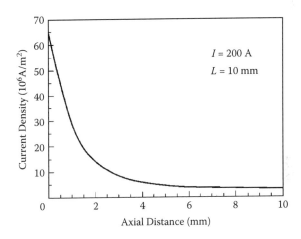

FIGURE 11.3
Axial current density in the arc axis. (From Wu C S, Ushio M, and Tanaka M, *Comput. Mater. Sci.*, 7: 308–314, 1997.)

velocity for 100- and 200-A arcs at an arc length of 10 mm. The model predicts a maximum velocity of 184 m/s for the 100-A arc and 235 m/s for the 200-A arc. As the welding current increases, the arc pressure at the electrode tip also increases. Consequently, the axial velocity in the arc column also increases. For examining the flow pattern of the shielding gas, the streamlines represented by the flow rate are shown in Figure 11.7. In the cathode region, a part of the shielding gas flowing from the nozzle is drawn into the arc due to the electromagnetic pumping force. The induced flow enters the arc core from the fringes, and as it approaches the anode, it is forced to turn around in the r-direction.

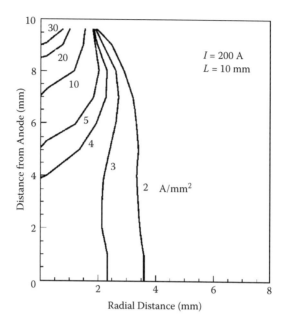

FIGURE 11.4
Isolines of current density. (From Wu C S, Ushio M, and Tanaka M, *Comput. Mater. Sci.*, 7: 308–314, 1997.)

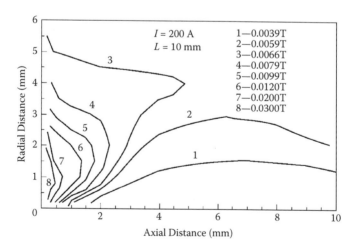

FIGURE 11.5
Azimuthal magnetic field in teslas. (From Wu C S, Ushio M, and Tanaka M, *Comput. Mater. Sci.*, 7: 308–314, 1997.)

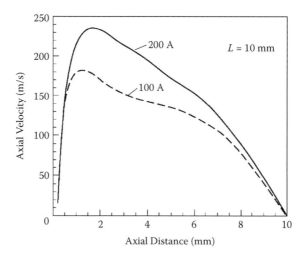

FIGURE 11.6

Axial velocity in the arc axis. (From Wu C S, Ushio M, and Tanaka M, *Comput. Mater. Sci.*, 7: 308–314, 1997.)

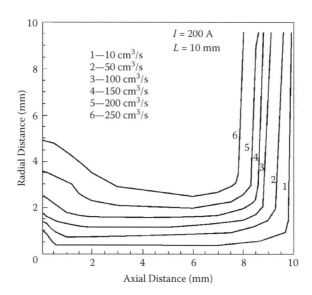

FIGURE 11.7

Streamlines in the arc. (From Wu C S, Ushio M, and Tanaka M, *Comput. Mater. Sci.*, 7: 308–314, 1997.)

The predicted isotherms for 10-mm arcs at 100 and 200 A are shown in Figures 11.8 and 11.9, respectively, along with the experimental measurements of Hsu et al.[173] It can be revealed that the calculated results are in good agreement with the measurement. Figure 11.10 shows the predicted temperatures along the centerline corresponding to Figures 11.8 and 11.9.

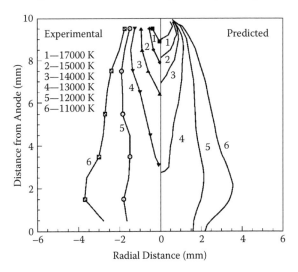

FIGURE 11.8
Isotherms in the 100-A arc. (From Wu C S, Ushio M, and Tanaka M, *Comput. Mater. Sci.*, 7: 308–314, 1997.)

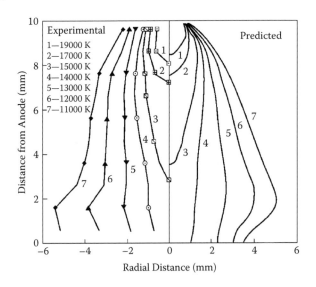

FIGURE 11.9
Isotherms in the 200-A arc. (From Wu C S, Ushio M, and Tanaka M, *Comput. Mater. Sci.*, 7: 308–314, 1997.)

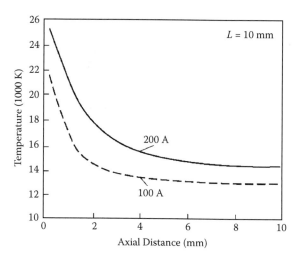

FIGURE 11.10
Temperature in the arc axis. (From Wu C S, Ushio M, and Tanaka M, *Comput. Mater. Sci.*, 7: 308–314, 1997.)

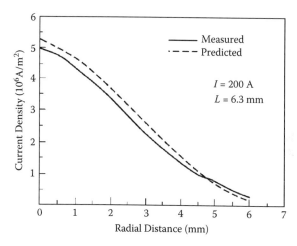

FIGURE 11.11
Current density at the anode surface. (From Wu C S, Ushio M, and Tanaka M, *Comput. Mater. Sci.*, 7: 308–314, 1997.)

The two curves have almost the same shape, but the absolute values differ by 2,000 K.

Figures 11.11 and 11.12 show a comparison of the theoretical predictions with the experimental measurements of Nestor[187] for the radial distributions of the current density and the heat flux at the anode, respectively. Again, the predictions agree closely with the experimental values. Because the heat flux into the anode is mainly determined by the electron flow, the heat flux distribution has a similar trend as the corresponding current density distribution.

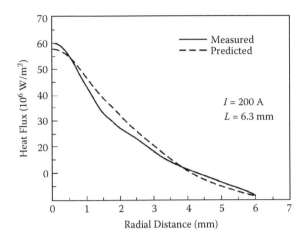

FIGURE 11.12
Heat flux at the anode surface. (From Wu C S, Ushio M, and Tanaka M, *Comput. Mater. Sci.*, 7: 308–314, 1997.)

11.3 Modeling of the Arc Anode Region

For direct current electrode negative (DCEN) welding, there is a thin transformation layer between the arc plasma and the anode surface; this is referred to as the *anode boundary layer*.[190] Within the anode boundary layer, there exists a larger temperature gradient because the temperature changes from over 10,000 K in the plasma to about 2,000 K at the anode surface. Various transport phenomena take place in this layer, which determines directly the magnitude and distribution of both heat and current density on the workpiece surface.

In general, an arc may be divided into a flow-affected region that represents the main body of the arc (arc column), an anode region, and a cathode region. In spite of extensive efforts in the arc physics, there is still a lack of basic understanding of the arc electrode regions of high-intensity arcs and of the associated electrode phenomena.[182,196] The proximity of the electrode gives rise to extremely steep gradients of the plasma properties, which render experimentation even more difficult in this generally hostile environment. Achieving effective utilization and exploitation of the welding arcs requires a thorough understanding of the plasma properties and its physical processes in the proximity of the anode surface.

The *anode region* is generally defined as that part of the arc discharge that contains the surface of the anode, the sheath in front of the anode, and the boundary layer, which makes the connection to the arc column.[192] Although there has been some research of the anode region of electric arcs, a consistent theory is still lacking that would, for example, predict the sign and the magnitude of the anode fall in arcs. Some investigators focused on the numerical simulation of fluid flow and heat transfer inside the weld pool by taking the arc action on the weld pool as boundary conditions (Gaussian distributed heat and current density).[30,73] Others mainly studied the arc column but did not deal with the anode region.[176,183] In this section, the transport mechanisms in the anode region are numerically simulated.

11.3.1 Anode Boundary Layer

As shown in Figure 11.13, the anode region is divided into three subzones: the anode boundary layer, the presheath, and the sheath. In the boundary layer, the presence of the relatively cold anode is felt. The anode boundary layer is characterized by steep gradients of temperature and particle densities. The boundary layer thickness is on the order of 10^{-1} mm, which is much larger than the particle mean free path length λ_m (on the order of 10^{-3} mm). The main feature of the boundary layer is that the ionized gas in the layer may be treated as a continuum and as a true plasma. Very close to the anode (on the order of one electron mean free path), the usual continuum approach is no longer valid. The ionized gas in the presheath and in the sheath may not be treated as a continuum because the total thickness of the presheath and the sheath is equal to one electron mean free path. In the presheath, the neutrality of the plasma is maintained. In the Debye sheath, the neutrality is broken, and a sharp potential drop occurs. The sheath is formed immediately in front of the anode accommodating the transition from electrical conduction in the plasma to metallic conduction in the anode. The thickness of the sheath is on the order of the Debye length λ_D (on the order of 10^{-5} mm). Compared to the arc length, the thickness of the anode boundary layer is so thin that it may be treated as a one-dimensional problem.

11.3.2 Governing Equations

Due to its small thickness, the anode region may be treated as a one-dimensional region. For the purpose of solving the conservation equations, the plasma parameters are assumed to vary only in the direction

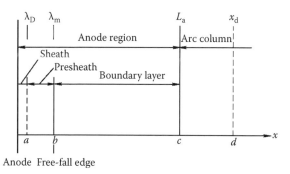

FIGURE 11.13
Schematic of the anode region.

perpendicular to the anode surface. In the boundary layer, the continuum approach and the condition of charge neutrality are valid. Since the electron temperature T_e may be higher than the temperature (T_i) of the heavy particles (atoms and ions), electrons and heavy particles are regarded as two separate fluids coexisting in the plasma.

In a three-component system (electrons, atoms, and ions), the electron flux may be written as

$$\vec{\Gamma}_e = -\mu_e n_e \vec{E} - \frac{\mu_e k_e T_e}{e} \nabla n_e - \frac{\mu_e k_e n_e}{e} \nabla T_e \tag{11.21}$$

where n_e is the electron number density, μ_e is the electron mobility, $\vec{E}$ is the electric field strength, k_e is the Boltzmann constant, and e is the elementary charge. The electron flux is driven by the potential gradient ($\vec{E} = -\nabla\phi$), the electron density gradient, and the electron temperature gradient.

The ion flux may be expressed by

$$\vec{\Gamma}_i = \mu_i n_e \vec{E} - \frac{\mu_i k_e T_h}{e} \nabla n_e - \frac{\mu_i k_e n_e}{e} \nabla T_h \tag{11.22}$$

where $n_e \approx n_i$ is the ion number density, and μ_i is the ion mobility. The condition of charge neutrality is valid in the boundary layer; thus, there is the relation $n_e \approx n_i$. The forces driving the ion flux are the potential gradient, the ion density gradient, and the ion temperature gradient.

In a steady-state situation without macroscopic mass flow, the particle conservation equation of the electrons becomes

$$\nabla \cdot \vec{\Gamma}_e = \dot{n}_e \tag{11.23}$$

where $\dot{n}_e$ is the net electron production rate. Under the same conditions, the particle conservation equation for the ions is

$$\nabla \cdot \vec{\Gamma}_i = \dot{n}_e \tag{11.24}$$

Since electron and ion production occurs in pairs, the net ion production rate equals that of the electrons.

In a plasma, the net electron production rate is the difference between the ionization rate by electron impact and the three-body recombination rate. In terms of the recombination coefficient γ_e and species composition, the net electron production rate is given by

$$\dot{n}_e = \gamma_e n_a \left[\left(\frac{n_e^2}{n_a} \right)_{equil} - \frac{n_e^2}{n_a} \right] \tag{11.25}$$

where $\left(n_e^2 / n_a \right)_{equil}$ is the function of T_e given by the Saha equation

$$\left(\frac{n_e^2}{n_a}\right)_{equil} = \frac{2Z_i}{Z_a}\left(\frac{2\pi m_e k_e T_e}{h_e^2}\right)^{3/2} \exp(-\frac{\varepsilon_i}{k_e T_e}) \tag{11.26}$$

where Z_i is the partition function of the ion, Z_a is the partition function of the neural atom, m_e is the mass of the electron, h_e is Plank's constant, n_a is the number density of neutral atoms, and ε_i is the ionization potential. The number density n_a is related to the pressure p by the expression

$$n_a = \frac{p}{k_e T_h} - n_e\left(1+\frac{T_e}{T_h}\right) \tag{11.27}$$

The current density is given by

$$\vec{J} = \vec{J}_e + \vec{J}_i \tag{11.28}$$

where $\vec{J}_e$ and $\vec{J}_i$ represent the electron and the ion contribution, respectively. According to Eqs. (11.21) and (11.22), the current density may be expressed as

$$\vec{J}_e = -e\vec{\Gamma}_e, \quad \vec{J}_i = e\vec{\Gamma}_i \tag{11.29}$$

Based on Eqs. (11.23) and (11.24), there is the following relation:

$$\nabla\cdot\left(\vec{\Gamma}_i - \vec{\Gamma}_e\right) = 0, \vec{\Gamma}_i - \vec{\Gamma}_e = \text{constant} \tag{11.30}$$

Thus, in the boundary layer

$$\vec{J} = \vec{J}_e + \vec{J}_i = e\left(\vec{\Gamma}_i - \vec{\Gamma}_e\right) = \text{constant} \tag{11.31}$$

In a steady-state situation without macroscopic mass flow, the energy conservation equation of the electrons may be expressed as

$$\nabla\cdot\left(\lambda_e \nabla T_e\right) + \left(\frac{5}{2} + \frac{e\phi_d}{k_e\sigma_e}\right)\frac{J}{e} k_e \nabla T_e + JE = \left(\frac{5}{2}k_e T_e + \varepsilon_i\right)\dot{n}_e$$

$$+ \frac{3m_e}{m_i} k_e\left(T_e - T_h\right)n_e\vec{v}_{ei} \tag{11.32}$$

where λ_e is the electron thermal conductivity, ϕ_d is the thermal diffusion coefficient of the electrons, σ_e is the electrical conductivity, m_i is the mass of the ion, and $\vec{v}_{ei}$ is the average collision frequency between electrons and ions. The terms on the left-hand side of Eq. (11.32) contain several energy inputs in the sequence, that is, heat transfer by the pure conduction, the transport of enthalpy due to the random thermal energy of electrons and the Thomson effect, and the internal energy dissipation due to Joule heating. The first term on the right-hand side of Eq. (11.32) represents the energy used for the production of electrons by ionization, while the second term represents the energy losses by elastic collisions between electrons and heavy particles. Here, only collisions between electrons and ions have been taken into consideration since in the anode boundary layer the collision frequency between electrons and ions is much larger than that between electrons and atoms.

For heavy particles, the joule heating term due to the ion current may be neglected. The energy equation of the heavy particles becomes

$$-\nabla \cdot \left(\lambda_h \nabla T_h \right) + \frac{3m_e}{m_i} k_e \left(T_e - T_h \right) n_e \overline{v}_{ei} = 0 \tag{11.33}$$

where λ_h is the thermal conductivity of heavy particles.

The main unknowns in the set of equations are n_e, T_e, T_h, and E. With these key variables, all other quantities of interest can be computed. There are also four main differential equations: Eqs. (11.23), (11.24), (11.32), and (11.33). The other equations are auxiliary relations.

11.3.3 Boundary Conditions

To solve the system of equations, appropriate boundary conditions must be specified. Because the thickness of the anode boundary layer L_a is an additional unknown, the boundary conditions cannot be specified at the interface between the arc column and the anode boundary layer. In this section, the boundary conditions are specified at a place lying in the arc column and taken from the solutions of the arc plasma region. The distance from this place to the anode surface is a little bigger than L_a. The differential equations are solved starting from this place and then entering the anode boundary layer soon.

As shown in Figure 11.13, at $x = x_d$, (1) the heavy particle temperature and electron temperature are identical, $T_e = T_h = (T)_{x_d}$; (2) the temperature gradients of the heavy particle and electron are identical, $dT_e/dx = dT_h/dx = (dT/dx)_{x_d}$; (3) the electron number density is $n_e = (n_e)_{x_d}$; (4) the gradient of electron number density is $dn_e/dx = (dn_e/dx)_{x_d}$; (5) the current density is $J = (J)_{x_d}$; and (6) the electric field strength is $E = (E)_{x_d}$. Among these six parameters, electron number density and its gradient are both calculated from the Saha equation at $(T)_{x_d}$ and from $(dT/dx)_{x_d}$, while others are determined by the model of arc plasma region introduced in Section 11.2.

Very close to the anode (on the order of one electron mean free path λ_m), the usual continuum approach is no longer valid. Thus, the calculation about the boundary layer should be stopped at the free-fall edge (point b in Figure 11.13) because $x_b = \lambda_m$. An additional boundary condition at the free-fall edge should be specified since the thickness of the anode boundary layer L_a is unknown. The ion temperature in the vicinity of the anode surface is still a problem for further study,[197,198] and there are no specific data for reference. Thus, an adjustable parameter is defined as

$$\theta_b = \frac{T_{hb}}{T_a} \tag{11.34}$$

where T_{hb} is the heavy particle temperature at the free-fall edge, and T_a is the anode surface temperature.

11.3.4 Solution Methods

The set of conservation equations, that is, Eqs. (11.23), (11.24), (11.32), and (11.33), represents four second-order differential equations, including highly nonlinear, nonequilibrium thermodynamic, and transport properties. By applying the Runge-Kutta procedure, these differential equations are solved starting from the boundary conditions and proceeding step by step toward the anode. The value of key variables n_e, T_e, T_h, and E are determined in the boundary layer. For each step, the nonequilibrium composition at that particular location is determined first, and then the thermodynamic and transport properties are calculated. This procedure continues until the supplementary condition $T_h = \theta_b T_a$ may be recovered, and the variation of the electron number density over one electron mean free path reaches the same order of magnitude as the electron number density itself. At this point, the continuum approach is no longer valid. The thickness of the anode boundary layer is defined as the distance from this point to the location where the difference $T_e - T_h$ becomes less than 10% of the electron temperature T_e. The procedure allows the determination of the boundary layer thickness in addition to the values of plasma properties.

11.4 Analysis Results of the Anode Boundary Layer

As mentioned, based on the model of fluid flow and heat transfer in an argon arc plasma, the values of T_e, T_h, dT_e/dx, dT_h/dx, J, and E at a location lying in the arc column can be determined. The value of n_e and dn_e/dx can be calculated from the Saha equation at T_e and from dT_e/dx. These values are used as the boundary conditions. For a free-burning argon arc at 150 A and 1 atmosphere pressure, the corresponding boundary conditions are specified as follows:

At $x = x_d$:

$$T_e = T_h = 1.26 \times 10^4 \ \text{K}$$

$$\frac{dT_e}{dx} = \frac{dT_h}{dx} = 4.5 \times 10^5 \ \text{K/m}$$

$$n_e = 8.82 \times 10^{23} \ \text{1/m}^3$$

$$\frac{dn_e}{dx} = 2.38 \times 10^{25} \ \text{1/m}^4$$

$$J = 3.2 \times 10^6 \ \text{A/m}^2$$

$$E = 2.75 \times 10^2 \ \text{V/m}$$

After solving the governing equations and boundary conditions, we obtain the plasma properties in the anode boundary layer.[199] Figures 11.14, 11.15, and 11.16 are the distributions of electron temperature, electron number density, and electric field strength in the anode boundary layer, respectively. It can be seen that the value of parameter θ_b, that is, the ratio of the heavy particle temperature at the free-fall edge to the anode surface temperature, has a marked effect on the distribution of plasma properties in the boundary layer.

Table 11.2 demonstrates the influence of the parameter θ_b on plasma scale lengths and potential drop in the boundary layer. The boundary layer thickness L_a, the electron mean free-path λ_m, the Debye length λ_D, and the potential drop $\varphi_{\lambda m} - \varphi_{La}$ in the boundary layer are all computed from the solution

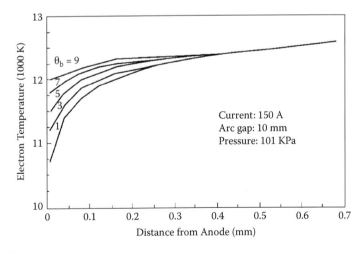

FIGURE 11.14
Distribution of electron temperature in the anode boundary layer. (From Wu C S, Ushio M, and Tanaka M, *Comput. Mater. Sci.*, 15: 302–310, 1999.)

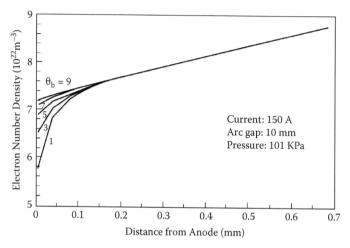

FIGURE 11.15
Distribution of electron number density in the anode boundary layer. (From Wu C S, Ushio M, and Tanaka M, *Comput. Mater. Sci.*, 15: 302–310, 1999.)

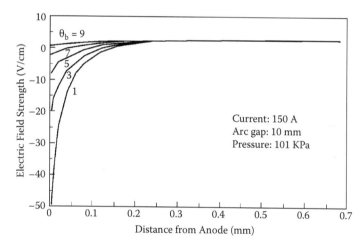

FIGURE 11.16
Electric field strength distribution in the anode boundary layer. (From Wu C S, Ushio M, and Tanaka M, *Comput. Mater. Sci.*, 15: 302–310, 1999.)

of the governing equations of the anode boundary layer. The results indicate that the boundary layer thickness decreases as the value of θ_b increases. A bigger value of θ_b results in smaller temperature gradients for electrons and ions, which makes the deviation between the electron temperature and the heavy particle temperature emerge at a point nearer the free-fall edge so that the boundary layer thickness decreases. The computed results show that the boundary layer thickness is two orders of magnitude larger than the electron mean free path and four orders of magnitude larger than the Debye length.

TABLE 11.2

Influence of the Ratio $\theta_b = T_{hb}/T_a$ on Plasma Scale Lengths in the Anode Boundary Layer (150-A Current, 10-mm Arc Gap, 101-kPa Pressure)

θ_b	L_a (mm)	λ_m (mm)	λ_D (mm)
1	0.287	0.237×10^{-2}	0.298×10^{-4}
3	0.266	0.228×10^{-2}	0.286×10^{-4}
5	0.247	0.225×10^{-2}	0.282×10^{-4}
7	0.235	0.224×10^{-2}	0.281×10^{-4}
9	0.161	0.223×10^{-2}	0.280×10^{-4}

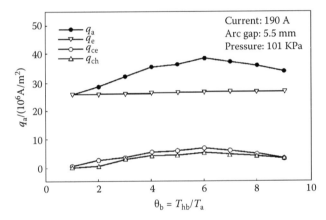

FIGURE 11.17

The effect of θ_b on the heat flux on the anode. (From Wu C S, Ushio M, and Tanaka M, *Comput. Mater. Sci.*, 15: 302–310, 1999.)

TABLE 11.3

Comparison of Calculated and Measured Anode Maximum Heat Intensity q_a

Current (A)	Arc Length (mm)	Pressure (kPa)	θ_b	Predicted q_a (10^6 W/m²)	Measured q_a (10^6 W/m²)
190	5.5	101	6	38.62	40.0
100	5.5	101	6	26.24	27.5

The temperature of heavy particles at the free-fall edge is unknown, and we have to estimate it. Through evaluating the heat flux on the anode surface, as shown in Figure 11.17, the effect of θ_b on the anode heat flux is quantitatively analyzed. In Table 11.3, a comparison is made for the calculated heat flux at the anode and the measured results from Reference 117. It shows that if θ_b is equal to 6, that is, the heavy particle temperature at the free-fall edge takes a value of 6,000 K, the calculated results agree well with the experimental measurements. Thus, the calculated results shown in the following are under the condition of $\theta_b = 6$.

Figure 11.18 shows that the electron temperature and the heavy particle temperature are identical at the arc column side. They deviate increasingly from each other toward the anode. The heavy particle temperature drops from 12,600 K in the arc column to 6,000 K at the free-fall edge, while the electron temperature decreases from 12,600 to 11,700 K. The electron temperature maintains sufficiently high value over the entire thickness of the anode boundary layer to ensure the required electrical conductivity for the passage of the electric current.

Figure 11.19 demonstrates the distribution of the electric field strength and the electrical potential in the boundary layer. The electric field, starting from a small positive value at the arc column side, decreases inside the boundary

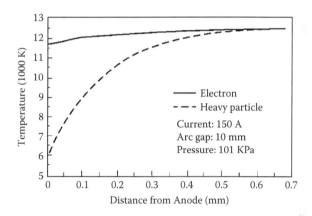

FIGURE 11.18
The temperature of electron and heavy particles. (From Wu C S, Ushio M, and Tanaka M, *Comput. Mater. Sci.*, 15: 341–345, 1999.)

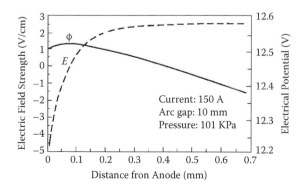

FIGURE 11.19
The electric field strength and potential in the anode boundary layer. (From Wu C S, Ushio M, and Tanaka M, *Comput. Mater. Sci.*, 15: 341–345, 1999.)

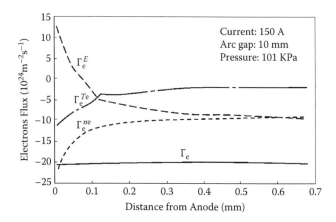

FIGURE 11.20
The electron flux and its components. (From Wu C S, Ushio M, and Tanaka M, *Comput. Mater. Sci.*, 15: 341–345, 1999.)

layer and turns to negative. As it approaches the anode, the electrical potential first increases and then falls. The decrease of the potential signifies that the electric field in the boundary layer is opposing the motion of the electrons and accelerating the positive ions. The net potential drop in the whole boundary layer is slightly positive.

The electron flux and its components are shown in Figure 11.20. The curve denoted by Γ_e^E represents the flux component driven by the electric field, expressed by the first term on the right-hand side of Eq. (11.21). The second and third terms of this equation are represented by the curves $\Gamma_e^{n_e}$ and $\Gamma_e^{T_e}$, respectively. The temperature and electron density gradients push the electrons toward the anode. The electric field drives the electrons toward the anode first. But, as it approaches the anode slightly, it drives the electrons away from the anode. Although the electric field strength is very high near the free-fall edge, this opposing force is overbalanced by the rapidly increasing gradients of the electron density and temperature close to the free-fall edge, which keep the electrons flowing toward the anode. It is obvious that the electron density gradient plays a more important role in driving the electron flux.

Figure 11.21 shows the ion flux and its components. The electric field drives the ions away from the anode first. But, as it approaches the anode slightly, it turns to push the ions toward the anode. The important driving forces for the ion flux are the ion temperature gradient and ion density gradient.

Equation (11.20) states that the heat flux transferred to the anode consists of three components: the heat brought by the electron flux Q_e, the heat caused by convection from the plasma Q_c, and the heat radiated from the plasma Q_r. The values of these three components need to be determined according to the plasma properties at the free-fall edge. For determining the component Q_e, there exists a problem whether the anode fall ϕ_f should be taken into

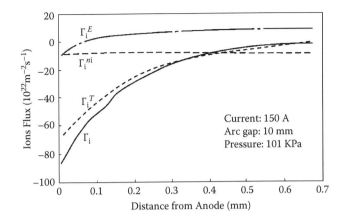

FIGURE 11.21
The ion flux and its components. (From Wu C S, Ushio M, and Tanaka M, *Comput. Mater. Sci.*, 15: 341–345, 1999.)

consideration or not. If $\phi_f \geq 0$, its contribution equal to the kinetic energy $I\phi_f$ (*I* is the arc current) needs to be considered. But, if $\phi_f < 0$, there is no contribution to the energy that arrives at the anode surface.

The potential difference in the boundary layer (ϕ_{fb}) and the potential difference in the sheath (ϕ_{fs}) are included in ϕ_f, which can be calculated by the model of the anode boundary layer. For the available models in the literature, some predicted a slightly negative ϕ_f,[200] some gave all positive values of ϕ_f,[201] and some resulted in large negative values of ϕ_f.[202] They all could not correlate the magnitude and sign of anode fall ϕ_f with the process conditions.

The sheath potential ϕ_{fs} may be determined by the following equation:[203]

$$\phi_{fs} = -\frac{k_e T_{eb}}{e} \ln\left[\frac{(1-\alpha_b)\Gamma_{eb}}{\Gamma_{ib}}\right] \tag{11.35}$$

where T_{eb}, Γ_{eb}, and Γ_{ib} are the electron temperature, the electron flux, and the ion flux at the free-fall edge, respectively, α_b is the percentage of electrons entering the lattice of anode. They all can be determined from the calculations of the anode boundary layer described in Section 11.2. As shown in Figure 11.22, when $\alpha_b < 0.958$, $\phi_{fs} < 0$; when $\alpha_b = 0.958$, $\phi_{fs} = 0$; when $\alpha_b > 0.958$, $\phi_{fs} > 0$. Here, by introducing an adjustable parameter α_b, the correlation between the anode fall and the process conditions is established. The calculation shows that the predicted anode fall is around the range of −3.0–3.8 V, which is in agreement with the practical situation.[204]

We note that the main factors affecting the heat transferred to the anode are as follows: θ_b and α_b, the values of the electron temperature, current density, viscosity, density, thermal content at the free-fall edge, the work function of the anode material, and anode fall. As demonstrated in Figure 11.23, when

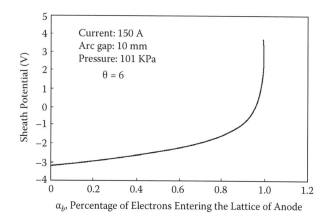

FIGURE 11.22

The effect of α_b on the sheath potential ϕ_{fs}. (From Wu C S, Ushio M, and Tanaka M, *Comput. Mater. Sci.*, 15: 341–345, 1999.)

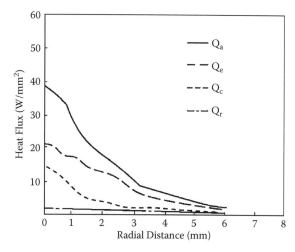

FIGURE 11.23

The heat flux and the anode and its components. (From Wu C S and Gao J Q, *Comput. Mater. Sci.*, 24: 323–327, 2002.)

the arc current is 120 A and arc length is 4.5 mm, the energy carried by the electrons, the convection, and the radiation from the plasma constitute 68.9%, 28.3%, and 2.8% of the heat on the anode surface, respectively.[205]

It should be noted that although some advances in welding arc physics and anode boundary layer have been made,[194,199,203–204] but there are still unsolved problems that need further deep investigation.

Nomenclature

A_d: Transverse cross section area of wire
A_R: Transverse cross section area of weld reinforcement
A_w: Transverse cross section area of weld
a: Thermal diffusivity
a_f, a_h, a_r: Distribution parameters of heat source
b: Damping coefficient of drops
b_c: Thermal loss coefficient of thin plate
b_h: Distribution parameter of heat source
B: Magnetic flux density
B_θ: Component of magnetic flux density
B_i: Biot number
c_h: Distribution parameter of heat source
C_p: Specific heat
C_1, C_2, C_3, C_4: Constants in equation of free surface deformation
d_w: Wire diameter
E: Electric field strength
$F(i,j,t)$: Fluid volume fraction
F_0: Fourier number
F_m: Electromagnetic force
F_v: Body force
F_x, F_y, F_z: Components of body force
F_r, F_z: Components of electromagnetic force
g: Acceleration of gravity
H: Workpiece thickness
H_D: Projected length of drop in its z-axis
H_m: Heat content of molten metal
H_{dr}: Heat content of droplets
H_{pv}: Heat content of weld pool
H_v: Latent of liquid–gas phase change
h_c: Heat transfer coefficient between workpiece and ambient environment
h_e: Plank's constant
I: Welding current
J: Current density
J_a: Current density at anode surface
J_r, J_z: Radial and axial components of current density, respectively
K: Centralization coefficient of heat flux
k: Thermal conductivity
k_d: Centralization coefficient of current density in droplet
k_e: Boltzmann's constant
k_{seg}: Equilibrium constant of segregation

L_e: Wire extension
L_m: Latent of fusion
L_q: Latent of vaporization
m_e: Electron mass
m_i: Ion mass
m_D: Droplet mass
n_3: Dimensionless operating parameter
n_a: Number density of atoms
n_e: Number density of electrons
n_i: Number density of ions
P: Pressure in fluids
P_a: Arc pressure
p: Pressure inside a drop
p_d: Droplet impact
Q: Effective power of arc
Q_0: Arc power
Q_{vd}: Droplet heat content delivered into weld pool
Q_{vh}: Inner heat generation term
q: Heat flux
q_v: Power density of volumetric heat source
q_{vo}: Maximum value of power density of volumetric heat source
r: Radius
R: Distance
R_D: Drop diameter at detachment
R_g: Gas constant
r: Radial distance
r_D: Drop radius
r_H: Radius of arc heating spot
r_w: Wire radius
S_m: Melting rate of wire
T: Temperature
T_0: Initial temperature of workpiece
T_f: Ambient temperature
T_L: Liquidus temperature
T_m: Melting point
T_s: Solidus temperature
T_a: Temperature of anode surface
T_e: Temperature of electrons
T_h: Temperature of ions
t: Time
U_a: Arc voltage
U, V, W: Fluid velocity components in rectangular coordinates
U^*, V^*, W^*: Fluid velocity components in body-fitted coordinates
u, v, w: Fluid velocity components
V_D: Droplet volume

V_φ: Work function of electrons
v_D: Droplet velocity impinging on the pool surface
v_R: Remnant droplet velocity
v_0: Welding speed
W_e: Wire extension
W_h: Nozzle standoff distance
x, y, z: Rectangular coordinates
x^*, y^*, z^*: Body-fitted coordinates
α: Heat exchange coefficient
α_b: Percentage of electrons entering the lattice of anode
α_{cr}: Heat loss coefficient due to convection and radiation
β: Thermal expansion coefficient
δt: Time step
$\delta x, \delta y, \delta z$: Interval distance of grid
ΔT: Temperature difference
ΔH_d: Heat content difference between droplet and weld pool
ε: Radiance
ε_i: Ionization potential
φ: Electrical potential
φ_d: Thermal diffusion coefficient of electrons
φ_f: Anode fall
γ: Surface tension
γ_e: Recombination coefficient
λ: Thermal conductivity
λ_D: Deby length
λ_e: Thermal conductivity of electrons
λ_h: Thermal conductivity of heavy particles
λ_m: Mean free path of electrons
Γ_s: Surface oversaturation parameter
η: Weld thermal efficiency
η_m: Melting efficiency
ϕ: Shape function of top surface of weld pool
μ: Viscosity
μ_e: Electron mobility
μ_i: Ion mobility
μ_m: Magnetic permeability
θ: Dimensional temperature
θ_b: Ratio of heavy particle temperature at free-fall edge to anode surface temperature
θ_s: Weighted factor
ρ: Density
σ_e: Electrical conductivity
σ_j: Distribution parameter of current density
σ_q: Distribution parameter of heat density
σ_s: Stefan-Boltzmann's constant

ϖ: Vorticity
ω: Overrelaxation factor
ψ: Shape function of bottom surface of weld pool
Ψ: Stream function

NOTES: (1) The physical meaning and definition of every symbol is given at the appropriate position when it first appears in the text of this book. (2) Only main symbols are listed in this table.

References

1. Oreper G M, Eagar T W, and Szekely J. Convection in arc weld pool. *Weld. J.*, 1983, 62(11): 307s–312s.
2. Rosenthal D. Mathematical theory of heat distribution during welding and cutting. *Weld. J.*, 1947, 20(5): 220s–234s.
3. Rosenthal D. The theory of moving source of heat and its application to metal treatments. *Trans. ASME*, 1946, 43(11): 849–866.
4. Rykalin H H. *Calculation of weld thermal processes*. Translated by H S Zhuang and B Y Xu. Beijing: China Machinery Press, 1958.
5. Grong Ø. *Metallurgical modeling of welding* (2nd edition). London: Institute of Materials, 1997.
6. Eagar T W and Tsai N S. Temperature fields produced by traveling distributed sources. *Weld. J.*, 1983, 62(12): 346s–355s.
7. Nguyen N T, Ohta A, Matsuoka K, et al. Analytical solutions for transient temperatures of semi-infinite body subjected to 3-D moving heat source. *Weld. J.*, 1999, 78(8): 265s–274s.
8. Nguyen N T, Mai Y W, Simpson S, et al. Analytical approximate solution for double ellipsoidal heat source in finite thickness plate. *Weld. J.*, 2004, 83(3): 82s–93s.
9. Wu C S, Xu G X, Li K H, et al. Analysis of double-electrode gas metal arc welding. In *Proceedings of the 5th International Conference on Trends in Welding Research*, ed. David S A, ASM International, 2006, 813–817.
10. Dilthey U and Roosen S. Computer simulation of thin sheet gas-metal-arc welding. In *Theoretical prediction in joining and welding* (ed. M. Ushio), JWRI, Osaka, Japan, 1996, 133–153.
11. Radaj D, Sudnik W, Erofeew W, et al. Modeling of laser beam welding with complex joint geometry and inhomogeneous material. In *Mathematical modeling of weld phenomena 5* (ed. H. Cerjak), Institute of Materials, London, 2001, 645–669.
12. Pardo E and Weckman D C. Prediction of weld pool and reinforcement dimensions of GMA welds using a finite-element model. *Metall. Trans. B*, 1989, 20(12): 937–947.
13. Tekriwal P and Mazumder J. Finite element analysis of three-dimensional transient heat transfer in GMA welding. *Weld. J.*, 1988, 67(7): 150s–156s.
14. Kumar S and Bhaduri S C. Three-dimensional finite element modeling of gas metal arc welding. *Metall. Trans. B*, 1994, 25(6): 435–441.
15. Oreper G M and Szekeley J. Heat and fluid flow phenomena in weld pools. *J. Fluid Mech.*, 1984, 147, 53–79.
16. Kou S and Sun D K. Fluid flow and weld penetration in stationary arc welds. *Metall. Trans. A*, 1985, 16(2): 203–213.
17. Oreper G M, Szekely J, and Eagar T W. The role of transient convection in the melting and solidification in arc weld pools. *Metall. Trans. B*, 1986, 17(12): 735–744.
18. Oreper G M and Szekeley J. A comprehensive representation of transient weld pool development in spot welding operations. *Metall. Trans. A*, 1987, 18(7): 1325–1332.

19. Thompson M E and Szekely J. The transient behavior of weld pools with a deformed free surface. *Int. J. Heat Mass Transfer*, 1989, 32(6): 1007–1019.
20. Zacharia T, David S A, and DebRoy T. Weld pool development during GTA and laser beam welding of type 304 stainless steel. *Weld. J.*, 1989, 68(12): 499s–519s.
21. Zacharia T, David S A, and DebRoy T. Modeling of interfacial phenomena in welding. *Metall. Trans. B*, 1990, 21(6): 600–603.
22. Choo R T C, Szekely J, and Westhoff R C. Modeling of high-current arcs with emphasis on free surface phenomena in the weld pool. *Weld. J.*, 1990, 69(9): 346s–361s.
23. Tsai M C and Kou S. Electromagnetic-force-induced convection in weld pools with a free surface. *Weld. J.*, 1990, 69(6): 241s–246s.
24. Zacharia T, David S A, and Vitek J M. Effect of evaporation and temperature-dependent material properties on weld pool development. *Metall. Trans. B*, 1991, 22(4): 233–241.
25. Zacharia T, David S A, Vitek J M, et al. Computational modeling stationary gas-tungsten-arc weld pools and comparison to stainless steel 304 experimental results. *Metall. Trans. B*, 1991, 22(4): 243–257.
26. Choo R T C and Szekely J. The effect of gas shear stress on Marangoni flows in arc welding. *Weld. J.*, 1991, 70(9): 223s–233s.
27. Choo R T C and Szekely J. Vaporization kinetics and surface temperature in a mutually coupled spot gas tungsten arc weld and weld pool. *Weld. J.*, 1992, 71(3): 77s–93s.
28. Choo R T C, Szekely J, and David S A. On the calculation of the free-surface temperature of gas-tungsten-arc weld pools from first principles: Part II. Modeling the weld pool and comparison with experiments. *Metall. Trans. B*, 1992, 23(6): 371–378.
29. Kanouff M and Greif R. The unsteady development of a GTA weld pool. *Int. J. Heat Mass Transfer*, 1992, 35(4): 967–979.
30. Domey J, Aidum D K, Ahmadi G, et al. Numerical simulation of the effect of gravity on weld pool shape. *Weld. J.*, 1995, 74(8): 263s–268s.
31. Lee S Y and Na S J. A numerical analysis of molten pool convection considering geometric parameters of cathode and anode. *Weld. J.*, 1997, 76(11): 484s–497s.
32. Chen Y, David S A, Zacharia T, et al. Marangoni convection with two free surfaces. *Numer. Heat Transfer*, Part A, 1998, 33: 599–620.
33. Ko S H, Farson D F, Choi S K, et al. Mathematical modeling of the dynamic behavior of gas tungsten arc weld pools. *Metall. Mater. Trans. B*, 2000, 31(12): 1465–1473.
34. Ko S H, Choi S K, and Yoo C D. Effects of surface depression on pool convection and geometry in stationary GTAW. *Weld. J.*, 2001, 80(2): 39s–45s.
35. Wang Y, Shi Q, and Tsai H L. Modeling of the effects of surface-active elements on flow pattern and weld penetration. *Metall. Mater. Trans. B*, 2001, 32(2): 145s–161s.
36. Tanaka M, Terasaki H, Ushio M, et al. A unified numerical modeling of stationary tungsten-inert-gas welding process. *Metall. Mater. Trans. A*, 2001, 33(7): 2043–2052.
37. Kou S and Wang Y H. Computer simulation of convection in moving arc weld pools. *Metall. Trans. A*, 1986, 17(12): 2271–2277.
38. Kou S and Wang Y H. Weld pool convection and its effects. *Weld. J.*, 1986, 65(3): 63s–70s.

39. Zacharia T, Eraslan A H, Aidum D K, et al. Three-dimensional transient model for arc welding process. *Metall. Trans. B*, 1989, 20(10): 654–659.
40. Zacharia T, Eraslan A H, and Aidum D K. Modeling of non-autogenous welding. *Weld. J.*, 1988, 67(1): 18s–27s.
41. Zacharia T, Eraslan A H, and Aidum D K. Modeling of autogenous welding. *Weld. J.*, 1988, 67(3): 53s–62s.
42. Wu C S, Chen D H, and Wu L. Numerical simulation of fluid flow and heat transfer in TIG weld pool. *Trans. China Weld. Inst.*, 1988, 9(4): 263–269.
43. Wu C S and Tsao K C. Modelling the three-dimensional fluid flow and heat transfer in a moving weld pool. *Eng. Comput.*, 1990, 7(3): 241–248.
44. Wu C S, Cao Z N, and Wu L. Numerical analysis of three-dimensional fluid flow and heat transfer in TIG weld pool with full-penetration. *Acta Metall. Sinica*, 1992, 28(10): B428–432.
45. Wu, Cao, and Wu. Numerical analysis of three-dimensional fluid flow and heat transfer in TIG weld pool. *Acta Metall. Sinica*, 1993, 6(2): 130–136.
46. Wu C S and Dorn L. Computer simulation of fluid dynamics and heat transfer in full-penetrated TIG weld pools with surface depression. *Comput. Mater. Sci.*, 1994, 2: 341–349.
47. Zacharia T, David S A, Vitek J M, et al. Surface temperature distribution of GTA weld pools on thin-plate 304 stainless steel. *Weld. J.*, 1995, 74: 353s–362s.
48. Cao Z N, Wu C S, and Wu L. Modeling full-penetrated weld pool in TIG welding. *Trans. China Weld. Inst.*, 1996, 17(1): 62–70.
49. Joshi Y, Dutta P, Schupp P E, et al. Nonaxisymmetric convection in stationary gas tungsten arc weld pools. *ASME J. Heat Transfer*, 1997, 119: 164–172.
50. Li Z Y and Wu C S. Analysis of the transport phenomena in the interfacial region between TIG arcs and weld pools. *Comput. Mater. Sci.*, 1997, 8: 243–250.
51. Cao Z N, Zhang Y M, and Kovacevic R. Numerical dynamic analysis of moving GTA weld pool. *ASME J. Manuf. Sci. Eng.*, 1998, 120: 173–198.
52. Wu C S and Zheng W. Analysis of fluid flow and heat transfer in a moving pulsed TIG weldpool. *Int. J. Joining Mater.*, 1997, 9: 166–170.
53. Wu C S, Zheng W, and Wu L. Numerical simulation of TIG weld pool behavior under the action of pulsed current. *Acta Metall. Sinica*, 1998, 34(4): 416–422.
54. Wu C S, Zheng W, and Wu L. Modeling the transient behavior of pulsed current tungsten-inert-gas weldpools. *Model. Simul. Mater. Sci. Eng.*, 1999, 7: 15–23.
55. Hirata Y, Asai Y, Takenak K, et al. 3-D numerical model predicting penetration shape in GTA welding. *Mater. Sci. Forum*, 2003, 426–432: 4045–4050.
56. Zhao P C, Wu C S, and Zhang Y M. Numerical simulation of dynamic characteristics of weld pool geometry with step changes of welding parameters. *Model. Simul. Mater. Sci. Eng.*, 2004, 12: 765–780.
57. Wu C S, Zhao P C, and Zhang Y M. Numerical simulation of transient 3-D surface deformation of full-penetrated GTA weld pool. *Weld. J.*, 2004, 83(12): 330s–335s.
58. Zhao P C, Wu C S, and Zhang Y M. Modelling the transient behaviors of a fully-penetrated gas-tungsten arc weld pool with surface deformation. *Proc. Instn. Mech. Eng., Part B: J. Eng. Manuf.*, 2005, 219: 99–110.
59. Tsao K C and Wu C S. Fluid flow and heat transfer in GMA weld pools. *Weld. J.*, 1988, 67(3): 70s–75s.

60. Wu C S. Computer simulation of three-dimensional convection in traveling MIG weld pools. *Eng. Comput.*, 1992, 9(5): 529–537.

61. Kim J W and Na S J. A study on the three-dimensional analysis of heat and fluid flow in gas metal arc welding using boundary-fitted coordinates. *AMSE J. Eng. Industry*, 1994, 116: 78–85.

62. Wu C S and Dorn L. The influence of droplet impact on metal inert gas weld pool geometry. *Acta Metall. Sinica*, 1997, 33(7): 774–780.

63. Ushio M and Wu C S. Mathematical modelling of 3-D heat and fluid flow in a moving GMA weld pool. *Metall. Mater. Trans. B*, 1997, 28(6): 509–516.

64. Wu C S and Sun J S. Modelling the arc heat flux distribution in GMA welding. *Comput. Mater. Sci.*, 1998, 9: 397–402.

65. Wu C S and Sun J S. Determining the distribution of the heat content of filler metal droplet transferred into GMA weldpools. *Proc. Instn. Mech. Eng., Part B: J. Eng. Manuf.*, 1998, 212: 525–531.

66. Fan H G and Kovacevic R. Droplet formation, detachment, and impingement on the molten pool in gas metal arc welding. *Metall. Mater. Trans. B*, 1999, 30: 791–801.

67. Ohring S and Lugt H J. Numerical simulation of a time-dependent 3D GMA weld pool due to a moving arc. *Weld. J.*, 1999, 78(12): 416s–424s.

68. Sun J S and Wu C S. Modeling the weld pool behaviors in GMA welding. *Int. J. Joining Mater.*, 1999, 11(4): 158–166.

69. Sun J S and Wu C S. The effect of welding heat input on the weldpool behavior in MIG welding. *Sci. China* (Series E), 2002, 45(3): 292–299.

70. Wang Y and Tsai H L. Impingement of filler droplets and weld pool dynamics during gas metal arc welding. *Int. J. Heat Mass Transfer*, 2001, 44: 2067–2080.

71. Wang Y and Tsai H L. Effects of surface active elements on weld pool fluid flow and weld penetration in gas metal arc welding. *Metall. Mater. Trans. B*, 2001, 32(6): 501–515.

72. Kim C H, Zhang W, and DebRoy T. Modeling of temperature field and solidified surface profile during gas-metal arc fillet welding. *J. Appl. Phys.*, 2003, 94: 2667–2669.

73. Zhang W, Kim C H, and DebRoy T. Heat and fluid flow in complex joints during gas metal arc welding—part I. Numerical model of fillet welding. *J. Appl. Phys.*, 2004, 95: 5210–5219.

74. Hsu Y F and Rubinsky B. Two dimensional heat transfer study on the key hole plasma arc process. *Int. J. Heat Mass Transfer*. 1988, 31(7):1409–1421.

75. Keanini R G and Rubinsky B. Three dimensional simulation of the plasma arc welding process[J]. *Int. J. Heat Mass Transfer*, 1993, 36(13): 3283–3298.

76. Fan H G and Kovacevic R. Key hole formation and collapse in plasma arc welding. *J. Phys. D: Appl. Phys.*, 1999, 32(22): 2902–2909.

77. Wang H G, Wu C S, and Zhang M X. Finite element method analysis of temperature field in keyhole plasma arc welding. *Trans. China Weld. Inst.*, 2005, 26(7): 47–53.

78. Wu C S, Wang H G, and Zhang M X. Numerical analysis of transient development of temperature field in keyhole plasma arc welding. *Acta Metall. Sinica*, 2006, 42(3): 311–316.

79. Waszink J H and Graat L H J. Experiment investigation of the force acting on a drop of weld metal. *Weld. J.*, 1983, 62(4):108s–116s.

80. Kim Y S and Eagar T W. Analysis of metal transfer in gas metal arc welding. *Weld. J.*, 1993, 72(6): 269s–277s.
81. Lancaster J F. *The physics of welding*. Oxford, UK: Pergamon Press, 1984.
82. Allum C J. Metal transfer in arc welding as a varicose instability: II. Development of a model for arc welding. *J. Phys. D: Appl. Phys.*, 1985, 18: 1447–1468.
83. Allum C J. Metal transfer in arc welding as a varicose instability: I. Varicose instability in a current carrying liquid cylinder with surface charge. *J. Phys. D: Appl. Phys.*, 1985, 18: 1431–1446.
84. Watkins A D, Smartt H B, and Johnson J A. A dynamic model of droplet growth and detachment in GMAW. In *Proceeding of the 4th international conference*, Gatlinburg, TN, 5–8 June 1995.
85. Yang S Y. Project transfer control by additional mechanical force in MIG/MAG. PhD thesis, Harbin Institute of Technology, Harbin, China, 1998.
86. Joo T M, Yoo C D, and Lee T S. Effects of welding conditions on molten drop geometry in arc welding. *Weld. J.*, 1996, 75(4): 62s–76s.
87. Simpson S W and Zhou Peiyuan. Formation of molten droplets at a consumable anode in an electric welding arc. *J. Phys. D: Appl. Phys.*, 1995, 28: 1594–1600.
88. Haidar J and Lowke J. Prediction of metal droplet formation in arc welding. *J. Phys. D: Appl. Phys.*, 1996, 29(6): 2951–2960.
89. Choi S K, Yoo C D, and Kim Y S. Dynamic simulation of metal transfer in GMAW, Part I: globular and spray transfer modes. *Weld. J.*, 1998, 77(1): 28s–44s.
90. Choi S K, Yoo C D, and Kim Y S. The dynamic analysis of metal transfer in pulsed current gas metal arc welding. *J. Phys. D: Appl. Phys.*, 1998, 31: 207–215.
91. Shaw R. *The dripping faucet as a model chaotic system, the science frontier express*. Aerial Press, Santa Cruz, CA 1984.
92. Jae H C, Jihye L, and Choong D Y. Dynamic force balance model for metal transfer analysis in arc welding. *J. Phys. D: Appl. Phys.*, 2001, 34: 2658–2664.
93. Chen M A. Numerical analysis of dynamic metal transfer in GMAW. Postdoctoral report, Shandong University, Jinan, China, 2003.
94. Jones L A, Eagar T W, and Lang J H. A dynamic model of drops detaching from a gas metal arc welding electrode. *J. Phys. D: Appl. Phys.*, 1998, 31: 107–123.
95. Chen M A and Wu C S. Numerical analysis of dynamic process of metal transfer in GMAW. *Acta Metall. Sinica*, 2004, 40(11): 1227–1232.
96. Wu C S, Chen M A, and Li S K. Dynamic analysis of droplet growth and detachment in GMAW. *Chinese J. Mech. Eng.*, 2006, 42(2): 76–81.
97. Wu C S, Chen M A, and Li S K. Analysis of droplet oscillation and detachment in active control of metal transfer. *Comput. Mater. Sci.*, 2004, 31: 147–154.
98. Andrews J G and Atthey D R. Hydrodynamic limit to penetration of a material by a high-power beam. *J. Phys. D: Appl. Phys.*, 1976, 9: 2181–2194.
99. Dowden J, Postacioglu N, Davis M, et al. A keyhole model in penetration welding with a laser. *J. Phys. D: Appl. Phys.*, 1987, 20: 36–44.
100. Kaplan A. A model of deep penetration laser welding based on calculation of the keyhole profile. *J. Phys. D: Appl. Phys.*, 1994, 27: 1805–1814.
101. Lampa C, Kaplan A, Powell J, et al. An analytical thermodynamic model of laser welding. *J. Phys. D: Appl. Phys.*, 1997, 30: 1293–1299.
102. Lankalapalli K N, Tu J F, and Gartner M. A model for estimating penetration depth of laser welding processes. *J. Phys. D: Appl. Phys.*, 1996, 29: 1831–1841.
103. Colla T J, Vicanek M, and Simon G. Heat transfer in melt flowing past the keyhole in deep penetrating welding. *J. Phys. D: Appl. Phys.*, 1994, 27: 2035–2040.

104. Kroos J, Gratzke U, and Simon G. Towards a self-consistent model of the keyhole in penetration laser beam welding. *J. Phys. D: Appl. Phys.*, 1993, 26: 474–480.

105. Ducharme R, Willians K, Kapadia P, et al. The laser welding of thin metal sheets: an integrated keyhole and weld pool model with supporting experiments. *J. Phys. D: Appl. Phys.*, 1994, 27: 1619–1627.

106. Solana P and Ocana J L. A mathematical model for penetration laser welding as a free-boundary problem. *J. Phys. D: Appl. Phys.*, 1997, 30: 1300–1313.

107. Sudnik W, Radaj D, and Erofeew W. Computerized simulation of laser beam welding, modeling and verification. *J. Phys. D: Appl. Phys.*, 1996, 29: 2811–2817.

108. Sudnik W, Radaj D, and Erofeew W. 1996. Computerized simulation of laser beam welding, modeling and verification. *J. Phys. D: Appl. Phys.*, 1996, 29: 2811–2817.

109. Pecharapa W and Kar A. Effects of phase changes on weld pool shape in laser welding. *J. Phys. D: Appl. Phys.*, 1997, 30: 3322–3329.

110. Lee J Y, Ko S H, Farson D F, et al. Mechanism of keyhole formation and stability in stationary laser welding. *J. Phys. D: Appl. Phys.*, 2002, 35; 1570–1576.

111. Quigley M B C, Richards P H, Swift-Hook D T, et al. Heat flow to the workpiece from a TIG welding arc. *J. Phys. D. Appl. Phys.*, 1973, 6(18): 2250–2258.

112. Wilkinson J B and Milner D R. Heat transfer from arcs. *Br. Weld. J.*, 1960, 7(2): 115–128.

113. Guan Q, Peng W X, Liu J D, et al. Measurement-based calculation of effective efficiency of welding heat sources. *Trans. China Weld. Inst.*, 1982, 3(1),10–24.

114. Giedt W H, Tallerico L N, and Fuerschbach P W. GTA welding efficiency: calorimetric and temperature field measurement. *Weld. J.*, 1989, 68(1): 28s–32s.

115. Christensen N, Davies V, and Gjermunden K. Distribution of temperatures in arc welding. *Br. Weld. J.*, 1965, 12(12): 54–57.

116. Niles R W and Jackson C E. Weld thermal efficiency of the GTAW process. *Weld. J.*, 1975, 54(1): 25s–32s.

117. Glickstein S S and Friedman E. Temperature transients in gas tungsten-arc weldments. *Weld. Rev.*, 1983, 62(5): 72–75.

118. Glickstein S S and Friedman E. Weld modeling applications. *Weld. J.*, 1984, 63(9): 38–42.

119. Tsai N S and Eagar T W. Distribution of the heat and current flux in gas tungsten arcs. *Metall. Trans. B*, 1985, 16(12): 841–846.

120. Knorovsky G A and Fuerschbach P W. Calorimetry of pulsed versus continuous gas tungsten arc welds. *Proceedings of the international conference on advances in welding science and technology*, (ed. S A David), ASM International, Metals Park, OH, 1986, 393–400.

121. Smart H B, Stewart J A, and Einerson C J. *Heat transfer in gas tungsten arc welding. ASM metals/materials technology series*, no. 8511-011, 1-14, Metals Park, OH: ASM International, 1986.

122. Goldak J, Chakravarti A, and Bibby M. A new finite element model for welding heat sources. *Metall. Trans. A*, 1984, 15: 299–305.

123. Croft D R. *Finite difference calculation of heat transfer*. Translated by F L Zhang. Beijing: Metallurgy Industry Press, 1982.

124. Kong X Q. *Applications of finite element method in heat transfer*. Beijing: Science Press, 1981.

125. Fudan University, Mathematics Department. *Selected explanation of finite element method*. Beijing: Science Press, 1981.

126. ESI Group. *SYSWELD Manual*. SYSTUS International, Rungls, Cedex, France 2000.

127. Turner J S. *Buoyancy effects in fluid*. Cambridge, UK: Cambridge University Press, December 1979.

128. Choo R T C and Szekely J. The possible role of turbulence in GTA weld pool behavior. *Weld. J.*, 1994, 73(2): 25s–31s.

129. Hong K, Weckman D C, Strong A B, et al. Vorticity based turbulence model for thermofluids modelling of welds. *Sci. Technol. Weld. Joining*, 2003, 8(5): 313–326.

130. American Society for Metals. *ASM handbook of engineering mathematics*. Materials Park, OH: ASM International, 1983, 607–609.

131. Choi M, Greif R, and Salcudean M. A study of heat transfer during arc welding with applications to pure metals or alloys and low or high boiling temperature materials. *Numer. Heat Transfer*, 1987, 11: 477–489.

132. Sahoo P, Debroy T, and McNallan M J. Surface tension on binary metal surface active solute systems under conduction relevant to welding metallurgy. *Metall. Trans. B*, 1988, 19: 483–491.

133. Wu C S. *Numerical analysis of welding thermal processes*. Harbin, China: Harbin Institute of Technology Press, 1990.

134. Patankar S V and Spalding D B. A calculating procedure for heat, mass and momentum transfer in three-dimensional parabolic flows. *Int. J. Heat Mass Transfer*, 1972, 15: 1787–1806.

135. Patankar S V. *Numerical heat transfer and fluid flow*. New York: McGraw-Hill, 1980.

136. Doormaal J P Van and Raithby G D. Enhancement of the SIMPLE method for predicting incompressible fluid flows. *Numer. Heat Transfer*, 1984, 7: 147–163.

137. Raithby G D and Schneider G E. Elliptic systems: finite difference methods II. In *Handbook of numerical heat transfer* (ed. W J Minkowycz, E M Sprrow, R H Pletcher, and G E Schneider). New York: Wiley, 1988, 241–289.

138. Sheng Y, Shoukri M, Sheng G, et al. A modification to the SIMPLE method for buoyancy-driven flows. *Numer. Heat Transfer*, Part B, 1988, 33: 65–78.

139. Tao W Q. *Numerical heat transfer*. Xi'an, China: Xi'an Jiaotong University Press, 1987.

140. Patankar S V and Baliga B R. Computation of heat transfer and fluid flow, Course notes—a university short course for engineers and scientists. University of Minnesota, Minneapolis, MN, 16–19 November 1981.

141. Jakobson H T and Chajar A J. Numerical solutions of heat conduction and simple fluid flow problems. *Int. Comm. Heat Mass Transfer*, 1987, 14: 67–69.

142. Ando H and Tanigawa M. *Welding arc phenomena*. Translated by Y X Shi and R H Peng R H. Beijing: China Machinery Press, 1985.

143. Press W H. *Numerical recipes*. Cambridge, UK: Cambridge University Press, 1986.

144. Lesenwich A. Control of melting rate and metal transfer in gas shielded metal arc welding. *Weld. J.*, 1958, 37(4): 343s–353s.

145. Choi S, Kim Y S, and Yoo C D. Dimensional analysis of metal transfer in GMA welding. *J. Phys. D: Appl. Phys.* 1999, 32: 326–334.

146. Smart H B and Einerson C J. A model for heat and mass input control in GMAW. *Weld. J.*, 1993, 72(5): 217s–229s.

147. Essers W G and Walter R. Heat transfer and penetration mechanisms with GMA and plasma-GMA welding. *Weld. J.*, 1981, 60(2): 37s–42s.
148. Chen C, et al., A two-dimensional transient model for convection in laser melted pool. *Metall. Trans. A*, 1984, 15(12): 2175–2184.
149. Zacharin T, David S A, Vitek J M, et al. Heat transfer during Nd:YAG pulsed laser welding and its effect on solidification structure of austenitic stainless steels. *Metall. Trans. A*, 1989, 20(5): 957–967.
150. Ketter R L and Pravel S R. *Modern methods of engineering computation*. New York: McGraw-Hill, 1969.
151. Liguo E, Wu L, and Zhang J. A new system for metal transfer study-DIA-1 dynamic image analyzer. *China Weld.*, 1993, 2(1): 55–62.
152. Ma T Y. *Computational fluid dynamics*. Beijing: Beijing Institute of Astronautics Press, 1986.
153. Goldak J, Bibby M, Moore J, et al. Computer modeling of heat flow in welds. *Metall. Trans. B*, 1986, 17(5): 587–600.
154. Patankar S V. *Numerical computation of heat transfer and fluid flow*. Translated by Z. Zhang. Beijing: Science Press, 1984.
155. Heiple C R and Burgardt P. Welding, brazing, and soldering In *ASM handbook*, Vol. 6. Metals Park, OH: ASM International, 1993, 19–24.
156. American Welding Society. *Welding handbook*, Vol. 1. Beijing: China Machinery Press, 1985.
157. Nishiguchi K. *The fundamentals and recent development in the field of arc and plasma welding processes*. Osaka, Japan: Textbook of Osaka University, 1982.
158. Chandel R S. Wire melting rate in mild steel MIG welding. *Metal Construct.*, 1988, 5: 214–216.
159. Heiple C R and Burgardt P. Effect of SO_2 shielding gas additions on GTA weld shape. *Weld. J.*, 1985, 64(6): 159s–162s.
160. Metcalfe J C, Quigley M B C. Heat transfer in plasma-arc welding. *Weld. J.*, 1975, 54(3): 99s–103s.
161. Zhu Y F, Zhang H, Dong C L, and Shao Y C. Keyhole plasma arc welding technology. *Astronaut. Manuf. Technol.*, 2002, 2: 22–26.
162. Dong C L, Wu L, and Shao Y C. The history and present situation of keyhole plasma arc welding. *Chinese Mech. Eng.*, 2000, 11: 577–581.
163. Zhang Y M and Zhang S B. Observation of keyhole during plasma arc welding. *Weld. J.*, 1999, 78(2): 53s–58s.
164. Dong H G, Gao H M, and Wu L. Numerical calculation of heat conduction in stationary PAW. *Trans. China Weld. Inst.*, 2002, 23: 24–26.
165. Sun J S and Wu C S. Numerical simulation of heat transfer process in double-sided arc (PAW + TIG) welding. *Acta Metall. Sinica*, 2003, 39(5): 499–504.
166. Nehad A K. Enthalpy technique for solution of Stefan problems: application to the keyhole plasma arc welding process involving moving heat source. *Int. Comm. Heat Mass Transfer*, 1995, 22: 779–790.
167. Wu C S, Wang H G, and Zhang Y M. A new heat source model for keyhole plasma arc welding in FEM analysis of the temperature profile. *Weld. J.*, 2006, 85(12): 284s–291s.
168. Zhang Y M and Zhang S B. Keyhole double-sided arc welding increasing weld joint penetration. *Weld. J.*, 1998, 77(6): 57–61.
169. Zhang Y M, Zhang S B, and Jiang M. Keyhole double-sided arc welding process. *Weld. J.*, 2002, 81(11): 248s–255s.

170. Wu C S, Sun J S, and Zhang Y M. Numerical simulation of dynamic development of keyhole in double-sided arc welding. *Model. Simul. Mater. Sci. Eng.*, 2004, 12: 423–442.

171. Wu C S, Gao J Q , Liu X F, et al. Vision-based measurement of weld pool geometry in constant-current gas tungsten arc welding. Proc. Instn. Mech. Eng., Part B: *J. Eng. Manuf.*, 2003, 217: 879–882.

172. Wu C S, Gao J Q, and Zhang M. Sensing weld pool geometrical appearance in gas-metal arc welding. Proc. Instn. Mech. Eng., Part B: *J. Eng. Manuf.*, 2004, 218: 813–818.

173. Hsu K C, Etemadi K, and Pfender E. Study of the free-burning high-intensity argon arc. *J. Appl. Phys.*, 1983, 54(3): 1293–1301.

174. Hsu K C and Pfender E. Two-temperature modeling of the free-burning high-intensity arc. *J. Appl. Phys.*, 1983, 54(8): 4359–4366.

175. McKelliget J and Szekely J. Heat transfer and fluid flow in the welding arc. *Metall. Trans. A*, 1986, 17(7): 1139–1148.

176. Lee S-Y and Na S-J. Analysis of TIG welding arc using boundary-fitted coordinates. Proc. Instn. Mech. Eng., Part B: *J. Eng. Manuf.*, 1995, 209: 153–164.

177. Lowke J J, Morrow R, and Haidar J. A simplified unified theory of arcs and their electrodes. *J. Phys. D: Appl. Phys.*, 1997, 30: 2033–2042.

178. Lowke J J, Kovitya P, and Schmidt H P. Theory of free-burning arc columns including the influence of the cathode. *J. Phys. D: Appl. Phys.*, 1992, 25: 1600–1606.

179. Zhu P, Lowke J J, and Morrow R. A unified theory of free burning arcs, cathode sheaths and cathodes. *J. Phys. D: Appl. Phys.*, 1992, 25: 1221–1230.

180. Zhu P, Lowke J J, Morrow R, et al. Prediction of anode temperatures of free burning arcs. *J. Phys. D: Appl. Phys.*, 1995, 28: 1369–1376.

181. Dowden J and Kapadia P. Plasma arc welding: a mathematical model of the arc. *J. Phys D: Appl. Phys.*, 1994, 27: 911–913.

182. Eagar T W. An iconoclast's view of the physics of welding—rethinking old ideas. In *Recent trends in welding science and technology TWE'89*, (ed. David S A, ASM International, Materials Park, OH) 1989, 341–346.

183. Jonsson P G, Eagar T W, and Szekely J. Heat and metal in gas metal arc welding using argon and helium. *Metall. Trans. B*, 1995: 26: 383–395.

184. Nemchinsky V A. Plasma flow in a nozzle during plasma arc cutting. *J. Appl. Phys.*, 1998, 31: 3102–3107.

185. Fan H G, Na S-J, and Shi Y W. Mathematical model of arc in pulsed current gas tungsten arc welding. *J. Appl. Phys.*, 1997, 30: 94–102.

186. Choo R T C, Szekely J, and Westhoff R C. On the calculation of the free surface temperate of gas-tungsten-arc weld pools from first principles: Part I. Modeling the welding arc. *Metall. Trans. B*, 1992, 23: 357–369.

187. Nestor O H. Heat intensity and current density distributions at the anode of high current, inert gas arcs. *J. Appl. Phys.*, 1962, 33(5): 1638–1648.

188. Rosenhow W M and Hartnett J P. *Handbook of heat transfer*. New York: McGraw-Hill, 1973, 8–126.

189. Guile A E, Hilton M A, and McLelland I A. Arc current distribution in "continuous high-speed" anode tracks. *J. Phys. D: Appl. Phys.*, 1975, 8: 964–970.

190. Boulos M I, Fauchais P, and Pfender E. *Thermal plasmas*, Vol. 1. New York: Plenum Press, 1994, 6–43.

191. Mitchner M and Kruger C H Jr. *Partially ionized gases*. New York: Wiley, 1973, 146–155.

192. Sanders N A and Pfender E. Measurement of anode falls and anode heat transfer in atmospheric pressure high intensity arcs. *J. Appl. Phys.*, 1984, 55: 714–722.

193. Ushio M, Fan D, and Tanaka M. Contribution of arc plasma radiation energy to electrodes. *Trans. JWRI*, 1993, 22: 201–207.

194. Wu C S, Ushio M, and Tanaka M. Analysis of the TIG welding arc behavior. *Comput. Mater. Sci.*, 1997, 7: 308–314.

195. Hsu K C. A self-consistent model for the high intensity free-burning argon arc. Ph.D. thesis, University of Minnesota, Minneapolis, 1982.

196. Zacharia T, Vitek J M, Goldak J A, et al. Modeling of fundamental phenomena in welds. *Model. Simul. Mater. Sci. Eng.*, 1995, 3: 265–288.

197. Emmert G A, Wieland R M, Mense A T, et al. Electric sheath and pre-sheath in a collisionless finite ion temperature plasma. *Phys. Fluids*, 1980, 23: 803–812.

198. Senda I. The sheath potential formation in the presence of a hot plasma-flow. *Phys. Plasmas*, 1995, 2: 6–13.

199. Wu C S, Ushio M, and Tanaka M. Modeling the anode boundary layer of high-intensity argon arcs. *Comput. Mater. Sci.*, 1999, 15: 302–310.

200. Bose T K. One-dimensional analysis of the wall region for a multiple-temperature argon. *Plasma Chem. Plasma Proc.*, 1990, 10: 189–206.

201. Nemchinsky V A and Peretts L N. Calculating near-anodal layer of a heavy-current arc of high-pressure. *Thuanal Tekhnich. Fiziki*, 1977, 47: 1868–1875.

202. Dinulescu H A and Pfender E. Analysis of the anode boundary-layer of high intensity arcs. *J. Appl. Phys.*, 1980, 51: 3149–3157.

203. Wu C S, Ushio M, and Tanaka M. Determining the sheath potential of gas tungsten arcs. *Comput. Mater. Sci.*, 1999, 15: 341–345.

204. Tanaka M, Ushio M, and Wu C S. One-dimensional analysis of the anode boundary layer in free-burning argon arcs. *J. Phys. D: Appl. Phys.*, 1999, 32(5): 605–611.

205. Wu C S and Gao J Q. Analysis of the heat flux distribution at the anode of a TIG welding arc. *Comput. Mater. Sci.*, 2002, 24: 323–327.

Index

Printed and bound by CPI Group (UK) Ltd, Croydon, CR0 4YY

23/10/2024

01777667-0017